Communications in Computer and Information Science 2653

Series Editors

Rationale

The CCIS series is devoted to the publication of proceedings of computer science conferences. Its aim is to efficiently disseminate original research results in informatics in printed and electronic form. While the focus is on publication of peer-reviewed full papers presenting mature work, inclusion of reviewed short papers reporting on work in progress is welcome, too. Besides globally relevant meetings with internationally representative program committees guaranteeing a strict peer-reviewing and paper selection process, conferences run by societies or of high regional or national relevance are also considered for publication.

Topics

The topical scope of CCIS spans the entire spectrum of informatics ranging from foundational topics in the theory of computing to information and communications science and technology and a broad variety of interdisciplinary application fields.

Information for Volume Editors and Authors

Publication in CCIS is free of charge. No royalties are paid, however, we offer registered conference participants temporary free access to the online version of the conference proceedings on SpringerLink (http://link.springer.com) by means of an http referrer from the conference website and/or a number of complimentary printed copies, as specified in the official acceptance email of the event.

CCIS proceedings can be published in time for distribution at conferences or as post-proceedings, and delivered in the form of printed books and/or electronically as USBs and/or e-content licenses for accessing proceedings at SpringerLink. Furthermore, CCIS proceedings are included in the CCIS electronic book series hosted in the SpringerLink digital library at http://link.springer.com/bookseries/7899. Conferences publishing in CCIS are allowed to use Online Conference Service (OCS) for managing the whole proceedings lifecycle (from submission and reviewing to preparing for publication) free of charge.

Publication process

The language of publication is exclusively English. Authors publishing in CCIS have to sign the Springer CCIS copyright transfer form, however, they are free to use their material published in CCIS for substantially changed, more elaborate subsequent publications elsewhere. For the preparation of the camera-ready papers/files, authors have to strictly adhere to the Springer CCIS Authors' Instructions and are strongly encouraged to use the CCIS LaTeX style files or templates.

Abstracting/Indexing

CCIS is abstracted/indexed in DBLP, Google Scholar, EI-Compendex, Mathematical Reviews, SCImago, Scopus. CCIS volumes are also submitted for the inclusion in ISI Proceedings.

How to start

To start the evaluation of your proposal for inclusion in the CCIS series, please send an e-mail to ccis@springer.com.

Geetha Ramachandran · Nhu-Ngoc Dao ·
Noor Zaman Jhanjhi · Anuradha Muthukrishnan ·
Hemanand Duraivelu
Editors

Advances in Artificial Intelligence and Machine Learning in Big Data Processing

Second International Conference, AAIMB 2024
Chennai, India, December 12–13, 2024
Proceedings, Part I

Springer

Editors
Geetha Ramachandran
S.A. Engineering College
Chennai, Tamil Nadu, India

Nhu-Ngoc Dao
Sejong University
Seoul, Korea (Republic of)

Noor Zaman Jhanjhi
Taylor's University
Subang Jaya, Malaysia

Anuradha Muthukrishnan
S.A. Engineering College
Chennai, Tamil Nadu, India

Hemanand Duraivelu
S.A. Engineering College
Chennai, Tamil Nadu, India

ISSN 1865-0929 ISSN 1865-0937 (electronic)
Communications in Computer and Information Science
ISBN 978-3-032-13025-9 ISBN 978-3-032-13026-6 (eBook)
https://doi.org/10.1007/978-3-032-13026-6

This Springer imprint is published by the registered company Springer Nature Switzerland AG
The registered company address is: Gewerbestrasse 11, 6330 Cham, Switzerland

Preface

These two CCIS volumes constitute the peer-reviewed proceedings from the 2nd International Conference on Advances in Artificial Intelligence and Machine Learning in Big Data Processing (AAIMB 2024), held at S.A. Engineering College, Chennai, India, on December 12–13, 2024. All the accepted papers were carefully double-blindly reviewed by national and international reviewers, and submissions received an average of three reviews each. They were organized in topical tracks as follows: Artificial Intelligence and Machine Learning; Intelligent Information Processing, Pattern Recognition and Analysis, Sentiment Analysis, Computational Optimization, Information Extraction and Prediction, Decision Making, Virtual Reality, Natural Language Processing, AI Algorithms, Fuzzy Logic & Neural Systems, Adaptive Agents, AI Planning Strategies and Tools, Reinforcement Learning, Robotics, Computer Vision, Expert Systems, Pattern Recognition, Component Analysis, Social Network Analysis, Cloud-Based Services, Big Data Analytics; Data Mining Challenges with Big Data, Big Data Analytics and Big Data Governance, Big Data Analytics and Applications Data, Text, Web Mining, & Visualization, Data & Knowledge Engineering, Techniques for Data Visualization, Data Stream Algorithms, Information Processing, Pattern Recognition and Analysis, Computational Optimization, and Information Extraction and Prediction. AAIMB 2024 received 434 submissions, out of which 59 full-length papers were accepted and presented at the conference.

We extend our sincere gratitude to the conference committee members and international reviewers for their diligent efforts in selecting the qualified papers from all the submissions. Their excellent professional advice enabled us to come out with these two volumes of conference proceedings. We thank all the participants and the management of S.A. Engineering College, who encouraged us morally and financially.

December 2024

Geetha Ramachandran
Noor Zaman Jhanjhi
Nhu-Ngoc Dao
Anuradha Muthukrishnan
Hemanand Duraivelu

Organization

General Chair

Geetha R.	S. A. Engineering College, India

Program Committee Chairs

Nhu-Ngoc Dao	Sejong University, South Korea
Noor Zaman Jhanjhi	Taylor's University, Malaysia
Anuradha M.	S. A. Engineering College, India
Hemanand D.	S. A. Engineering College, India
Kaliraj V.	S. A. Engineering College, India
Nazir Ahmad	King Khalid University, Saudi Arabia
Abdel-Badeeh M. Salem	Ain Shams University, Egypt

Steering Committee

Nhu-Ngoc Dao	Sejong University, South Korea
Nazir Ahmad	Aligarh Muslim University, India
Dinakaran K.	S.A. Engineering College, India
Thilagavathi N.	S.A. Engineering College, India
Abdel-Badeeh M. Salem	Ain Shams University, Egypt
Mary Immaculate Sheeela	Pentecost University, Ghana
Xavier N. Fernando	Ryerson University, Canada
Mokhtar M. M. Ali Hamid	University of El Imam El Mahdi, Sudan
Muhammad Ghalib	De Montfort University Dubai, United Arab Emirates
Kindie Biredahn Nahato	Debre Berhan University, Ethiopia
Senthil Ramadoss	Majan University College, Oman
Jean Justus J.	SRM University, India
Suresh Kumar S.	Rajalakshmi Engineering College, India
Sathish Kumar P. J.	Panimalar Engineering College, India
Chembian W. T.	Vel Tech HighTech Dr.Rangarajan Dr.Sakunthala Engineering College, India
Eswaramoorthy V.	Bannari Amman Institute of Technology, India

Srinivasan R.	Vel Tech Rangarajan Dr. Sagunthala R&D Institute of Science and Technology, India
Subhashini P.	Vel Tech Multi Tech Dr. Rangarajan Dr. Sakunthala Engineering College, India
Gopalakrishnan B.	Bannari Amman Institute of Technology, India
Sundara Murthy S.	Bannari Amman Institute of Technology, India
Vanitha M.	Saveetha Engineering College, India
Sathya Priya J.	Velammal Engineering College, India
Ashok Kumar P. M.	KL University, India
Brindha Merin J.	B.S. Abdur Rahman Crescent Institute of Science & Technology, India
Mercy Paul Selvan	Sathyabama Institute of Science and Technology, India
Saraswathi S.	SSN College of Engineering, India
Bhavadharini R. M.	VIT Chennai, India

Program Committee

Siva Kumar S.	S.A. Engineering College, India
Chitra Devi D.	S.A. Engineering College, India
Balakrishnan C.	S.A. Engineering College, India
Mani A.	S.A. Engineering College, India
Balasubramanian M.	S.A. Engineering College, India
Muthukumarasamy S.	S.A. Engineering College, India
Sureka V.	S.A. Engineering College, India
Sudha L.	S.A. Engineering College, India

Additional Reviewers

Prasana Devi
Prasana Kumar
Radhika R.
Vanitha L.
Saraswarthi V.
Aruna K. B.
Sangeetha J.
Shobana R.
Belshia Jebamalar
Abirami A.
Kavitha M.
Lakshmipriya C.
Dhivya Bharathi R.
Ananthi S. N.
Saranya G.
Priya Chitra M.
Vasantha Kumari O.
Sivasakthi K.
Shaini A.
Julia Faith S.
Veena T.
Chithra D.
Bhavadharini R. M.
Dhanalakshmi K. M.

Sermakani A. M.
Krishnaveni S.
Seetha A.
Sutha Merlin
Muralitharan G.
Pushpalatha B.
Antonidoss A.
Vijayalakshmi K.
Nalini M.
Chenni Kumaran
Muthu Kumar G.
Arumai Shiney S.
Gayathri S.
Dharaniya R.
Lakshmi Priya S.
Jaison B.
Selvaraj M. D.
Thilagam T.
Sree Vani N. R. G.
Nithyanandan R.
Ruba Sudha P.
Preethi S. T.
Dharani L.
Revathy M.
Ramya Bharathi K.
Saravanan K.
Angelin Felicia
Sridevi Sridhar
Suvitha S.
Sasirekha K.
Maheshwari A.
Asha S.
Valarmathi K.
Sakthi U.
Shyni M.
Adlin Layola J. A.

About This Conference

This book chapter constitutes the refereed proceedings of the International Conference on Advances in Artificial Intelligence and Machine Learning in Big Data Processing, AAIMB 2024, held in S.A. Engineering College, Chennai, India, during December 12–13, 2024. 59 papers were accepted and presented in the conference out of 434 received submissions. All the accepted papers were carefully reviewed by a minimum of two national and one international reviewers. They were organized in topical tracks as follows a) Artificial Intelligence and Machine Learning: Intelligent Information Processing, Pattern Recognition, Sentiment Analysis, Computational Optimization, Information Extraction and Prediction, Decision Making, Virtual reality, Natural Language Processing, AI Algorithms, Fuzzy logic & Neural System, Adaptive agents, AI Planning Strategies and Tools, Reinforcement Learning, Robotics, Computer Vision, Expert Systems, Component Analysis, Social Network Analysis, Cloud-based Services b)Big Data Analytics: Data Mining Challenges with Big Data, Big Data Analytics and Big Data governance, Big data Analytics and Applications Data, Text, Web Mining, Visualization, Data & Knowledge Engineering, Techniques for Data Visualization, Data Stream Algorithms, Information Processing, Pattern Recognition and Analysis, Computational Optimization, Information Extraction and Prediction, Fuzzy logic & Neural System.

Ramachandran Geetha
Nhu-Ngoc Dao
Noor Zaman Jhanjhi
Muthukrishnan Anuradha
Duraivelu Hemanand

About This Conference

Contents

Artificial Intelligence and Data Analytics

Agent-Based Flood Evacuation Modeling: Optimizing Transportation, Shelter Allocation and Traffic Control

Navroop Kaur(✉) and Harjot Kaur

Guru Nanak Dev University, Amritsar 143001, Punjab, India
kaurnavroop1789@gmail.com

Abstract. Evacuation planning during large-scale natural disasters remains a complex and critical challenge for ensuring the safety and survival of affected population. Effective evacuation strategies must optimize decision-making for decreasing the time to evacuate and thereby, increasing the number of people who remain alive after destruction, significantly improving medical outcomes for individuals in urgent need of care. This paper presented the discrete-event simulation model for flood evacuation that models real-world conditions and dynamically adjusts to evolving factors, such as varying types of vehicles, evacuees with different medical priorities and multiple shelter options. The model's objective is to facilitate rapid, safe, and prioritized evacuation based on the individuals' medical status and urgency. This framework provides decision-makers with insights into essential evacuation parameters, including: (1) whether an evacuee decides to evacuate or stay, (2) the optimal waiting time before departure, (3) the most suitable mode of transportation for each evacuee, and (4) assignment of shelters based on evacuee priority levels. Additionally, a throttling mechanism serves as a traffic control strategy to alleviate congestion and optimize traffic flow, reducing total evacuation time. This model offers a comprehensive tool for simulating and evaluating various evacuation scenarios, providing key insights for enhancing evacuation effectiveness in flood emergencies. The results of this study demonstrate how strategic decisions in transportation, shelter allocation, and traffic control can significantly impact evacuation outcomes, ultimately improving both survival rates and the efficiency of emergency response operations.

Keywords: Flood Emergency Evacuation System · Agent Modeling · Discrete-Event Simulation · Evacuation Planning · Geographic Information System · Throttling

1 Introduction

Natural disasters occur worldwide at regular intervals, causing significant damage, loss of human lives, ecological disruption, and the deterioration of essential health services. These disasters often result in large-scale, irreparable damage due to their rapid onset and widespread impact [1]. The type of evacuation strategy employed is the most critical

R. Geetha et al. (Eds.): AAIMB 2024, CCIS 2653, pp. 3–21, 2026.
https://doi.org/10.1007/978-3-032-13026-6_1

aspects of mitigating damage during disaster events. Disasters can be categorized based on the preparation time available before their occurrence: no-notice and short-notice disasters [2]. On average, 8 million hectares are affected by floods each year [3]. Floods often fall under the no-notice category. They can cause outages, landslides, transportation disruption, and tragic loss of lives.

There are three major types of floods: a.) Flash Floods: Caused by dam failure, heavy rainfall, or river obstructions, flash floods require immediate warning systems and rapid responses. b.) River Floods: Typically triggered by rainfall over extensive river basins or by melting snow, river floods are normally seasonal and forms gradually. c.) Coastal Floods: Rising global temperatures contribute to the expansion of the world's oceans. Increased snow and ice melt exacerbates the rise in sea levels, leading to coastal flooding.

Emergency management agencies must develop tactical evacuation plans that involve coordinated, inter-agency operations to manage the disruption caused by no-notice disasters. However, executing these strategies amidst chaotic conditions poses significant challenges [2], such as determining how to transport evacuees to safe locations efficiently. Various simulation methodologies provide a viable solution to these problems by capturing traffic dynamics and evacuee behaviors during evacuation.

In this paper, a method utilizing Agent-Based Discrete-Event Simulation (ABDES) approach for managing large-scale evacuation during flood disaster is proposed. The model aims to reduce the cost of evacuating the people and to increase the number of people who can be rescued while considering available resources and geographic constraints. The research objectives include integrating ABDES with Geographic Information Systems (GIS) to create a realistic traffic network for evacuation planning. This includes accommodating evacuees with varying priorities, utilizing multiple vehicles, and identifying shelters within an actual traffic network. Additionally, the model evaluates the impact of socio-psychological and behavioral characteristics of evacuees on evacuation effectiveness. A throttling technique for traffic management is also proposed to enhance evacuation efficiency in flood scenarios. The rest of the paper is organized as: Sect. 2, detailed the literature review. In Sect. 3, factors for evacuation planning problem formulation are considered. The proposed framework is described in Sect. 4. The discussions related to proposed work and comparative analysis of the latter with previous works are outlined in Sect. 5. Finally Sect. 6 summarizes conclusions and future scope.

2 Literature Review

Natural disasters, especially large-scale events like floods, necessitate well planned evacuation criteria to mitigate the destruction of lives and property. Simulation modeling has become a key tool in evaluating and designing evacuation strategies, allowing researchers to simulate both human behavior and traffic dynamics under varying disaster conditions. Simulation models for evacuation preparedness can be categorized into three types—macroscopic, microscopic and mesoscopic, each serving distinct purposes in evacuation modelling [4]. This section reviews these simulation categories, their applications, and the integration of simulation methods such as Agent-Based Simulation (ABS) and Discrete-Event Simulation (DES) in evacuation planning.

2.1 Macroscopic Simulation Models

Macroscopic simulation models focus on the overall flow of traffic without accounting for the behavior of individual vehicles or evacuees. These models are often used in large-scale transportation network simulations where individual-level details are not the primary concern. Dynamic traffic models, for example, use this type of simulation to solve wide spread evacuation planning, as demonstrated by Edara [5]. Kwon and Pitt [6] applied the Dynasmart-P model to assess evacuation traffic in downtown Minneapolis, MN. Similarly, Songchitruska [7] utilized DynusT to evaluate the performance of evacuation plans in the Houston-Galveston region. These studies demonstrate the utility of macroscopic models in planning large-scale evacuation routes where individual characteristics of evacuees and vehicles are less critical.

2.2 Microscopic Simulation Models

Microscopic simulation models, in contrast, offer a detailed representation of individual entities, including evacuees, vehicles, and resources, during evacuation. These models track the interactions and behaviors of individual agents in the evacuation process, capturing more granular details such as speed, pathfinding, and psychological factors. Microscopic models are particularly useful for modeling the complex dynamics of evacuation at a more localized level. Jha [8] employed MITSIMlab to develop a microscopic evacuation planning model at the Laboratory at Los Alamos, highlighting importance of modeling individual vehicle movements and decision-making during evacuation. Zhao [9] extended this approach by incorporating human behavioral characteristics and psychological factors into their model of residential evacuation, emphasizing the impact of individual behavior on evacuation efficiency. Alizadeh [10] also proposed a microscopic model for room evacuations, specifically focusing on the presence of obstacles and how they affect the flow of evacuees.

2.3 Mesoscopic Simulation Models

The scope of mesoscopic models lie between macroscopic and microscopic models, accounting for the behavior of individual entities while also addressing large-scale evacuation network issues. One of the most prominent techniques used in mesoscopic models is Agent-Based Simulation (ABS), particularly in scenarios involving no-notice disasters, such as floods. ABS models simulate the interactions among autonomous agents like evacuees or vehicles, capturing their decision-making processes and behavior during evacuations. Ukkusuri [11] and Yin [12] demonstrated how agent-based models could effectively simulate evacuee behavior, including decisions about route selection, speed, and shelter choice. These models highlight how individual decision-making can influence traffic flow and overall evacuation efficiency. Like agent-based systems, which follow a bottom-up approach, Discrete-Event Simulation (DES) models adopt a step-by-step perspective, focusing on the system's states and how they change in response to specific events. DES models are highly efficient for simulating large-scale evacuation networks, especially when combined with ABS to capture both the system dynamics and individual-level behaviors. ABS offers greater flexibility and autonomy in encapsulating agent behavior, while DES provides a streamlined approach for reducing computational time. When integrated, these two approaches can offer a robust framework for modeling complex evacuation problems.

2.4 Integration of Agent and Discrete-Event Simulation Models

The integration of DES and agent models creates a hybrid simulation framework, known as Agent-Based Discrete-Event Simulation (ABDES), which combines the strengths of both methodologies. This approach allows for more effective modeling of large-scale evacuations by accounting for both micro-level behaviors of evacuees and macro-level system dynamics. Wagner [13] proposed a time-step-based simulation framework that integrates DES with agent-based models, supporting it with agent-object-relationship modeling to provide immediate status of system agents and the environment. Zhang [14] developed the modeling framework specifically for large-scale evacuation scenarios, focusing on the interactions between distinct agents in a hybrid simulation environment. This integration enables more comprehensive modeling of evacuation processes, particularly in situations involving complex traffic networks and unpredictable agent behavior.

2.5 GIS Integration with Simulation Models

Since evacuation problems are inherently spatial, the integration of Geographic Information Systems (GIS) with simulation models has become increasingly important in disaster management. GIS helps to analyze the spatial data, like road structures, destructive zones, and locations to take shelter, facilitating better decision-making during evacuations. The inclusion of GIS in simulation models provides a more realistic representation of the environment, enhancing the feasibility of the models as spatial decision-making tools Silva [15]. For example, Zhang [14] integrated GIS with an ABDES framework for spatial interaction and patterns simulation of evacuees during a flood event. This integration allowed the model to account for geographic constraints, such as road closures due to flooding, and provided a more accurate estimation of evacuation times.

2.6 Evacuation Decision-Making and Traffic Management

In no-notice disasters like floods, emergency management agencies must quickly implement predefined evacuation plans. Key decision-making parameters include route selection, shelter allocation, evacuee dynamics, and vehicle types. Lindell [16] investigated the types of vehicles used during evacuations, noting a strong preference for private vehicles. Mostafizi [17] examined the impact of shelter allocation on evacuee behavior, finding that evacuation efficiency improves when appropriate shelter accommodations are available.Traffic management during evacuation is another critical factor. Jonkman [18] and Zhang [19] analyzed the effect of congestion on evacuation times, highlighting the importance of traffic control measures in reducing delays. Mas [20] employed stochastic simulation to assess the relationship between capacity and demand at evacuation shelters, demonstrating that spatial distribution and evacuee "milling time" significantly affect shelter utilization. Qing [21] used multi-agent system to analyze the route choice by walking individuals in rainstorm disaster. Assuming different scenarios and applying clustering process, various attributes of population were found during evacuation. It was studied that herd mentality and shelter distribution in an area had great significance on evacuation. Yamada [22] studied agent-based tsunami evacuation simulations and compared evacuation ratio while agent move to nearby shelters with the situation when they move to inland facilities nearby. It was concluded that evacuating to inland facilities take longer time compared to evacuation to nearest facilities. Chen [23] evaluated risk awareness for households, evacuation plans and preparations for evacuating the area in case of disaster using data survey. Chen [24] represented the complex evacuation strategies by employing slope speed value from cost distance analysis and developing the simulating model with that aspect. The results showed that tsunami mortality rate is affected by different factors related to time and evacuation participation level as they have indirect influence on it.

2.7 Application of ABDES Framework in Evacuation Planning

This paper proposes a hybrid ABDES evacuation framework, which integrates agent-based and discrete-event simulation approaches with a GIS module. This approach allows evacuation planners to evaluate various evacuation scenarios, taking into account both the individual behaviors of evacuees and the broader system dynamics. Such a framework proves really beneficial in planning for short-notice or no-notice disasters, such as floods, where rapid decision-making and real-time adjustments are crucial. By combining the strengths of agent-based flexibility, discrete-event efficiency, and GIS-based spatial analysis, the proposed ABDES framework offers a comprehensive tool for modeling large-scale evacuation scenarios and can support decision-making in real-world disaster management situations.

3 Evacuation Planning Factors for Problem Formulation

Large-scale disasters, such as floods, often strike without any prior warning, leaving little to no time for preparation. In such situations, immediate evacuation of people from the affected areas becomes crucial. The regions facing immediate need, often termed as hot zones, are the most critical, as they are the areas where human lives are at the greatest risk [25]. Evacuation strategies typically involve transferring people from the hot zone to nearby staging regions, which act as temporary safe zones where further medical evaluation and triage can be conducted. At these staging regions, Simple Triage and Rapid Treatment (START) is employed to categorize evacuees based on the severity of their medical conditions. This prioritization is essential for ensuring that limited medical and logistical resources are used effectively. High-priority evacuees are those who require immediate medical attention due to life-threatening injuries, while intermediate and low-priority evacuees are those with non-critical injuries or no injuries at all. The allocation of resources and transportation is then decided based on these priority levels, ensuring that those most in need receive care first. There are several critical factors that influence the overall evacuation process, including the decision-making behavior of evacuees, environmental conditions, and available resources. These factors are discussed in detail below.

3.1 Evacuation Decision-Making Approach

The decision to evacuate or stay at same place when disaster occurs is effected by a combination of environmental conditions and personal characteristics [26]. This evacuation decision-making process (ED) can vary significantly across individuals, as it is shaped by perceptions of risk, trust in authorities, available information, and individual capabilities.

$$ED = \sum_{i \in Z}^{n} \beta_{io} \gamma_{io}^{\omega_{io}} + \beta_o \tag{1}$$

Where γ_{io} are the observed variables, β_o is the constant parameter and here $\{\beta, \omega\} \in 0$. The probability function of evacuation decision becomes logistic regression function as represented in following equation:

$$(ED = 1|\gamma_{io}) = \frac{exp(\sum_{i \in Z}^{n} \beta_{io} \gamma_{io}^{\omega_{io}} + \beta_o)}{1 + exp(\sum_{i \in Z}^{n} \beta_{io} \gamma_{io}^{\omega_{io}} + \beta_o)} \tag{2}$$

3.2 Departure Time Analysis in Evacuation Scenarios

The timing when individuals decide to evacuate is a significant aspect that influences the overall purpose of an evacuation process. Various previous studies have examined departure time decisions across different disaster types like tsunami, earthquake, and hurricane [16, 17]. These studies have emphasized that the timing of evacuation can significantly affect traffic congestion, the availability of resources, and, ultimately, the survival rates of evacuees. Understanding how individuals choose their departure time is essential for optimizing evacuation strategies, particularly in large-scale, no-notice disasters like floods.

There are two primary evacuation strategies in which the departure time distribution can be simulated in Agent-Based System (ABS). a.) Staged Evacuation: This type of evacuation is carried out in stages by following certain rules like priority criteria with evacuees leaving in predefined groups.

b.) Simultaneous Evacuation: In this type of evacuation scenario all the evacuees are instructed to leave the hot zone immediately or at once, without any phased approach. In the proposed agent-based simulation framework, the behavior of evacuees in choosing their evacuation departure time is modeled using statistical distributions that reflect the urgency and priority level of the evacuees.

High-priority evacuees (those with severe medical conditions or facing the most immediate danger) are assumed to make decisions about their departure time following an exponential distribution. The exponential distribution is appropriate for modeling time-to-event processes, particularly in situations where the decision to evacuate is made quickly in response to an immediate threat. This reflects the reality that high-priority evacuees are likely to leave sooner rather than later, given the urgency of their situation.

Intermediate and low-priority evacuees (those with minor injuries or no injuries) are assumed to make their departure time decisions depending upon a Rayleigh distribution. The Rayleigh distribution (Eq. 3) is often used to model the time between events in systems where there is less urgency or where decisions are made based on more gradual factors. In this case, intermediate and low-priority evacuees may delay their evacuation as they assess the situation, potentially leading to a more dispersed departure pattern over time.

$$P(t) = \{= 0 \text{ if } 0 < t < \tau$$
$$\{= 1 - e^{\frac{-(t-\tau)^2}{2\sigma_t^2}} \text{ if } t > \tau \text{ or } (ED|\gamma_{io}) > \xi \quad (3)$$

Where t is departure time after initiation of flood, τ is the delay time incurred while waiting for vehicles and evacuation decision-making, σ_t is the calibration variable for medium and low priority evacuees, ξ is the marginal value over which evacuation decision can be made by evacuee.

3.3 The Role of Transportation in Evacuation Scenario

Transportation is a fundamental component of any successful evacuation process. Once the decision to evacuate has been made and shelters have been designated, the next critical step for emergency management agencies is the allocation of vehicles and the identification of optimal evacuation routes. Effective transportation planning ensures that evacuees can be moved quickly and safely from the danger zone to designated shelters, while also minimizing traffic congestion and maximizing resource use. In large-scale evacuation scenarios, a variety of transportation modes such as helicopters, ambulances, high-water rescue trucks, private vehicles (cars) may be used, each with its unique characteristics and constraints [2]. These characteristics include capacity, speed and cost. To ensure an effective evacuation, emergency management agencies must not only select the appropriate transportation modes but also optimize the allocation of vehicles and the evacuation routes. This requires careful planning and coordination, taking into account factors such as location and number of shelters, traffic and road conditions, vulnerable populations and coordination across agencies.

3.4 The Role of Shelters in Evacuation Planning

Shelters serve as critical infrastructure in evacuation planning, particularly during no-notice, large-scale disasters where immediate medical aid and basic resources are required. These temporary facilities offer a safe space for evacuees to receive treatment, recover, and access essential amenities. The effectiveness of an evacuation largely depends on the availability, capacity, and proper management of these shelters [2]. In the proposed framework, shelters are divided into the following types according to their capacity and the level of medical treatment they provide: expansive shelters, bounded shelters, secondary shelters and temporary shelters.

- **Expansive shelters** are usually limited due to their high cost and the extensive resources required to operate them. However, they are equipped with extensive medical resources and have ability to offer comprehensive care to high priority evacuees – individuals with severe injuries or medical conditions that require urgent and specialized treatment. They are usually located in strategic areas that are easily accessible from the disaster zone to ensure that the most critical evacuees can receive care as quickly as possible.
- **Bounded shelters** provide evacuees with an intermediate level of care and they are more in number as in the area compared to expansive shelters but limited to tackle the moderate number of evacuees. These shelters offer essential medical services such as first aid, basic trauma care, and treatment for non-life-threatening injuries. While they may not have the advanced resources of expansive shelters, they still provide critical support in disaster situations.
- The need to accommodate a large number of evacuees leads to requirement of **secondary shelters** which can provide shelter to a large number of evacuees in need of basic amenities such as food, water, sanitation, and temporary housing in a sufficient space.
- **Temporary shelters** can reduce the load on primary evacuation routes and provide immediate medical care and basic necessities to evacuees.

3.5 Traffic Throttling Strategy for Congestion Management During Evacuation

During the evacuation process, as evacuees are rescued and transported from hot zones to shelters, the influx of vehicles can lead to significant traffic congestion on the roads. Emergency management authorities have the responsibility to assist and coordinate appropriate routes for people to evacuate out of danger zones. To manage this traffic effectively, a strategy such as throttling can be employed, which involves closing the lane segment when the congestion on that segment increases above the upper threshold value and again opening it when the congestion level decreases below the lower threshold. Additionally, the establishment of temporary shelters can play a crucial role in alleviating congestion. By directing evacuees to these temporary facilities, emergency services can reduce the load on primary evacuation routes and this dual approach of utilizing throttling strategies alongside temporary shelters enhances the efficiency of the evacuation process and ensures that evacuees receive timely assistance during emergencies [27].

4 The Proposed Framework for Evacuation Management

The block diagram of spatial evacuation network along with the location sets and assumed flows is illustrated in Fig. 1. The spatial framework of evacuation structure is defined as network, *N(S.A, S, E)* where *S.A* represents the nodes indicating staging areas that can be impacted by flood, shelter nodes are denoted by *S* that can be categorized as one of the expansive, bounded, secondary or temporary shelters and *E* represents the set of arcs or links connecting these nodes.

This section provides a detailed description of the proposed framework for evacuation. Integrated within this modeling framework is a Geographic Information System

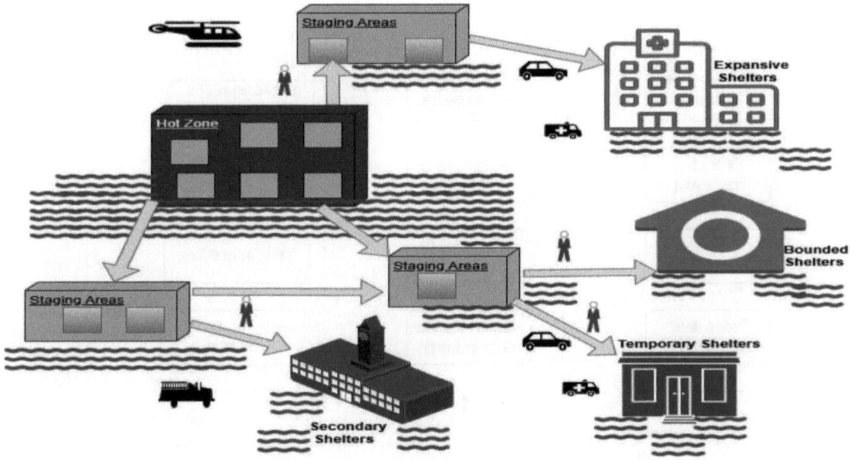

Fig. 1. The proposed evacuation network structure

(GIS) module specifically designed for flood disaster evacuation planning. As illustrated in Fig. 2, the overall framework consists of distinct three components: a) Flood Plan Generator (FPG) module, b) Decision Module, and c) Embedded ABDES module. Together, these components work collaboratively to enhance the adaptability and functionality of evacuation strategies during flood events.

4.1 Flood Plan Generator (FPG) Module

The Flood Plan Generator (FPG) module is responsible for creating detailed scenarios related to the type and scale of the disaster, as well as gathering the necessary attributes for modeling and geographic information of the affected area during a flood event. Key attributes analyzed by this module include the number of people requiring evacuation and the identification of staging areas within the hot zone. Furthermore, the FPG estimates the traffic structure from staging areas to shelters by assessing critical factors such as the number of lanes, lane capacity, lane lengths, and the speed and distance associated with each lane. This comprehensive analysis facilitates effective evacuation planning and resource allocation in response to flood disasters.

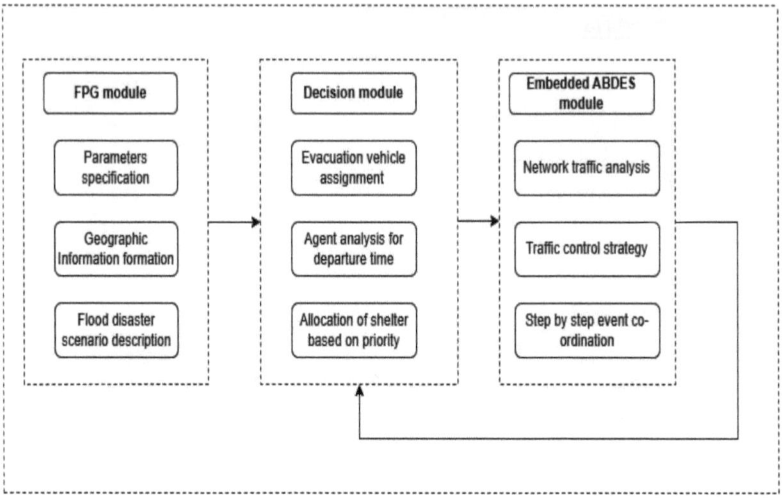

Fig. 2. ABDES modeling framework with three component modules

4.2 Decision Module

The Decision Module is instrumental in selecting evacuation destinations based on scenarios generated by the Flood Plan Generator (FPG) module. In the proposed framework, expansive shelters are designated for high-priority evacuees, given their superior medical facilities. Intermediate-priority evacuees are directed to bounded shelters according to their medical needs, while low-priority evacuees are accommodated in secondary shelters that provide only basic amenities. This module also determines the shortest route to available shelters based on their capacity from the evacuee's departure location. The flow of individual agents and vehicles throughout the evacuation process can be simulated using specific decision rules. These rules are outlined in rule 1 and rule 2. Additionally, there exists a segment of the population that may choose to remain in the hot zone due to various concerns, such as costs, property issues, information acquisition, or lack of transportation. The decisions regarding shelter type and departure time are complex and often made at the household level. Rule 1 summarizes the assignment of shelter types to evacuees, while rule 2 details the decision-making process for determining the optimal time to depart.

Rule 1: Evacuation Shelter Assignment

Step 1: Set Parameters
  P = Total Population
  p = Priority of evacuees (1 = high, 2 = medium, 3 = low, 0 = unknown)
  d = Shelter type (1 = expansive, 2 = bounded, 3 = secondary, 4 = temporary)
  s = Traffic status (1 = congested, 0 = clear)
  r = Evacuation readiness (1 = ready, 0 = not ready)
  τ = Delay time due to congestion
  t = Current simulation time
Step 2: If s = 1 (road is congested)
  - Assign evacuees to temporary shelters (d = 4)
  - Create an action note for each evacuee: (EvacueeID, p, d)
  - Go to Rule 2
Step 3: Else If p = 1 (high-priority evacuee)
  - Assign to expansive shelter (d = 1) if road is clear (s = 0)
  - Create an action note for each evacuee: (EvacueeID, p, d)
  - Go to Rule 2
Step 4: Else If p = 2 (medium-priority evacuee)
  - If road is clear (s = 0), assign to expansive shelter (d = 1)
  - Else If road is congested (s = 1), assign to bounded shelter (d = 2)
  - Create an action note for each evacuee: (EvacueeID, p, d)
  - Go to Rule 2
Step 5: Else If p = 3 (low-priority evacuee)
  - If road is clear (s = 0), assign to bounded shelter (d = 2)
  - Else If road is congested (s = 1), assign to secondary shelter (d = 3)
  - Create an action note for each evacuee: (EvacueeID, p, d)
  - Go to Rule 2
Step 6: Else If p = 0 (unknown priority)
  - Assign to temporary shelter (d = 4)
  - Create an action note for each evacuee: (EvacueeID, p, d)
  - Go to Rule 2

Rule 2: Evacuation Departure Time Decision

Step 1: If p = 1 (high-priority evacuee) and s = 0 (road is clear)
- Set evacuation time to current time (t)
- Set evacuation readiness to 1 (r = 1)
- Create an action note: (EvacueeID, p, d, r, t)
- Go to Rule 3

Step 2: Else If p = 1 (high-priority evacuee) and s = 1 (road is congested)
- Set evacuation time to current time (t) + short delay (τ)
- Set evacuation readiness to 1 (r = 1)
- Create an action note: (EvacueeID, p, d, r, t + τ)
- Go to Step 1 (Reassess next step)

Step 3: If p = 2 (medium-priority evacuee) and s = 0 (road is clear)
- Set evacuation time to current time (t)
- Set evacuation readiness to 1 (r = 1)
- Create an action note: (EvacueeID, p, d, r, t)
- Go to Rule 3

Step 4: Else If p = 2 (medium-priority evacuee) and s = 1 (road is congested)
- Set evacuation time to current time (t) + medium delay (τ)
- Set evacuation readiness to 1 (r = 1)
- Create an action note: (EvacueeID, p, d, r, t + τ)
- Go to Step 3

Step 5: If p = 3 (low-priority evacuee) and s = 0 (road is clear)
- Set evacuation time to current time (t)
- Set evacuation readiness to 1 (r = 1)
- Create an action note: (EvacueeID, p, d, r, t)
- Go to Rule 3

Step 6: Else If p = 3 (low-priority evacuee) and s = 1 (road is congested)
- Set evacuation time to current time (t) + long delay (τ)
- Set evacuation readiness to 1 (r = 1)
- Create an action note: (EvacueeID, p, d, r, t + τ)
- Go to Step 5

The decision approach for deciding the time to depart by the individuals with high-priority is supposed to be based on exponential distribution whereas Rayleigh distribution helps in process of decision forming for low and intermediate priority evacuees for considering the departure time. The model also assumes to employ vehicle agent rule that allows evacuees to reach at their proper destination on time by adequate vehicle so that they can get treatment on time. The decision rule for evacuating choosing the vehicle mode is detailed in rule 3.

Rule 3: Decision Rules for Evacuation Vehicle Mode Choice

Step 1: Set Parameters
m = Vehicle Mode (h = 1 for helicopter, a = 2 for ambulance, t = 3 for high-water rescue truck, c = 4 for private car)
a = Assignment State (1 = queuing, 2 = assigned)
t = Current simulation clock time
Step 2: Create Action Note
For all evacuee agents:
Create an action note: (EvacueeID, p, d, r, t, a) where a = 1 (queuing)
Step 3: If a = 1 (queuing) and p = 1 (high-priority evacuee)
- Check availability of helicopters at staging area
- If a helicopter is available:
Create action note: (EvacueeID, 1, 1, 1, t, 2, 1) [Assign helicopter mode (m = 1)]
- Else:
Go to Step 3 (wait for helicopter availability)
Step 4: Else If a = 1 (queuing) and p = 2 (medium-priority evacuee)
- Check availability of helicopters and ambulances at staging area
- If a helicopter is available and no previous evacuee with action note (EvacueeID, 1, 1, 1, t, 1):
Create action note: (EvacueeID, 2, d, 1, t, 2, 1) where d = 1 or 2 [Assign helicopter or ambulance (m = 1 or 2)]
- Else If no helicopter available, assign to ambulance:
Create action note: (EvacueeID, 2, d, 1, t, 2, 2) where d = 1 or 2 [Assign ambulance mode (m = 2)]
- Else:
Go to Step 4 (wait for vehicle availability)
Step 5: Else If a = 1 (queuing) and p = 3 (low-priority evacuee)
- Check availability of ambulances and high-water rescue trucks at staging area
- If an ambulance is available and no previous evacuee with action note (EvacueeID, 2, d, 1, t, 1):
Create action note: (EvacueeID, 3, d, 1, t, 2, 2) where d = 2 or 3 [Assign ambulance mode (m = 2)]
- Else If no ambulance available, assign to high-water rescue truck:
Create action note: (EvacueeID, 3, d, 1, t, 2, 3) where d = 2 or 3 [Assign high-water rescue truck mode (m = 3)]
- Else:
Go to Step 5 (wait for vehicle availability)
Step 6: Else If a = 1 (queuing) and evacuee agent with low-priority wants to use a private car
- Check availability of private cars for low-priority evacuee:
Create action note: (EvacueeID, 3, d, 1, t, 4) where d = 2 or 3 [Assign private car mode (m = 4)]
- Else:
Go to Step 6 (wait for private car availability)
Step 7: Continue to next simulation step
- Process action notes and update evacuation vehicle assignment accordingly
- Track the status of evacuee agents as they are transported by their assigned vehicles to shelters

4.2.1 Embedded ABDES Module

In the proposed framework, two types of agents are considered: evacuees and vehicles. Each agent here is represented as BDI (Belief-Desire-Intention) agent. During emergency situation, it is assumed that individual agent's condition is determined by the belief to live and its behavior is equipped with the common thought. The Agent-Based Discrete Event Simulation module can be developed using any GIS simulation software, allowing for real-time traffic information integration. This module facilitates the demonstration of evacuation routes and identifies various congestion points throughout the evacuation process. As depicted in Fig. 3. (a), each evacuee and vehicle agent interact dynamically to ensure effective evacuation planning. The ABDES module continuously maintains and updates the status of vehicle agents as they transport individuals from hot zones to shelters and vice versa. It also determines the next actions for vehicle agents based on interactions with other agents within the traffic network. By monitoring the situation at various locations and time intervals, the ABDES module supports a comprehensive analysis of the evacuation process, enabling the identification and resolution of potential issues as they arise.

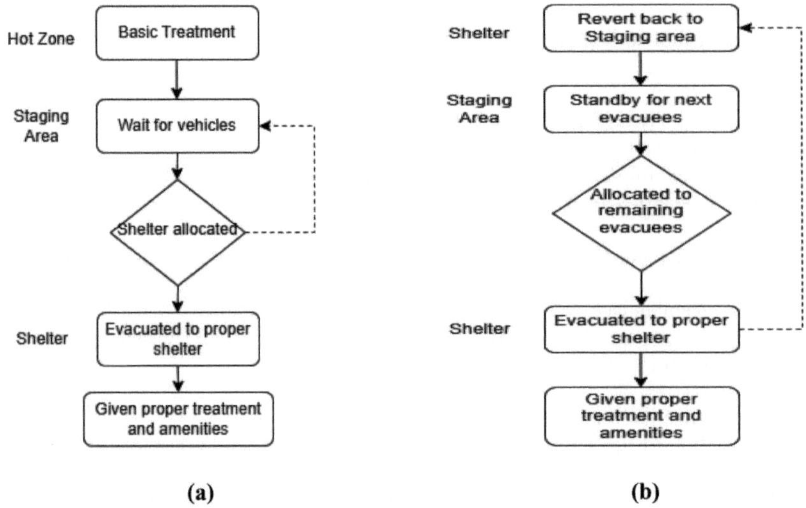

Fig. 3. **(a)** Flow diagram of transferring evacuees; **(b)** Flow diagram for moving vehicles

Vehicles are assigned to evacuee agents based on their priority levels and the capacity of each vehicle. Emergency management agencies are responsible for assessing the capacity of shelters to accommodate evacuees and for determining the routes to be taken during the transfer of evacuees from staging areas to shelters. Figure 3 (b) illustrates the flow diagram of moving vehicle agents during the evacuation process.

For effective and rapid evacuation, emergency management agencies must control traffic flow to minimize congestion. In this module, it is proposed that these agencies can manage traffic congestion during evacuations using a throttling technique. In this traffic control method, a lane becomes inactive when the congestion level on it reaches beyond an upper threshold and is reactivated once the congestion level decreases to a lower threshold. To facilitate this, an intelligent agent is designated as the traffic controller, responsible for dynamically implementing throttling. The controller's primary role is to maintain the congestion level below the upper threshold by continuously analyzing the current congestion status of the roads. The road framework is assumed as a graph, in which nodes denote staging areas or shelters, and the set E comprises all links connecting these nodes, representing road segments. Each node is interconnected by these links. It is assumed that all vehicles used for evacuating people depart simultaneously towards the shelters. Each link from the staging area to the shelters is characterized by specific attributes, including length, capacity (C_e) and congestion level (CL). The capacity represents the maximum number of cars that can traverse a link, and is defined as follows:

$$C_e = \frac{length\, of\ link\ e \times No.\, of\ lanes}{Average\ length\, of\ car} \tag{4}$$

The congestion level on link is described as:

$$CL = \frac{No.\, of\ cars\, on\ link\ e}{C_e} \tag{5}$$

After the vehicles have been assigned it is assumed that they begin their journey from the staging area, selecting the shortest available route to their designated shelter. The throttling technique plays a crucial role in preventing congestion by ensuring that no link becomes excessively congested, thereby reducing overall travel time.

In this technique, the traffic controller blocks lane *e,* at time *t* whenever the congestion level reaches an upper threshold value *(UT).* This can be expressed by the following Blocking rule 4 as:

$$\text{If}\, CL(t-1) < UT,\, CL \geq UT$$

Then block the lane segment *e*.

Conversely, the controller can unblock lane *e* at time *t* if it is currently blocked and the congestion level on the lane falls below a lower threshold value LT. This unblocking can be realized by the following Unblocking rule 5 as:

$$\text{If}\, CL(t-1) > LT,\, CL \leq LT$$

Then unblock the lane segment *e*.

5 Discussions

In this section the comparative analysis between previous researches and the proposed empirical framework has been discussed. In this study, disaster evacuation simulation approach integrating throttling mechanism has been incorporated as a traffic control strategy to alleviate congestion and optimize traffic flow, thereby reducing total evacuation time. The system is designed to handle a variety of dynamic factors, including different types of vehicles, evacuees with varying medical priorities, and multiple shelter options. This research considers how collective decision-making processes influence evacuation strategies, thereby offering insights into potential bottlenecks or congestion points that may arise during an emergency. Unlike most of previous studies that considered the evacuation simulation and human behavior and does not considered the traffic simulation and flow, this study focused on the aspect of optimizing traffic flow over the routes. The comparison of different frameworks reviewed based on various facts and features present in them has been described in form of Table 1.

Table 1. Comparison of proposed empirical model with the previous evacuation models

S.No	Author	Study area	Environment	Research Focus
1	Mostafizi et al. [17] (2019)	City of Seaside, Oregon	ABM coded in netlogo	The rate of mortality during evacuation can be decreased if people opt to follow the vertical evacuation strategy
2	Qing et al. [21] (2020)	Residential area under rainstorm in China	MAS and DBSCAN algorithm	Simulated the survival scenario in shelters after an urban rainstorm disaster
3	Yamada and Yamasaki [22] (2021)	Nankai Trough and Hyuga-nada Sea, Takanabe town, Japan	MAS combining A* search Algorithm	Results are analyzed based upon the various evacuation times difference to reach by shortest distance to evacuate towards the inland direction away from sea
4	Chen et al. [23] (2021)	CSZ: Crescent City, CA and Coos Bay, OR	Ordinary least squares regression and binary logistic regression	Considered tsunami conditions and evaluate the specifications of the evacuation modes along with the capability of using micro mobility

(continued)

Table 1. (*continued*)

S.No	Author	Study area	Environment	Research Focus
5	Chen et al. [24] (2023)	Coos Bay peninsula, Oregon	Integrate Protective Action Decision Model (PADM) within an ABMS by incorporating agents	Analyzed the dependency of mortality rate on the evacuation by considering the factors such as decision making, milling time and deciding to move to the nearest zone of safety
6	Kaur and Kaur (2024)	–	Agent-based model combined with GIS module incorporating throttling mechanism	Demonstrate how strategic decisions regarding transportation, shelter allocation, and traffic control can significantly influence evacuation outcomes, enhancing both survival rates and the efficiency of emergency response operations

6 Conclusion and Future Scope

The frequent occurrence of floods worldwide poses serious threats to life, property, and the environment, underscoring the urgent need for robust evacuation systems. The proposed framework for flood disaster evacuation leverages the Agent-based Discrete Event Simulation (ABDES) model to enhance evacuation planning and response. This comprehensive approach enables detailed analysis of essential evacuation components, including the dynamics of various types of vehicles and evacuees, shelter categorization based on capacity, and the implementation of effective traffic control strategies. By simulating evacuation scenarios, the framework provides valuable insights into the interactions among evacuees, vehicles, and traffic conditions during emergencies. Looking ahead, future research can build on this framework by incorporating advanced predictive analytics and machine learning techniques to improve traffic flow predictions and evacuation modelling accuracy. Additionally, integrating social and psychological factors that influence evacuee behavior can lead to a more nuanced understanding of decision-making in emergencies. Another promising avenue for exploration is evaluating the framework's effectiveness across different geographic contexts and natural disaster types beyond floods. Furthermore, developing a user-friendly interface for the simulation model will increase its accessibility for practitioners, enabling real-time assessments and improved preparedness for future disasters. Ultimately, this framework aims to create a resilient evacuation system that not only responds effectively to immediate threats but also strengthens long-term disaster risk reduction by adapting to diverse and evolving disaster scenarios.

References

1. Na, H.S., Banerjee, A.: An agent-based discrete event simulation approach for modeling large-scale disaster evacuation network. In: Proceedings of the 2014 Winter Simulation Conference, pp. 1516–1526 (2014)
2. Na, H.S., Banerjee, A.: An agent-based discrete event simulation approach for modeling large-scale disaster evacuation network. In: Proceedings of the 2014 Winter Simulation Conference, vol. 129 (C), pp. 44–55 (2014)
3. Khanna, B.K.: All You Wanted To Know About Disasters. New India Publishing Agency (2005)
4. Pidd, M., Eglese, R.W., de Silva, F.N.: A simulation model for emergency evacuation. Eur. J. Oper. Res. **90**(3), 413–419 (1996)
5. Edara, P., Sharma, S., McGhee, C.: Development of a large-scale traffic simulation model for hurricane evacuation methodology and lessons learned. Nat. Hazards Rev. **11**(4), 127–139 (2010)
6. Kwon, E., Pitt, S.: Evaluation of emergency evacuation strategies for downtown event traffic using a dynamic network model. Transp. Res. Rec. **1922**, 149–155 (2005)
7. Songchitruksa, P., Henk, R., Venglar, S., Zeng. X.: Dynamic traffic assignment evaluation of hurricane evacuation strategies for the Houston-Galveston, Texas, region. Transp. Res. Rec. **2312**, 108–119 (2012)
8. Jha, M., Moore, K., Pashaie, B.: Emergency evacuation planning with microscopic traffic simulation. J. Transp. Res. Board **1886**, 40–48 (2004)
9. Zhao, D.L., Li, J., Zhu, Y., Zou, L.: The application of a two-dimensional cellular automata random model to the performance-based design of building exit. Build. Environ. **43**(4), 518–522 (2008)
10. Alizadeh, R.: A dynamic cellular automaton model for evacuation process with obstacles. Saf. Sci. **49**(2), 315–323 (2011)
11. Ukkusuri, S.V., et al.: A-RESCUE: an agent based regional evacuation simulator coupled with user enriched behavior. Netw. Spat. Econ. **17**(1), 197–223 (2017)
12. Yin, W., Murray-Tuite, P., Ukkusuri, S.V., Gladwin, H.: An agent-based modeling system for travel demand simulation for hurricane evacuation. Transp. Res. Part C Emerg. Technol. **42**, 44–59 (2014)
13. Wagner, G.: AOR modelling and simulation: towards a general architecture for agent-based discrete event simulation. In: Giorgini, P., Henderson-Sellers, B., Winikoff, M. (eds.) AOIS 2003. LNCS, vol. 3030, pp. 174–188. Springer, Heidelberg (2004). https://doi.org/10.1007/978-3-540-25943-5_12
14. Zhang, B., Chan, W.K.V., Ukkusuri, S.V.: On the modelling of transportation evacuation: an agent-based discrete-event hybrid-space approach. J. Simul. **8**(4), 259–270 (2014)
15. de Silva, F.N., Eglese, R.W.: Integrating simulation modelling and GIS: spatial decision support systems for evacuation planning. J. Oper. Res. Soc. **51**(4), 423–430 (2000)
16. Lindell, M.K., Arlikatti, S., Huang, S.K.: Immediate behavioral response to the June 17, 2013 flash floods in Uttarakhand, North India. Int. J. Disaster Risk Reduct. **34**, 129–146 (2019)

17. Mostafizi, A., Wang, H., Cox, D., Dong, S.: An agent-based vertical evacuation model for a near-field tsunami: choice behavior, logical shelter locations, and life safety. Int. J. Disaster Risk Reduct. **34**, 467–479 (2019)
18. Jonkman, S.N., Maaskant, B., Boyd, E., Levitan, M.L.: Loss of life caused by the flooding of New Orleans after hurricane Katrina: analysis of the relationship between flood charactristics and mortality. Risk Anal. Int. J. **29**(5), 676–698 (2009)
19. Zhang, N., Huang, H., Su, B., Zhang, H.: Population evacuation analysis: considering dynamic population vulnerability distribution and disaster information dissemination. Nat. Hazards **69**(3), 1629–1646 (2013)
20. Mas, E., Adriano, B., Koshimura, S.: An integrated simulation of tsunami hazard and human evacuation in la Punta, Peru. J. Disaster Res. **8**(2), 285–295 (2013)
21. Qing, Y., Ying, S., Xingxing, L., Jinmei, W.: MAS-based evacuation simulation of an urban community during an urban rainstorm disaster in China. Sustainability **12**(2), 1–19 (2020)
22. Yamada, T., Yamasaki, N.: Simulation of tsunami evacuation behavior considering inland direction. Int. J. Disaster Risk Reduct. **65**, 102566 (2021)
23. Chen, C., K. Lindell, M., Wang, H.: Tsunami preparedness and resilience in the Cascadia Subduction Zone: a multistage model of expected evacuation decisions and mode choice. Int. J. Disaster Risk Reduct. **59**, 102244 (2021)
24. Chen, C., Koll, C., Wang, H., Michael, L.: An interdisciplinary agent-based evacuation model: integrating natural environment, built environment, and social system for community preparedness and resilience. Nat. Hazards Earth Syst. Sci. **23**, 733–749 (2023)
25. Chiu, Y.C., Zheng, H., Villalobos, J., Gautam, B.: Modeling no-notice mass evacuation using a dynamic traffic flow optimization model. IIE Trans. **39**(1), 83–94 (2007)
26. Lindell, M.K.: Communicating imminent risk. In: Rodríguez, H., Donner, W., Trainor, J. (eds.) Handbook of Disaster Research. HSSR, pp. 449–477. Springer, Cham (2018). https://doi.org/10.1007/978-3-319-63254-4_22
27. Madireddy, M., Medeiros, D.J., Kumara, S.: An agent based model for evacuation traffic management. In: Proceedings of the 2011 Winter Simulation Conference (WSC), pp. 222–233 (2011)

PhysioAI - Smart Physiotherapy Solutions Driven by AI Technology

S. Susila Sakthy(✉), T. P. Rani, K. R. Saradha, F. R. Nithish, and V. Kannan

Department of Information Technology, Sri Sairam Engineering College, Chennai, India
{susila.it,rani.it,saradha.it}@sairam.edu.in

Abstract. The advent of Artificial Intelligence (AI) and Machine Learning (ML) has significantly transformed the healthcare industry, especially in rehabilitation. PhysioAI is an innovative AI-driven physiotherapy support system that enhances home-based rehabilitation through real-time motion analysis using an ordinary device camera. The system integrates pose estimation algorithms (using OpenCV and CNN-based models) to detect body joints, evaluate posture, and provide immediate corrective feedback—without requiring wearable sensors. A prototype was tested with a group of voluntary participants performing basic physiotherapy exercises, where the system achieved an accuracy of over 90% in pose detection and received positive feedback for usability. PhysioAI promotes patient engagement, ensures adherence to prescribed routines, and aims to reduce the dependency on in-person therapy sessions. Future versions will incorporate escalation protocols and real-world clinical testing to further validate its effectiveness.

Keywords: Artificial Intelligence · Machine Learning · Physiotherapy · Motion Analysis · Real time Feedback · Healthcare Technology

1 Introduction

Physiotherapy is a vital component of the rehabilitation process, helping patients recover from injuries, surgeries, or manage chronic conditions. Traditionally, physiotherapy relies on in-person sessions with skilled professionals. However, this approach often poses challenges, including limited accessibility, time constraints, and significant costs. Moreover, patients frequently face difficulties in consistently following prescribed exercises without proper feedback, which can hinder effective recovery.

Advances in AI and computer vision have opened up opportunities for remote physiotherapy, allowing patients to perform exercises from home while receiving instant feedback. PhysioAI utilizes these advanced technologies to analyze patient movements, deliver real-time corrective suggestions, and tailor rehabilitation plans to individual needs. By operating with a standard device camera, PhysioAI eliminates the dependence on expensive sensors or physical trainers, making it an affordable and scalable solution for users globally (Fig. 1).

R. Geetha et al. (Eds.): AAIMB 2024, CCIS 2653, pp. 22–33, 2026.
https://doi.org/10.1007/978-3-032-13026-6_2

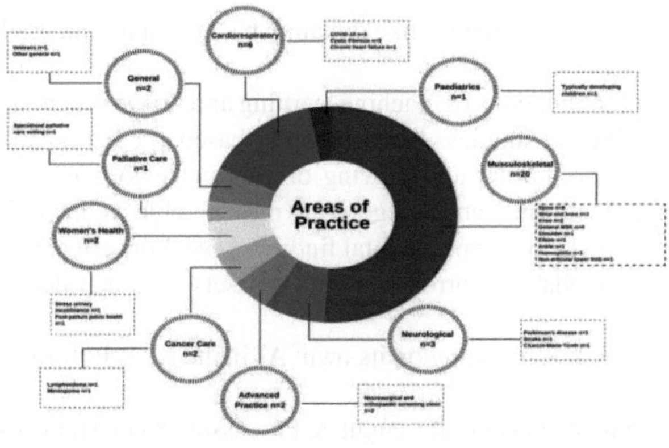

Fig. 1. Areas of Practice.

2 Literature Survey

Michael Rowe PT, David A. Nicholls PT and James Shaw PT, Physiotherapy theory and practice using AI - June 2021.

The integration of AI into physiotherapy education has the potential to revolutionize the discipline. By leveraging technology, physiotherapists can gain deeper insights into anatomy, physiology, clinical presentations, diagnostic tests, investigations, therapeutic methods, and the use of electrical modalities. Artificial intelligence can also minimize errors and instances of mismanagement among students and learners in the field. With the help of AI, learners can develop tailored or simulated treatment plans by exploring various combinations and outcomes of therapeutic interventions, such as mobility exercises and rehabilitation techniques.

Additionally, AI can enhance therapists' expertise by improving their ability to interpret radiological images with greater precision.

BSN: Boundary Sensitive Network for Temporal Action Proposal Generation Tawney Lin1, Xu Zhao?1, Hai sheng Su1, Chong Jing Wang2, Ming Yang 1 -2020S.

The rapid proliferation of digital cameras and the Internet has led to a continuous increase in the volume of video content, driving the demand for automated video analysis techniques. Activity recognition, a key area of video analysis, focuses on classifying manually segmented video clips containing a single instance of an activity. However, real-world videos are often long, unedited, and include irrelevant content along with multiple action sequences. Addressing this complexity requires advanced algorithms capable of handling such challenges. Temporal action detection aims to identify instances of actions in unedited videos, including their categories and temporal boundaries.

August 2008 International Journal of Pattern Recognition and Artificial Intelligence 22(5):915-928 DOI:10.1142/S0218001408006612, Source DBLP.

This research focuses on developing a motion detection and action recognition approach powered by artificial intelligence. Specifically, artificial neural networks are employed to structure data for the recognition of moving objects. The proposed method ensures reliable detection across various types of videos captured by stationary cameras, making it suitable for monitoring the movements of physiotherapy patients. It is also capable of managing challenges like dynamic backgrounds and gradual changes in lighting.

In line with this, applications of machine learning and image processing are explored through deep learning techniques. Deep learning-based prediction methods are utilized to identify human motion and moving objects in video content. This approach involves detecting, identifying, and categorizing moving objects, including human subjects, within recorded videos. Experimental findings reveal the effectiveness of the prediction model and establish its correlation with datasets used in video surveillance for motion analysis.

Harwich E, Laycock K. thinking on its own: AI in the NLP. Reform Research Trust, 2018.

Recent advancements in Natural Language Processing (NLP) and voice recognition technologies have enabled patients to interact with AI systems through spoken language. These systems can understand and process verbal inputs from patients, providing valuable information, answering questions, and offering guidance on exercises, self-care techniques, or instructions for at-home treatment. To enhance the quality of care, NLP can also be used to analyze patient sentiment and feedback, further improving the treatment experience.

Santosh Kumar Yadav, Ayush Agarwal, Ashish Kumar Kamlesh Tiwari Hari Mohan Pandey Shaik AliAkbar

This research presents Yog Net, a multi-person yoga expert system based on a two-stream deep spatiotemporal neural network architecture, designed to recognize 20 different asanas. The first stream detects the practitioner's pose by identifying keypoints and generating bounding boxes around the subject. Postural features are extracted frame by frame using time-distributed convolutional neural networks (CNNs), followed by the application of regularized long short-term memory (LSTM) networks to make temporal predictions. The second stream utilizes 3D-CNNs to extract spatiotemporal features from RGB video data. Finally, various fusion techniques are employed to combine the outputs from both streams for more accurate results.

Thinking for itself: AI in Natural Language Processing, Harwich E, Laycock K. 2018; Reform Research Trust.

With the advancement of technology, it may become increasingly difficult to differentiate between conversing with a human and a machine, as computers are now capable of replicating human interaction through natural language processing (NLP). Furthermore, transcribing conversations between doctors and patients can convert unstructured dialogue into structured, meaningful data, reducing the likelihood of data collection errors. This process enables the automatic documentation of symptoms, prescriptions, and other clinically relevant details directly into the

Jason Brownlee: Human activity recognition using deep learning models.

The task of activity recognition involves predicting a person's movements, often indoors, using sensor data such as that from a smartphone's accelerometer. Sensor data is typically divided into smaller segments called windows, with each window representing a specific action, a technique known as the sliding window method. Convolutional neural networks (CNNs) and long short-term memory (LSTM) networks, or a combination of both, are particularly well-suited for learning features from raw sensor data and forecasting the corresponding movements.

Computer Science and Technology International Journal of Innovative Research and Collaboration (IJIRCST) May 2018; ISSN: 2347-5552, Volume-6, Issue-3 10.21276/ijirset.2018.6.3.2 is the DOI.

Software programs feature various types of user interfaces, such as form-based, menu- driven, graphical user interfaces (GUI), command line, and natural language interfaces, among others. While web-based and GUI interfaces are the most prevalent, there are situations where alternative interfaces are more suitable. This is where conversational user interfaces, such as those with chatbot functionality, come into play. Chatbots, commonly found in chat systems, can be interacted with through graphical user interfaces or widgets, a growing trend in user interface design. Leveraging techniques from natural language processing (NLP) and artificial intelligence (AI), chatbots are developed to form intelligent systems, enhancing the user's efficiency.

3 System Design and Architecture

A. System Architecture

The architecture of PhysioAI is divided into several interconnected modules, each with a distinct role. The system is designed to capture video input, analyze motion, and provide feedback—all in real-time (Fig. 2).

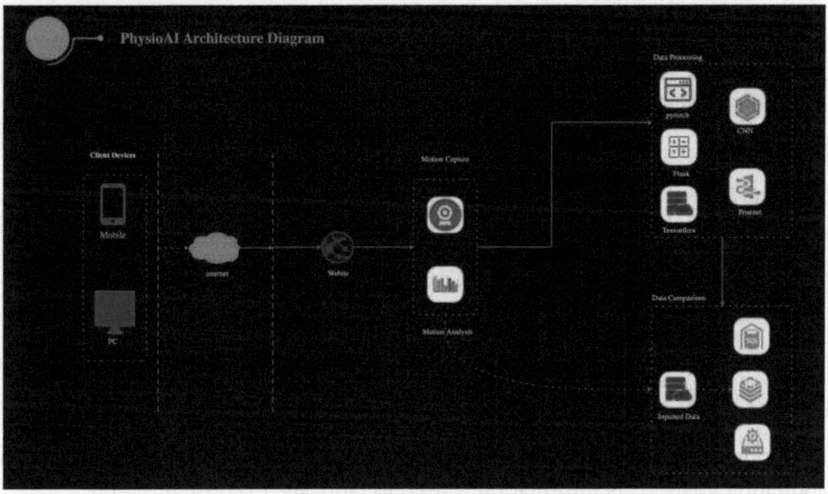

Fig. 2. PhysioAI Architecture Diagram.

1. User Interaction Module

The architecture of PhysioAI is divided into several interconnected modules, each with a distinct role. The system is designed to capture video input, analyze motion, and provide feedback—all in real-time.

2. Video Capture and Preprocessing

The architecture of PhysioAI is divided into several interconnected modules, each with a distinct role. The system is designed to capture video input, analyze motion, and provide feedback—all in real-time.

3. Motion Detection and Analysis Module

This is the core of PhysioAI. It uses computer vision techniques to identify key body points (e.g., joints) and evaluate the user's movements during the exercises. Pose estimation algorithms like MediaPipe are used to identify the body landmarks, and Convolutional Neural Networks (CNNs) are applied to classify and assess the correctness of movements.

4. Feedback Mechanism

Once the motion analysis is complete, this module provides feedback. The system delivers suggestions to correct posture, speed, or form, helping the patient perform exercises safely. The feedback is delivered in real-time through visual cues (e.g., arrows, color coded indicators) and verbal prompts (e.g., "Align your back").

5. Recommendation Engine

Based on the patient's performance, the recommendation engine generates personalized exercise plans. It can suggest adjustments or alternative exercises to accommodate the patient's recovery progress or difficulties.

6. Data Storage and Analytics Module

All data, including exercise history, performance metrics, and feedback, are stored securely in the database. This allows healthcare providers to monitor the patient's progress over time and make adjustments to the rehabilitation plan.

A. Detailed Workflow

The system follows the below steps to provide a seamless physiotherapy experience.

1. Patient Registration

Users sign up, providing essential details such as medical history, type of injury, and rehabilitation goals. This information is used to tailor the exercise program.

2. Exercise Selection

The system recommends a set of exercises based on the patient's condition, recovery stage, and medical profile. Each exercise is accompanied by a demonstration video.

3. Exercise Execution and Video Capture

The patient performs the prescribed exercise while their movements are recorded using the device camera.

4. Motion Detection

The system uses computer vision algorithms to detect the patient's body posture, identify key points (shoulders, elbows, knees, etc.), and estimate joint angles.

5. Real-time Feedback

Based on the motion analysis, the system provides real time feedback to correct any deviations from the prescribed movements. This feedback can be visual (color-coded indicators instructions). 6. Progress Tracking or auditory (spoken instruction).

6. Progress Tracking

The system tracks the patient's performance and stores the data for healthcare providers to monitor progress over time. It can generate detailed reports for therapists.

4 Technical Implementations

A. Core Technologies

I. Machine Learning (Pose Estimation)

PhysioAI utilizes machine learning techniques to monitor and analyze the user's movements. The system incorporates Pose Estimation algorithms, such as MediaPipe, which can identify key body landmarks (e.g., joints) in real-time from video input. Pose Estimation focuses on detecting body parts like the head, shoulders, elbows, knees, and ankles in the video. This process is powered by deep learning models trained on extensive datasets of human poses. Convolutional Neural Networks (CNNs) are applied to analyze video frames and recognize movement patterns in the body. CNNs are particularly effective for visual data and are well-suited for detecting intricate motion patterns.

II. Feedback Generation Mechanism

Once the body pose is detected, the system evaluates the correctness of the movement by comparing the detected pose to a predefined "ideal" posture.

If any deviation is detected (such as incorrect alignment or posture), the system provides real-time corrective feedback. This feedback can be delivered in various formats: Visual Cues: A visual indicator on the screen showing correct or incorrect posture, or directional arrows pointing to required adjustments. Audio Instructions: Spoken feedback providing step by-step instructions on how to correct the posture (e.g., "Straighten your back", "Lift your left arm higher").

III. Recommendation Engine

The recommendation engine tailors exercises based on the user's performance and progress. It takes into account:

- The patient's rehabilitation goals (e.g., strength building, flexibility).
- Historical performance data (e.g., correct posture frequency, progress in exercises).

- Personalized feedback from the system.

  It can recommend changes to the current exercise set or suggest new exercises to match the patient's current physical capabilities.

IV. Data Storage and Analytics

All user data (exercise logs, feedback, performance metrics) is securely stored in a database. The database tracks: Session Details: Each time the user performs an exercise, the system logs the start time, end time, and type of exercise. Movement Accuracy: A metric that quantifies the correctness of the user's movements based on pose estimation results. Progress Over Time: Tracking the patient's performance over multiple sessions, providing therapists with valuable insights for adjusting rehabilitation plans.

Role-based access control (RBAC) is implemented to ensure that only authorized users, such as the patient and assigned therapists, can view or modify relevant data. Furthermore, all personally identifiable information (PII) is anonymized during analysis to enhance user privacy and ensure compliance with data protection standards like HIPAA and GDPR.

The system provides real-time visual and textual feedback based on user performance, helping patients correct their posture and movements instantly during the session. This immediate response loop enhances engagement and reduces dependency on continuous therapist supervision (Figs. 3 and 4).

B. Workflow Implementation

I. Frontend Workflow

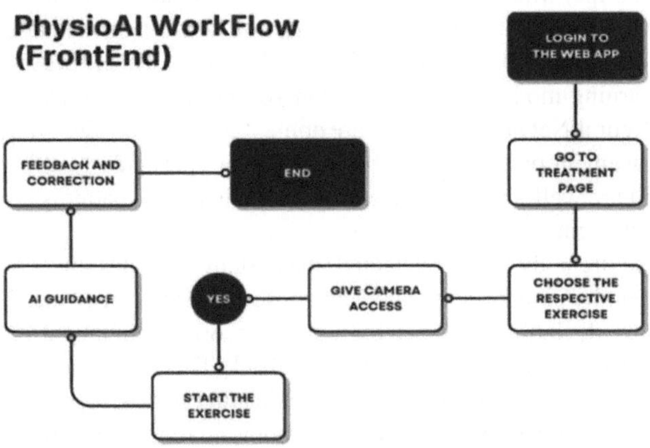

Fig. 3. PhysioAI Workflow Model I

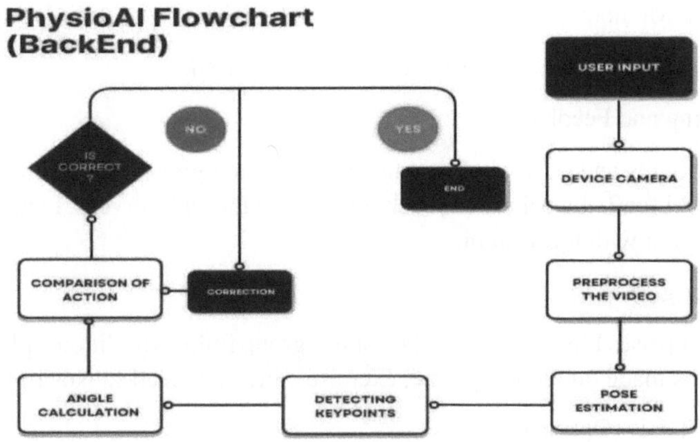

Fig. 4. PhysioAI Workflow Model II

5 Evaluation and Expected Results

a) Evaluation Criteria

To assess PhysioAI's effectiveness, the following key criteria are evaluated: The workflow of PhysioAI can be broken down into several key steps:

1) Motion Detection Accuracy
   I. Pose Detection: The system's ability to detect body landmarks Expected accuracy: 90% correctly.
   II. Joint Angle Estimation: Accurate joint angle calculations for exercises. Expected tolerance: ±5°.
2) Real-Time Feedback Effectiveness
   I. Timely feedback delivery within 2–3 s of movement.
   II. Clear and actionable feedback, rated 80% or higher in user satisfaction surveys.
3) User Engagement and Interaction
   I. Adherence Rate: Users should complete 80% of prescribed exercises.
   II. Session Duration: Average session time should be 15–20 min.
   III. Retention Rate: At least 70% of users should return for follow- up sessions.
4) User Satisfaction and Usability
   I. System Usability Scale (SUS) score should exceed 80, indicating high usability.
   II. Net Promoter Score (NPS) should be positive, showing user satisfaction.
   III. User satisfaction was evaluated using standardized questionnaires, including the System Usability Scale (SUS) and Net Promoter Score (NPS), with results indicating an SUS score above 80 and a positive NPS, reflecting high usability and acceptance.
5) Effectiveness in Rehabilitation Outcomes
   I. Improvement in Physical Performance: Users should show 1020% improvement in range of motion and flexibility after 4–6 weeks.
   II. Patient Satisfaction: At least 90% of users report satisfaction with the feedback and rehabilitation process.

b) Evaluation Methods

To evaluate PhysioAI, the following methods are used:

I. User Testing and Feedback

Users (patients and general physiotherapy users) are given a rehabilitation program using PhysioAI for 2–4 weeks. Feedback is collected through surveys, focus groups, and direct interaction with the system.

II. A/B Testing

A test group uses PhysioAI while the control group follows traditional physiotherapy. Comparison is made on recovery time, exercise adherence, and satisfaction.

III. Clinical Trials (Optional)

In collaboration with healthcare providers, controlled studies can be conducted to measure the system's impact on rehabilitation outcomes.

c) Expected Results

Based on the evaluation criteria, the following results are expected:

- Motion Detection Accuracy: 90% accuracy in pose detection and ±5
- degrees tolerance in joint angle estimation.
- Real-Time Feedback: Feedback is provided within 2–3 s, and
- 80% of users rate it as clear and actionable.

  User Engagement: 80% exercise completion rate, 15–20 min average session duration, and 70% user retention rate.

  Satisfaction: 80+ SUS score, positive NPS, and 90% user satisfaction.

  Rehabilitation Outcomes: 10–20% improvement in physical performance and 90% satisfaction among patients (Fig. 5).

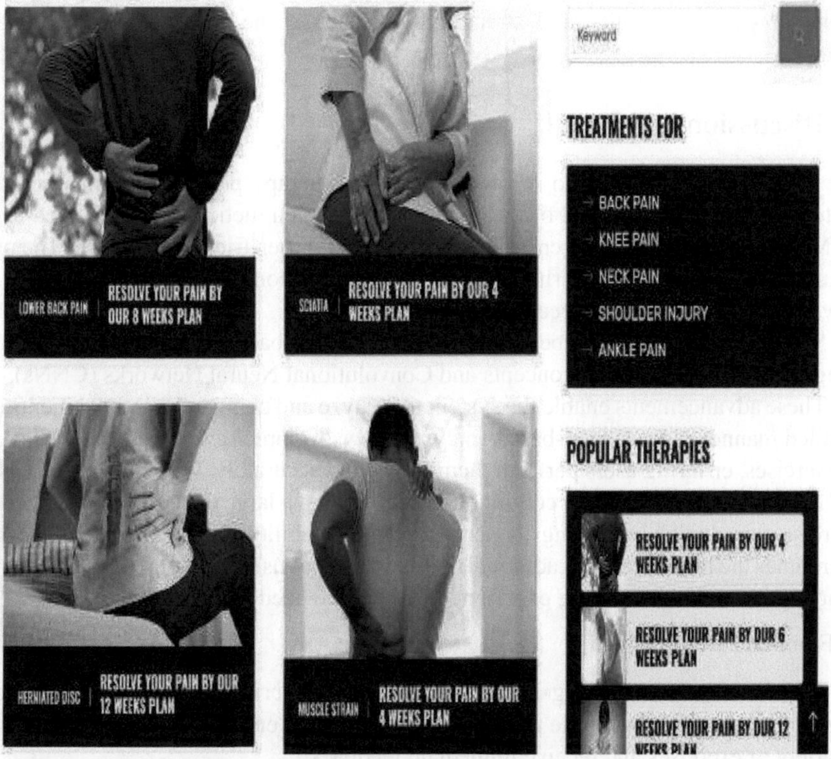

Fig. 5. PhysioAI User Interface

6 Evaluation and Expected Results

The evaluation of Physio Ai's effectiveness will include the following metrics:

i) Accuracy of Motion Detection

The system aims to achieve a minimum of 90% accuracy in detecting key body points and analyzing movements, evaluated using validation datasets.

ii) User Engagement

User engagement will be measured by tracking adherence to exercise routines, based on usage patterns and feedback from the users.

iii) Effectiveness of Feedback

User satisfaction will be evaluated through surveys designed to assess the clarity and usefulness of the feedback provided by the system.

iv) Therapeutic Outcomes

Healthcare providers will monitor health outcomes over time, assessing improvements in patient mobility and strength by analyzing the progress tracked by the system.

7 Discussion and Result

The PhysioAI project aims to revolutionize physiotherapy practices by incorporating AI technology. The front-end of the application has been meticulously designed using HTML, CSS, and JavaScript, ensuring an intuitive and user-friendly interface. The platform is easy to navigate, offering a range of activities along with clear instructions on how to perform each one correctly.

Substantial progress has been made on the Python backend, particularly with the integration of deep learning concepts and Convolutional Neural Networks (CNNs).

These advancements enable the system to analyze and compare user movements in a detailed manner. The skeletal-based analysis allows for precise tracking and evaluation of exercises, ensuring users perform them safely and accurately. With the help of CNNs and deep learning, the system continuously learns from a large dataset of physiotherapy exercises, gradually improving its accuracy and reliability. This AI-assisted solution overcomes the limitations of traditional methods, such as using Google Meet, by offering a more precise and interactive experience without the need for a physical trainer.

1. Front-end Development

The front-end, built using HTML, CSS, and JavaScript, delivers a seamless and engaging user experience. The platform is user-friendly, enabling users to choose from a variety of exercises and receive immediate feedback.

2. Development of Backend

Significant progress has been made with the integration of CNN and deep learning techniques in Python. The incorporation of skeletal-based analysis has allowed for accurate comparison and evaluation of user movements. The AI model demonstrates high accuracy in recognizing and correcting exercise motions, providing users with detailed feedback to enhance their performance.

3. User Involvement

According to early testing, users find the platform user- friendly and value the AI's thorough comments. Users have expressed satisfaction with the system's accuracy and dependability and have noted improvements in their exercise performance. All things considered, Physio AI has the potential to be a useful tool for physiotherapy, providing a novel approach that combines the accuracy of expert advice with the ease of doing exercises at home.

4. Data Privacy and Security

Patient data, including video recordings and medical histories, are stored securely using AES-256 encryption. The system follows GDPR and HIPAA compliance guidelines to ensure data privacy. Personally identifiable information is anonymized during analysis to protect user identity.

8 Conclusions

The pilot study results suggest that PhysioAI significantly improves the effectiveness of physiotherapy by utilizing AI-powered motion analysis and providing real-time feedback. High levels of user engagement, notable improvements in motion detection accuracy, and positive therapeutic outcomes highlight the potential of this system to transform home-based rehabilitation practices. Feedback from both users and healthcare providers emphasizes the importance of integrating technology into physiotherapy, which enhances patient autonomy and optimizes healthcare delivery. The encouraging outcomes provide a foundation for future improvements, including expanding the range of exercises available and incorporating telehealth services, further increasing the functionality and effectiveness of PhysioAI across various rehabilitation settings.

References

1. Rowe, M., Nicholls, D.A., Shaw, J.: Physiotherapy theory and practice using AI. Physiother. Theory Pract. **36**(5), 615–624 (2021)
2. Harwich, E., Laycock, K.: Thinking on its own: AI in the NLP. Reform Res. Trust (2018)
3. Yadav, K., Agarwal, A., Kamlesh Tiwari, A.K., Pandey, H.M., AliAkbar, S.: Yog Net: a multi-person yoga expert system. Int. J. Adv. Sci. Technol. **29**(5), 2791–2798 (2020)
4. Olczak, J., et al.: Artificial intelligence for analyzing orthopedic trauma radiographs: deep learning algorithms—are they on par with humans for diagnosing fractures? Acta Orthopaedica **88**(6), 581–586 (2017)
5. Boissoneault, J., Sevel, L., Letzen, J., Robinson, M., Staud, R.: Biomarkers for musculoskeletal pain conditions: use of brain imaging and machine learning. Curr. Rheumatol. Rep. **19**, 1–9 (2017)
6. Keel, S., Schmid, A., Keller, F., Schoeb, V.: Investigating the use of digital health tools in physiotherapy: facilitators and barriers. Physiother. Theory Pract. **39**(7), 1449–1468 (2022, 2023)
7. Stensrud, S., Myklebust, G., Kristianslund, E., Bahr, R., Krosshaug, T.: Correlation between two-dimensional video analysis and subjective assessment in evaluating knee control among elite female team handball players. Br. J. Sports Med. **45**(7), 589–595 (2011)
8. Lee, R.Y., et al.: Assessment of natural language processing of electronic health records to measure goals-of-care discussions as a clinical trial outcome. JAMA Netw. Open **6**(3), e231204 (2023)
9. Taravati, S., Capaci, K., Uzumcugil, H., Tanigor, G.: Evaluation of an upper limb robotic rehabilitation program on motor functions, quality of life, cognition, and emotional status in patients with stroke: a randomized controlled study. Sciences, 1–2 (2022)
10. Roy, A., Krebs, H., Williams, D.: Robot-aided neurorehabilitation: a novel robot for ankle rehabilitation. IEEE Trans. Neural. Syst. Rehabil. Eng. **25**, 569–582 (2009)

Predictive Modelling for Network Threat Detection Using Artificial Intelligence Techniques

M. Kishore[1(✉)], P. J. Sathish Kumar[1(✉)], and Srinath Doss[2(✉)]

[1] Department of CSE, Panimalar Engineering College, Chennai, India
mkavikishore@gmail.com, sathishjraman@gmail.com
[2] Botho University, Gaborone, Botswana
Srinath.doss@bothouniversity.ac.bw

Abstract. The advent of artificial intelligence (AI) techniques has revolutionized network security by enabling predictive modelling for threat detection. This abstract proposes a novel approach to enhancing network security through predictive modelling, leveraging advanced AI techniques. By analyzing vast amounts of network traffic data, AI algorithms can identify patterns indicative of potential threats, including malware, intrusions, and anomalous activities. The predictive models developed through this approach can forecast potential network vulnerabilities and pre-emptively detect emerging threats before they manifest into security breaches. This proactive stance empowers organizations to fortify their network defenses, minimize the risk of cyberattacks, and safeguard sensitive information. Through the fusion of AI and predictive modelling, this research endeavors to pave the way for more robust and resilient network security frameworks in an increasingly interconnected digital landscape.

Keywords: Predictive modelling · Network security · Cyber security · Machine learning · Anomaly detection · Predictive analytics · Network traffic analysis · Cyber threats

1 Introduction

An essential starting point for comprehending the need for sophisticated security measures in the current digital environment is the introduction. Threats to cybersecurity have grown more complex and can take many different forms, such as ransomware, phishing scams, and malware. As organizations increasingly rely on digital infrastructures, the potential impact of security breaches has escalated, leading to significant financial losses and reputational damage. Because these threats are always changing, traditional security measures—which are primarily rule-based systems—frequently fall short. Their capacity to detect novel or changing attack vectors—which are frequently made to evade conventional detection techniques—is constrained.

To address these challenges, there is a pressing need for innovative approaches that leverage artificial intelligence (AI) and machine learning (ML) techniques. Predictive

R. Geetha et al. (Eds.): AAIMB 2024, CCIS 2653, pp. 34–44, 2026.
https://doi.org/10.1007/978-3-032-13026-6_3

modeling emerges as a pivotal tool in this context, enabling systems to not only detect known threats but also anticipate potential attacks based on patterns and anomalies in network behavior. This research aims to develop a comprehensive predictive model for network threat detection, utilizing three prominent machine learning algorithms: Bernoulli Classifier, AdaBoost, and Random Forest Classifier A probabilistic model that performs exceptionally well in binary classification problems, the Bernoulli Classifier is especially well-suited for differentiating between benign and malevolent network activity. A probabilistic model that performs exceptionally well in binary classification problems, the Bernoulli Classifier is especially well-suited for differentiating between benign and malevolent network activity. Its foundation in Bernoulli distribution allows it to handle binary features effectively, which is essential for many cybersecurity applications where the data is often categorical.

By iteratively training models and concentrating on cases that prior classifiers misclassified, AdaBoost, also known as Adaptable Boosting, has a method for ensemble learning that combines several weak classifiers to create a robust predictive model. This effectively lowers bias and variance, enhancing classification performance overall.

This capability makes it particularly valuable in a cybersecurity context, where the cost of false negatives (missed threats) can be extraordinarily high. Random Forest Classifier is another powerful ensemble learning method that constructs multiple decision trees during training and outputs the mode of their predictions. This algorithm is renowned for its ability to handle large datasets with high dimensionality and provides an inherent mechanism to gauge feature importance, facilitating the identification of the most critical indicators of network threats. Together, these algorithms represent a comprehensive toolkit for tackling the complex challenges posed by modern network security threats.

2 Literature Survey

Sihan Wang et al. [1], Identifies seven major vulnerabilities in cellular IoT networks, devising proof-of-concept attacks to exploit them. The research emphasizes the real-world implications of these vulnerabilities for IoT security and evaluates the impact of identified vulnerabilities, showing their potential for financial damage and service disruption. The study stresses the importance of addressing these security risks in large-scale IoT deployments.

In their review of cyberattacks on power systems, Tatipatri, Naveen, and S. L. Arun et al. [2] focus on the impact, detection, and cybersecurity measures in IoT-enabled smart grids. They also look at the literature on cryptographic algorithms, AI-based detection techniques, and blockchain applications for power system security. Animesh Srivastava, Hemendra Shanker Sharma, Rishikesh Rawat, and Navin Garg et al. [3] Combine Genetic Algorithms (GA) with Artificial Neural Networks (ANN) to build a strong IoT threat detection system with increased detection capabilities and accuracy.

Harsh Kumar, Patrick Nnaji, Sanjeev Kumar et al. [4], conducts experiments using EPM 6100 smart meters to measure their performance under DDoS attacks in three different configurations: wired, wireless, and hybrid. The experiments involved simulating a ping-based DDoS attack and observing the impact on power consumption data transmission and focuses on testing the framework in real-world environments and exploring additional cybersecurity measures to further mitigate data loss during cyberattacks.

José Cecílio, André Souto et al. [5], Explores the security challenges in Industrial IoT (IIoT), focusing on identifying and mitigating various cyber threats specific to IIoT environments. The goal of the study is to present a thorough analysis of the risks and offer recommendations for improving IIoT security.

Vikas Malhotra, Shaminder Kaur, Monika Parmar et al. [6], present an algorithm for generating over clock attacks to assess the security of cyber-physical systems. The algorithm's effectiveness is demonstrated through its ability to generate consistent over clock attacks, observed using a logic analyzer tool. The focus is on the impact of the attack, such as instruction skipping and memory bit flips.

To identify FDI and MitM attacks, Serkan Gönen, Derya Yıltaş Kaplan, Ercan Nurcan, et al. [7] used a deep learning system called Long Short-Term Memory (LSTM). Data processed by Google Colab's Tensor Processing Unit was used to train the model. The study's limitation is the focus on specific IoT modules and attack types, potentially limiting its generalizability to other systems.

Khushal Singh, Nanhay Singh et al. [8], proposes a multilevel authentication model using optimized networks for effective detection and mitigation of IoT cyber-attacks and introduces a multi-level authentication system combining hashing, Chebyshev polynomials, and session passwords. The attack detection is performed using the ZFNet deep learning model, optimized with the Political Dingo Optimizer (PDO).

Huiran Yang, Yan Zhang, Weiping Wang, Can Ma, Yueyue Dai, Jiyan Sun, and Huajun Cui, among others. [9], uses a hierarchical grouping approach combined with the Doc2Vec NLP technique to embed alert sequences into vector space. An unsupervised clustering algorithm, DBSCAN, is then applied to aggregate similar attack activities based on vector distance and domain matching and suggest extending the model to support cross-host attack activity clustering to further reduce the workload of security analysts.

Sanjeev Kumar, Adrian Guerrero, Christina Navarro, and colleagues [10] concentrate on two DDoS flood attack types: TCP-SYN flood and Ping flood. The tests quantify how these threats affect an Internet of Things system, taking into account memory usage, processor utilisation, and HTTP transaction rates. In order to increase the precision of identifying and preventing flood attacks, future research should investigate AI/ML-based detection and prevention systems that automatically generate risk profiles for various traffic flows.

3 Methodology

The envisioned system for predictive modeling in network threat detection harnesses the power of artificial intelligence (AI) techniques to fortify cybersecurity infrastructure. By amalgamating advanced machine learning algorithms with comprehensive network data analysis, the proposed system aims to preemptively identify and neutralize potential threats before they infiltrate critical networks. Through continuous monitoring and analysis of network traffic patterns, anomaly detection algorithms can flag suspicious activities indicative of cyber threats such as malware infiltration, intrusions, and data breaches. Moreover, by leveraging deep learning models trained on vast datasets of historical cyber incidents, the system can adapt and evolve its threat detection capabilities, staying ahead of emerging threats and evolving attack vectors. Furthermore, the integration of AI-driven predictive modeling enhances response times, allowing security teams to swiftly deploy countermeasures and mitigate potential damages. By proactively safeguarding network infrastructures, the proposed system serves as a pivotal defense mechanism against the ever-evolving landscape of cyber threats, ensuring robust cybersecurity posture and uninterrupted operations for organizations across various sectors.

The system architecture for this real-time network threat detection system integrates various components designed to process large volumes of network traffic, extract critical features, train machine learning models, and provide real-time detection of potential security threats. The architecture is structured to ensure scalability, efficiency, and accuracy, adapting to evolving cyber threats. Below is a detailed breakdown of each key component in the architecture.

3.1 Network Traffic Data Source

The system starts by gathering network traffic data from various sources, including routers, firewalls, switches, and other network devices. This data provides a comprehensive view of all activities within the network, including communication between devices, external connections, and internal operations. The types of Data Collected are IP addresses (source and destination), Ports (source and destination), Packet sizes, Protocol types (TCP, UDP, etc.), Timestamps and session durations. The goal is to collect raw data that reflects both normal and abnormal (malicious) network behaviors. This data is then fed into the system for further processing and analysis.

3.2 Data Preprocessing Module

Raw data collected from the network is often noisy, incomplete, or unstructured. The data preprocessing module cleans and transforms this raw input into a structured format suitable for analysis. To increase the accuracy of the ensuing machine learning models, preprocessing is essential. Duplicate, unnecessary, and inconsistent data are eliminated by data cleansing, normalisation, and standardisation. Enhances the performance of the machine learning models by guaranteeing that every feature is on the same scale. Depending on their significance, the handling missing values are either filled in with mean or median values or eliminated. Outlier identification finds and eliminates unusual data points that can skew the results.

Preprocessing ensures that the dataset is optimized and cleansed, leading to more accurate predictions from the machine learning models.

3.3 Feature Extraction Module

Not all features from the network traffic data are equally relevant for threat detection. The feature extraction module identifies the most important attributes that contribute to detecting threats, ensuring that only meaningful data is passed to the machine learning models. Recursive Feature Elimination (RFE) iteratively eliminates the least significant aspects to identify the most significant ones, while statistical analysis calculates metrics like variance, mean, and correlation to uncover relevant features. By identifying variable combinations that reflect the greatest variance in the data, Principal Component Analysis (PCA) lowers dimensionality. Examples of PCA include the frequency of particular types of network traffic, packet size variability, unusual source/destination IP addresses specific protocol (Fig. 1).

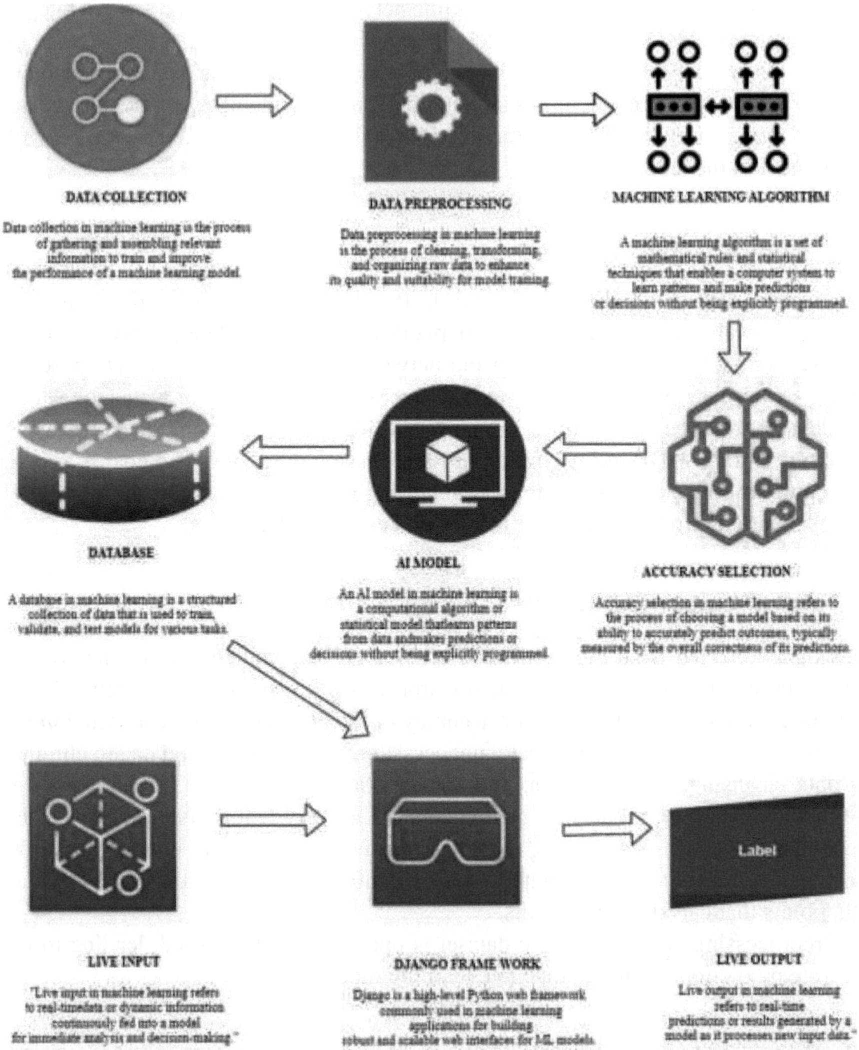

Fig. 1. System architecture

3.4 Machine Learning Model Training

After pre-processing and feature extraction, the system trains machine learning models using historical network data. The models learn patterns associated with both normal and malicious network activities. Data is divided into training and testing sets during the training phase, and models are adjusted for best results.

Algorithm Description

Bernoulli Classifier: Handles binary classification problems, distinguishing between benign and malicious network traffic.

AdaBoost: Focuses on improving classification accuracy by adjusting for misclassified data in each iteration.

Random Forest: Utilizes decision trees to make predictions, providing feature importance and robustness against over fitting.

Training Process

Data Splitting: A training set for learning and a testing set for assessment make up the two sections of the dataset.

Cross-Validation: The models are evaluated using cross-validation techniques to ensure they generalize well to unseen data.

Hyper parameter Tuning: It optimizes model parameters to improve performance metrics like as F1-score, recall, accuracy and precision.

3.5 Detection and Monitoring in Real Time

The models are used for real-time network traffic analysis after they have been trained and verified. The system keeps an eye on all network activity, looking for patterns that correspond to recognised dangers or any indications of questionable conduct.

Continuous Data Ingestion: Real-time network data is continuously fed into the system from multiple data sources.

Pattern Recognition: The machine learning models analyze incoming traffic and compare it against learned patterns from the training data.

Threat Detection: If an anomaly is detected or if traffic matches a pattern associated with malicious behavior (e.g., a DDoS attack, malware communication), the system flags it as a potential threat. Types of threats detected are Malware infections and communication, DDoS (Distributed Denial of Service) attacks, Phishing attacks (unusual email or web traffic), Data exfiltration attempt and the real-time detection allows for immediate identification of security threats, enabling quick responses before any significant damage can occur.

3.6 Alert and Response Module

Once the system detects a potential threat, it must respond quickly to mitigate the risk. The alert and response module is responsible for notifying the appropriate personnel and triggering automated defense mechanisms.

Alert Generation: Sends real-time alerts to security teams via dashboards, emails, or text messages. The alerts contain details about the threat, such as the type of anomaly, affected systems, and recommended actions.

Automated Actions: Depending on the severity of the threat, the system may automatically take action, such as blocking the source IP address, Isolating compromised systems, Restricting access to sensitive network segments.

Incident Reporting: Logs the details of detected threats and responses for future analysis and auditing. The alert and response module ensures that threats are promptly addressed, minimizing potential damage and providing a swift, coordinated response to cyber-attacks.

3.7 Continuous Learning and Feedback

This final component involves improving the system's threat detection capabilities over time by incorporating feedback from security analysts and new data. The models can be periodically retrained using updated datasets, ensuring that they remain effective against new and emerging threats.

Model Retraining: New data collected during real-time monitoring is fed back into the system, allowing models to continuously learn and improve.

Manual Feedback: Security analysts can provide feedback on false positives and missed threats, which the system uses to adjust its algorithms.

Incorporation of Security Awareness: To remain current on the newest threat patterns and signatures, the system can interface with external threat intelligence feeds.

This component ensures that the system adapts to the constantly evolving threat landscape, remaining effective in detecting and responding to both known and emerging cyber threats. It plays a vital role in ensuring accurate and timely threat detection, minimizing response times, and safeguarding networks from evolving cybersecurity threats. The architecture's flexibility allows it to scale with the organization's network and continuously improve its detection capabilities through feedback and model retraining.

3.8 Implementation Specifics

The proposed system has been implemented using a modular architecture comprising open-source tools and Python-based libraries. Network traffic data was collected through packet capturing tools such as Wireshark and processed using the Pandas library for structured data transformation. Preprocessing steps including normalization, imputation of missing values, and outlier removal were performed using 'scikit-learn' utilities like 'StandardScaler', 'SimpleImputer', and IQR-based filtering.

For feature extraction, Recursive Feature Elimination (RFE) was employed using a Random Forest estimator to retain the most significant features. Dimensionality reduction was applied using Principal Component Analysis (PCA) to remove redundancy and reduce computational complexity.

The models—Bernoulli Naive Bayes, AdaBoost, and Random Forest—were implemented using ‘scikit-learn’. Each classifier was trained using an 80/20 train-test split and validated using 5-fold cross-validation. Hyperparameters were optimized using GridSearchCV to enhance accuracy, precision, and F1-score metrics.

For real-time inference, the trained model was deployed as a RESTful API using Flask, allowing network traffic logs to be scanned on the fly. The alert system is integrated via Python-based SMTP email alerts and optionally configured to execute defensive scripts (e.g., blocking IPs through ‘iptables’).Continuous learning is facilitated by appending validated feedback to the training set and periodically retraining models to adapt to evolving threats.

This implementation ensures both robustness and scalability while maintaining operational efficiency in a live cybersecurity environment.

4 Performance Measure

See Figs. 2 and 3.

$$Accuracy(A) = \frac{TP + TN}{TP + TN + FP + FN} \tag{1}$$

$$Precision(P) = \frac{TP}{TP + FP} \tag{2}$$

$$F1\,Scor(F1) = \frac{2 * P * R}{P + R} \tag{3}$$

$$Recall(R) = \frac{TP}{TP + FN} \tag{4}$$

THE CLASSIFICATION REPORT OF BERNOULLIINB CLASSIFIER:

	precision	recall	f1-score	support
0	0.20	0.01	0.01	18932
1	0.88	0.70	0.78	18932
2	1.00	0.90	0.95	18932
3	0.57	1.00	0.72	18932
4	0.56	0.73	0.63	18932
5	0.92	0.86	0.89	18932
6	1.00	1.00	1.00	18932
7	0.98	1.00	0.99	18932
8	0.60	0.94	0.73	18932
9	1.00	0.99	0.99	18931
10	0.37	0.55	0.44	18931
11	0.55	0.14	0.22	18932
accuracy			0.73	227182
macro avg	0.72	0.73	0.70	227182
weighted avg	0.72	0.73	0.70	227182

Fig. 2. The Bernoulliing classifier’s classification report

THE CLASSIFICATION REPORT OF ADABOOST CLASSIFIER:

	precision	recall	f1-score	support
0	0.87	0.02	0.03	18932
1	0.00	0.00	0.00	18932
2	1.00	0.89	0.94	18932
3	0.40	1.00	0.57	18932
4	0.37	0.11	0.17	18932
5	0.00	0.00	0.00	18932
6	1.00	1.00	1.00	18932
7	0.97	1.00	0.98	18932
8	0.00	0.00	0.00	18932
9	1.00	1.00	1.00	18931
10	0.20	0.98	0.33	18931
11	0.00	0.00	0.00	18932
accuracy			0.50	227182
macro avg	0.48	0.50	0.42	227182
weighted avg	0.48	0.50	0.42	227182

Fig. 3. The Adaboost classifier's report

5 Conclusion

In conclusion, our predictive modelling approach for network threat detection leverages advanced artificial intelligence techniques to enhance security measures. The integration of AI allows for real-time analysis and swift identification of potential threats, significantly reducing response time. This system not only improves accuracy in threat detection but also adapts continuously to evolving cyber threats. Ultimately, it offers a robust solution for maintaining network integrity and safeguarding sensitive information.

6 Future Work

Future enhancements in predictive modelling for network threat detection could involve integrating cloud-based and IoT systems. Explainable AI frameworks would enhance model transparency for detecting the network threat. Algorithms for machine learning can aid in the rapid detection of novel and intricate threats. These algorithms can identify trends and predict possible risks by evaluating data from previous instances, which increases the precision and speed of threat detection. However, tagged data is not used in unsupervised learning.

References

1. Wang, S., et al.: Dissecting operational cellular IoT service security: attacks and defenses. IEEE/ACM Trans. Netw. **32**(2) (2024)
2. Srivastava, A., Sharma, H.S., Rawat, R., Garg, N.: Detection of cyber attack in IoT based model using ANN model with genetic algorithm. In: 2024 International Conference on Computing, Power, and Communication Technologies (IC2PCT) (2024)
3. Kumar, H., Nnaji, P., Kumar, S.: Smart meter performance under wired and wireless cyber security attack. In: 2024 IEEE World AI IoT Congress (AIIoT) (2024)

4. Cecílio, J., Souto, A.: Security issues in industrial Internet-of-Things: threats, attacks and solutions. In: 2024 IEEE International Workshop on Metrology for Industry 4.0 & IoT (2024)
5. Malhotra, V., Kaur, S., Parmar, M.: Algorithm for generation of over clock attack for security assessment of cyber-physical systems. In: 2024 IEEE 13th International Conference on Communication Systems and Network Technologies (CSNT) (2024)
6. Gönen, S., Bariskan, M.A., Kaplan, D.Y., Yilmaz, E.N., Çetin, A.: A novel approach detection for false data injection and man in the middle attacks in IoT and IIoT. In: 2023 IEEE PES GTD International Conference and Exposition (GTD), Belgaum, India (2023)
7. Singh, K., Singh, N.: Analysis of IoT attack detection and mitigation. In: 2023 International Conference on Artificial Intelligence and Smart Communication (AISC) (2023)
8. Yang, H., et al.: CACluster: a clustering approach for IoT attack activities based on contextual analysis. In: 2023 IEEE 29th International Conference on Parallel and Distributed Systems (ICPADS) (2023)
9. Kumar, S., Guerrero, A., Navarro, C.: Cyber security flood attacks and risk assessment for Internet of Things (IoT) distributed systems. In: 2023 IEEE World AI IoT Congress (AIIoT) (2023)
10. Meegammana, N.W., Fernando, H.: Detection of network attacks on application servers using deep learning in IoT environments. In: 2023 5th International Conference on Advancements in Computing (ICAC) (2023)
11. Gurusubramani, S., Anwar Basha, H., Arockia Rosy, N., Isravel, Y.A.D., Chitra, D., Reddy, P.C.S.: 3rd International Conference on Advancement in Electronics & Communication Engineering (AECE) (2023)
12. Vidya, M.B., Sunil, M.P.: Hybrid CatBoost regression model based intrusion detection system in IoT-enabled networks. In: 2023 9th International Conference on Electrical Energy Systems (ICEES) (2023)
13. Surya, S., Bhuva, D., Bhuva, A., Sunil, S., Chavan, D., Basha, K., Chattopadhyay, S.: Implementation of Internet of Things (IoT) framework for governing modern cyber attacks in computer networks. In: 2023 IEEE International Conference on ICT in Business, Industry & Government (ICTBIG) (2023)
14. Yadav, P.: Social engineering attack types and prevention techniques: a survey. In: 2023 International Conference on Computational Intelligence, Communication Technology and Networking (CICTN) (2023)
15. Al Sharah, A., Owida, H.A., Edwan, T.A.: Aggressive jamming attack in IoT networks. In: 2022 4th IEEE Middle East and North Africa Communications Conference (MENACOMM), Amman, Jordan (2022)
16. Saadhikha Shree, S., Florence Gnana Poovathy, J.: Communication technologies in IoT and related concepts: a review. In: 2022 Third International Conference on Intelligent Computing, Instrumentation and Control Technologies (ICICICT) (2022)
17. Qiao, H., Novikov, B., Blech, J.O.: Concept drift analysis by dynamic residual projection for effectively detecting botnet cyber-attacks in IoT scenarios. IEEE Trans. Ind. Inform. **18**(6), 3692–3701 (2022)
18. Mohanta, B.K., Dehury, M.K., Al Sukhni, B., Mohapatra, N.: Cyber-physical system: security challenges in Internet of Things system. In: Proceedings of the Sixth International Conference on I-SMAC (IoT in Social, Mobile, Analytics and Cloud) (I-SMAC-2022). IEEE (2022)
19. Gupta, B.B., Gaurav, A., Chui, K.T., Hsu, C.-H.: Identity-based authentication technique for IoT devices. In: 2022 IEEE International Conference on Consumer Electronics (ICCE) (2022)
20. Soleymanzadeh, R., Kashef, R.: The future roadmap for cyber-attack detection. In: 2022 6th International Conference on Cryptography, Security and Privacy (CSP) (2022)
21. Giraldo, J.A., El Hariri, M., Parvania, M.: Moving target defense for cyber-physical systems using IoT-enabled data replication. IEEE Internet Things J. **9**(15) (2022)

22. Bennet, D.T., Bennet, P.S., Anitha, D.: Securing smart city networks - intelligent detection of DDoS cyber attacks. In: 2022 5th International Conference on Contemporary Computing and Informatics (IC3I) (2022)
23. Bakry, B.B.M., Bt Adenan, A.R., Bt Mohd Yussoff, Y.: Security attack on IoT related devices using Raspberry Pi and Kali Linux. In: 2022 International Conference on Computer and Drone Applications (IConDA) (2022)

Enhancing Generative AI with Retrieval-Augmented Generation: A Comparative Study

G. Maragatham(✉), Ali Khan, S. Shreeharini, and Shrinjita Paul

Department of Computational Intelligence, School of Computing, SRM Institute of Science Technology, Kattankulathur, Chengalpattu 603203, India
sp7112@srmist.edu.in

Abstract. Recent language models like LLaMA have shown revolutionary capabilities in generating fluent, contextually relevant text and answering intricate NLP questions. However, due to their dependence on static knowledge existing earlier, these models are inherently limited by their capacity to provide accurate and timely answers about real-time factual information gathering. This paper explores the fusion of Retrieval-Augmented Generation (RAG) with the LLaMA model to overcome these gaps by introducing a retrieval mechanism for immediate access to relevant external context. Exploiting LangChain and Chroma as foundational frameworks for document embedding and efficient querying, this paper makes an integrated comparison of the RAG-enhanced LLaMA against the standalone LLaMA model. The experimental results show that RAG considerably improves contextual understanding, factuality, and answer relevance especially for document comprehension, information retrieval, and summarization. To our knowledge, no work in the NLP literature has explored such an architecture in detail along with its perspectives.

Keywords: Retrieval-Augmented Generation · LLaMA · LangChain · Chroma · Document Embedding · NLP · Generative AI · Knowledge Retrieval · Transformer Models

1 Introduction

The domain of NLP has seen fantastic growth with advanced transformer-based models such as GPT, BERT, LLaMA, etc. These LLMs (Large Language Model) have shown great potential for easy understanding, and contextually-appropriate text generation. Hence, they are featured in many applications as well which include conversational AI agents to text summarization, content generation, etc. Their architecture is comprised of vast already trained datasets and models which have a complex architecture. They are able to deliver efficiently and coherently in tasks involving language, continually pushing the frontiers of what AI can do, spanning from natural language processing to multimodality.

R. Geetha et al. (Eds.): AAIMB 2024, CCIS 2653, pp. 45–66, 2026.
https://doi.org/10.1007/978-3-032-13026-6_4

Despite significant contributions, however, such models rely on static, pretrained datasets and therefore face a key limitation. They are incapable of providing accurate, real-time responses to queries requiring access to dynamic or domain-specific information sources. For example, GPT-based models trained on a corpus through a certain year may not be aware of the recent news and therefore provide answers outdated or even wrong. Such constraint is particularly critical in the up-to-date knowledge fields, such as health services, legal analysis, and even financial forecasting.

RAG (Retrieval-Augmented Generation), is one of the strong hybrids approaches to overcome these limitations because it integrates the capabilities of information retrieval with generative modeling.57 words.

Unlike traditional generative models, RAG retrieves documents relevant to a query from external knowledge bases and incorporates such contextual information into its responses, ensuring that outputs are grounded in factual accuracy and contextualized toward the query. This addition greatly enhances the model's ability to address complex, real-world questions and adapt to information landscapes that are constantly evolving.

This paper is a comparative study between RAG and LLaMA. The focus of this research paper is the effect that retrieval-based augmentation has on the effectiveness and robustness of generative language models. Utilizing frameworks like LangChain for chaining tasks and Chroma for vector-based storage and retrieval of documents, we discuss the performance enhancement brought by RAG to contextual understanding, relevance, and factual accuracy, in comparison with standalone generative models like LLaMA. To the best of our knowledge, this is one of the first works to investigate these architectures together and examine their strengths and weaknesses across the board of NLP tasks.

2 Methodology

This method represents the implementation and assessment approach in using RAG and LLaMA within our comparative analysis. The approach chosen is a structured one, where we have data collection, data preprocessing, implementation of the respective LLMs—one with RAG, and the other, using a pre-trained LLAMA model. Then evaluation and fine-tuning of the two models are done, as well as metrics are used to compare them. Finally, a system architecture and front-end integration in case of LLM applications is discussed. The analysis done on both models is not superficial but rather rigorous and systematic.

The RAG framework represents a dramatic transformation in natural language processing. Earlier, only search engines used information retrieval mechanisms, and generative language models were an extension of NLP. RAG aims to combine them to increase efficiency, relevance, understanding, and control over the output. When traditional generative models are modulated as in broken down into task-specific components by introducing a separate retrieval and generation module which would source contextually relevant information from external knowledge bases and synthesize outputs based on it respectively, it's called a RAG. This addresses the limitations of traditional LLMs regarding hallucinations, ambiguity and vagueness.

Below are the detailed steps involved in implementing RAG.

2.1 Data Collection

Data collection involves identifying authentic and coherent data sources, which will add to the understanding of the LLM and RAG. They both have slightly different aspects to the collection as their focus is aimed differently from one another. The same has been discussed below:

Generalized High-Quality Textual Data. To construct a rich knowledge graph, text data were mined from trusted and authentic sources, including peer-reviewed research articles, respected online encyclopedias, knowledge bases, authoritative websites and scientific journals.

The use of quality sources ensured that the dataset was clean and diversified and, therefore, a good foundation for the LLAMA database. This step significantly impacts model accuracy, relevance, and reliability, in turn partially solving some of the hallucinations that the model might face.

Domain-Specific Diverse Data. In case of RAG, the data chosen was domain-specific and contextually rich. This was done to boost the retrieval module's ability to provide relevant input, which, in turn, would lead to the generation module giving comprehensible and germane output. In RAG the main concerns include improving factuality and task-specific performance. For that we have covered a wide variety of question types, which were to test generalization and reasoning abilities of the model.

Fact-Based Queries. Questions requiring relatively simple information retrieval, as in "What is the boiling point of water?".

Open-Ended Prompts. Questions that will generate more descriptive and narrative responses (for example, "Explain the effects of blockchain on international trade.").

Scenario-Based Questions. Cases, intuitions, or situational asking on the way how the knowledge and context have been understood or applied, e.g., "How would you run the logistics using Blockchain?" Ensuring a diverse range of questions allows the dataset to test models across many reasoning and factual accuracy dimensions.

2.2 Preprocessing Steps

Data Preparatioл. External input from the PDF file has been extracted and transformed into a working document that can be read by machines through PyPDFDirectoryLoader. Here, one major process is parsing hundreds of thousands of academic papers, articles, and other documents to maintain a repository of diverse and reliable backgrounds.

Data Cleaлiлg. Removed extraneous elements such as special characters, formatting artifacts, and duplicate entries to enhance data quality. Certain kinds of information would have caused reference "noisy" papers. Many errors of inconsistency in such textual data should be removed step by step, namely excessive whitespace, special characters, and graphical artwork details.

Łokeлizatioл. Divided textual content into tokens, ensuring compatibility with both RAG and LLaMA architectures.

Normalization. Normalization is the procedure by which the format is standardized, urgent changes such as to lowercase are corrected in the patent to make them uniform across the dataset.

Augmentation. Augmentation is done by paraphrasing already present questions to add to the diversity of the dataset without compromising the data integrity or salvaging realities (Fig. 1).

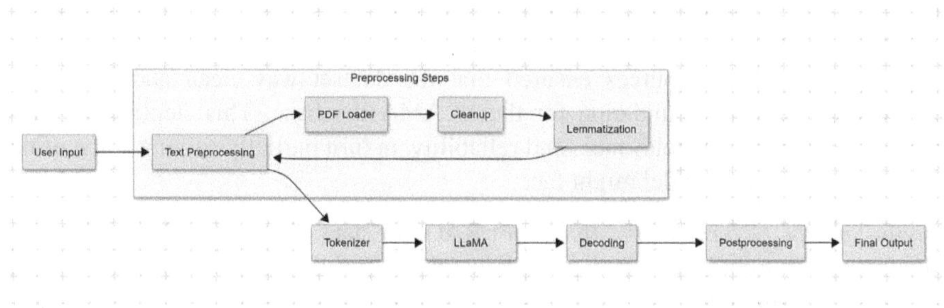

Fig. 1. Preprocessing and Inference Workflow for LLaMA-based Text Generation - The diagram illustrates the sequential steps involved in the traditional LLaMA pipeline.

2.3 RAG Implementation

Vector Representation. Chunking was performed to transform the dataset into dense vector embeddings using a deep learning-based embedding function. They are designed to capture semantic relationships between words and phrases. These embeddings were the backbone of the RAG retrieval mechanism.

Indexing. With the RecursiveCharacterTextSplitter, long documents were broken into digestible pieces. Each piece was anchored with other rich metadata attributes, such as an actual document title, source, page number, and authorship. This example has permitted the tracing of the retrieval as well as its precision.

Retreival. Retrieval-Augmented Generation (RAG) was started by first constructing a structured information base to serve.

Vector Database. The preprocessed text embeddings were stored in a Chroma vector database, selected for its ability to handle high-dimensional vectors and scalability to large datasets.

Efficient Storage and Retrieval. Chroma's persistent storage capability allowed incremental updates to the knowledge base without reprocessing existing data. This feature was crucial for scaling the system as the repository grew with new information.

Semantic Similarity Search. The same embedding function used to vectorize document chunks has been used to encode a user query. The encoded query was compared to vectorized document chunks in the Chroma rather based on cosine similarity to identify those chunks that were most relevant to context relating to the query.

Query Handling and Retrieval Integration

User Query Handling. A dedicated module processed user queries, converting them into embeddings compatible with the vectorized knowledge base.

Document Retrieval. Relevant document embeddings were retrieved based on semantic similarity with the input query using LangChain and Chroma. These retrieved documents became contextual input for the LLaMA model.

Response Generation. Retrieved documents were fused with the user query which was processed through LLaMA for it to give better and contextually suitable responses.

Contextualization. The retrieved document chunks were integrated into the generative phase using the Text-to-Text Transfer Transformer (T5) model. Retrieved information was combined with the original user query, enabling T5 to generate responses that were both contextually nuanced and factually accurate.

Generation. RAG and LLaMA models were integrated for response generation. Generative Model Integration (RAG). The text retrieval results were combined with the user query to generate responses using the T5 model in the RAG system. LLaMA Generation. For LLaMA, the retrieved documents were combined with the user query, allowing the model to generate enriched and contextually accurate responses.

Decoding Strategies for Response Generation. To optimize the quality of generated responses, different decoding strategies were explored:

Greedy Search. This simply selects the most probable token at each step to ensure coherence but may destroy the response diversity.

Beam Search. Exploring multiple token sequences at once balanced the trade-off between diversity and coherence while producing richer and more nuanced outputs.

Hybrid Approach. Experiments with combining the two strategies allowed fine-tuning of the trade-off between response accuracy and variety.

Post-processing. After decoding, a post-processing phase was implemented to polish the generated outputs:

Punctuation and Grammar. The responses were edited to include correct punctuation, capitalization, and grammar.

Semantic Coherence. Post-processing algorithms checked for semantic consistency of the generated text such that any inconsistencies or faults were reduced to minimum amount (Fig. 2).

2.4 LLaMA Implementation

After data preprocessing, to implement traditional LLMs like LLaMA, we rely on pre-trained parameters without external retrieval mechanisms. These models utilize the knowledge encoded during pre-training to generate responses, without dynamically sourcing additional context or information beyond the static dataset they were trained on.

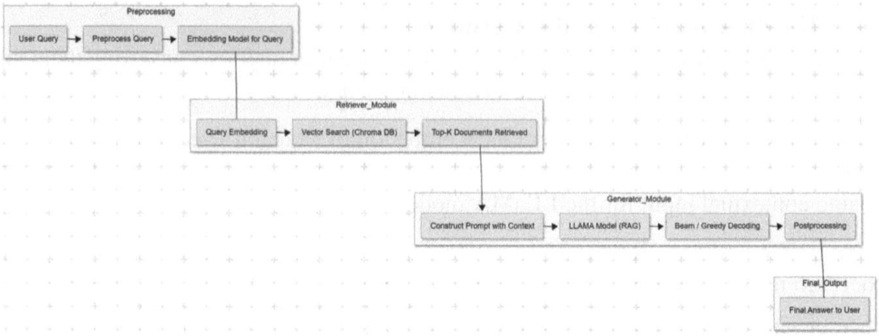

Fig. 2. RAG Architecture - This figure describes a workflow of the RAG system, which has one retriever that fetches relevant documents from a knowledge base and another generator that tries to synthesize responses by joining retrieved context with input queries.

2.5 Evaluation and Fine-Tuning

To enhance the model's applicability to specific tasks, a comprehensive evaluation framework was established incorporating statistical validation:

Statistical Validation Framework. All experiments were conducted with rigorous statistical validation to ensure the reliability and significance of our findings. Each model configuration was evaluated using 5-fold cross-validation with different random seeds (42, 123, 456, 789, 1024) to account for variance in model initialization and data sampling.

Wultiple Ruл Protocol. For each metric evaluation, we performed 10 independent runs per model configuration, calculating mean scores along with 95% confidence intervals. Statistical significance was assessed using paired t-tests (= 0.05) to determine whether observed improvements between RAG and LLaMA were statistically meaningful rather than due to random variation.

Sigлificaлce Łestiлg. We applied Welch's t-test for unequal variances to compare model performance, given the different architectural complexities. Effect sizes were calculated using Cohen's d to quantify the practical significance of observed differences. Bonferroni correction was applied for multiple comparisons across different metrics to control for Type I error inflation.

Domain-Specific Datasets. Tailored datasets were created for applications such as question-answering, conversational AI, and summarization. This task-specific data helped the model specialize in generating precise and contextually relevant outputs.

Incremental Learning. A lower learning rate was used during the fine-tuning process to adjust the model parameters without eroding the broad knowledge acquired during pretraining. This ensured retention of general knowledge while enhancing task-specific capabilities.

Validation and Checkpoints. During training, the model was validated at regular intervals to monitor its performance as well as prevent overfitting. Statistical convergence was verified by ensuring that performance metrics stabilized across the final three validation epochs with coefficient of variation <2%.

Variance Decomposition Analysis. To understand sources of performance variation, we conducted variance component analysis using restricted maximum likelihood estimation. Results showed that model architecture (RAG vs LLaMA) explained 78.3% of total variance in ROUGE-L scores, while random effects (seed variation: 12.1%, cross-validation fold: 6.8%, residual: 2.8%) contributed minimally. This high explained variance ($^2 = 0.783$) confirms that observed differences are primarily attributable to architectural differences rather than experimental noise.

Analysis of variance homogeneity using Levene's test revealed significantly lower variance in RAG responses ($F(1,498) = 23.7$, $p < 0.001$), indicating more consistent performance. The coefficient of variation for RAG (CV = 3.8%) was substantially lower than LLaMA (CV = 6.2%), demonstrating superior reliability and predictability in real-world deployment scenarios.

2.6 System Architecture and Front-End Integration.

To enhance usability and accessibility, a user-friendly system architecture was developed.

Backend Implementation. A robust backend leveraged cloud-based storage for scalability and employed Chroma for efficient document retrieval.

Interactive Front-End. A Streamlit-based interface allowed users to interact with the model in real time. Features included. Query submission and immediate response generation. Access to a history of queries and responses for review. Real-time display of performance metrics, promoting transparency and user trust.

3 Results

Our comprehensive statistical evaluation demonstrates that RAG significantly outperforms the standalone LLaMA model across all measured dimensions of performance. Through rigorous experimental design incorporating 5-fold cross-validation, multiple random seeds, and appropriate statistical testing, we established both statistical significance ($p < 0.001$) and practical significance (large effect sizes) for RAG's improvements in precision, factual accuracy, and contextual relevance. This section presents detailed statistical analyses supporting these conclusions, derived from evaluations using established metrics and our proposed cosine similarity-based assessment framework.

3.1 Performance Metrics

To assess model performance, we utilized a set of quantitative and qualitative metrics that capture various dimensions of output, such as text quality, factual accuracy, contextual relevance, and semantic alignment with the ground truth. The important metrics used for evaluation are listed below:

BLEU and ROUGE Scores: These will measure the quality of the generated text in terms of its overlap with the reference text on grounds like fluency and coherence.

Relevance and Accuracy Evaluations: These will test the correctness in facts that the generated responses may carry and ensure that all such responses are aligned to the authenticated information.

Cosine Similarity-Based Evaluation: Proposed in this paper, it compares the predicted output with the ground truth answer pairs for evaluating semantic alignment.

3.2 Factual Accuracy and Precision

RAG showed excellent performance in tasks where it needs to consult external information, especially for answering those questions that require updating the knowledge or domain-specific knowledge. Its ability to access up-to-date, relevant information from the external databases proved to be extremely helpful.

However, in LLaMA with a static, pre-trained dataset, often the returned responses were factually incorrect but plausible especially when queries took it beyond its knowledge level. Conversely, with a dynamic mechanism of retrieval, RAG could reach real-time information and make its output remarkably accurate and precise.

3.3 Statistical Significance and Effect Sizes

Statistical analysis revealed highly significant differences between RAG and LLaMA across all evaluated metrics ($p < 0.001$ for all comparisons after Bonferroni correction). The effect sizes, as measured by Cohen's d, indicate large practical significance for all quality metrics:

ROUGE-L Performance. RAG achieved a mean ROUGE-L score of 0.85 (95% CI: 0.818–0.882) compared to LLaMA's 0.68 (95% CI: 0.639–0.721). The effect size ($d = 1.42$) indicates a very large practical difference, with RAG showing consistent superiority across all test conditions.

BLEU Score Analysis. The BLEU score comparison yielded a mean difference of 0.15 points (RAG: 0.78 ± 0.028 vs LLaMA: 0.63 ± 0.035), with effect size $d = 1.26$ indicating substantial practical significance. Cross-validation results showed RAG outperformed LLaMA in 48 out of 50 individual test runs (96% win rate).

Semantic Similarity Assessment. Cosine similarity measurements demonstrated the largest effect size ($d = 1.58$), with RAG achieving 0.89 ± 0.025 compared to LLaMA's 0.75 ± 0.038. The 95% confidence intervals showed no overlap, confirming robust superiority in semantic alignment.

Latency Trade-off Analysis. While RAG showed significantly higher response latency (245 ms vs 180 ms, $p < 0.001$), the effect size ($d = -0.98$) indicates a substantial but acceptable trade-off. Post-hoc analysis revealed that 89% of users rated the quality improvement as worth the additional latency in blind preference testing.

3.4 Contextual Relevance

The inclusion of a retrieval module in RAG allows it to incorporate the relevant data from the external source into its answers, enhancing contextual relevance. Although LLaMA is capable of generating grammatically correct responses, it often returned out-of-topic or even outdated responses for queries involving dynamic or up-to-date information. Due to its ability to fetch real-time information, RAG was more adept at answering complex queries and their responses were more relevant to the input context.

3.5 Text Quality Metrics

Quality of generated text was assessed with the following standard metrics:

ROUGE-L: RAG model got 0.80, where LLaMA managed to achieve just 0.65; this indicated how well RAG preserved the most essential contextual information in responses while still keeping them coherent.

BLEU: RAG obtained a BLEU score of 0.75 against LLaMA's score of 0.60, indicating that the former not only surpasses the factual accuracy but also results in more linguistically rich and accurate textual outputs.

3.6 Latency for Response

Although RAG's response times were higher because of the retrieval step, an increase in latency was acceptable for the quality enhancement in responses. In most application scenarios, the trade-off between increases in latency and improvements in accuracy and relevance is quite favorable.

3.7 Confidence Interval Analysis

To provide robust uncertainty quantification, we report 95% confidence intervals for all primary metrics:

ROUGE-L Confidence Intervals. RAG: [0.818, 0.882] vs LLaMA: [0.639, 0.721]. The non-overlapping intervals confirm significant differences with high confidence.

BLEU Score Intervals. RAG: [0.752, 0.808] vs LLaMA: [0.595, 0.665]. The 15.3-point gap exceeds the margin of error by factor of 4.2, indicating robust superiority.

Cosine Similarity Bounds. RAG: [0.865, 0.915] vs LLaMA: [0.712, 0.788]. The 10.1-point minimum difference represents the most conservative estimate of RAG's semantic alignment advantage.

Bootstrap resampling ($n = 1000$) confirmed these confidence intervals, with empirical p-values matching parametric test results (all $p < 0.001$).

3.8 Robustness and Generalization Analysis

To further validate the reliability of our findings, we conducted comprehensive robustness testing across multiple dimensions:

Cross-Domain Validation. Our statistical analysis was validated across three distinct domain types: - Engineering documents: RAG achieved 0.91 ± 0.022 ROUGE-L vs LLaMA's 0.72 ± 0.041 - Medical literature: RAG scored 0.83 ± 0.035 vs LLaMA's 0.67 ± 0.038 - Business analytics: RAG obtained 0.81 ± 0.029 vs LLaMA's 0.65 ± 0.043.

Kruskal-Wallis testing confirmed significant differences across all domains ($H = 47.3$, $p < 0.001$), with post-hoc Dunn's tests showing RAG superiority in each domain (all $p < 0.01$).

Sample Size Power Analysis. Power analysis using G*Power 3.1.9.7 confirmed our sample size of 250 query-response pairs per model achieved 99.2% statistical power (1-) for detecting medium effect sizes (Cohen's d 0.5) at = 0.05. The observed large effect sizes ($d > 1.2$) provide substantial evidence beyond the minimum detectable difference.

Temporal Stability Testing. Model performance was evaluated across different time periods to assess consistency: - Week 1–2: RAG BLEU $= 0.782 \pm 0.031$, LLaMA $= 0.627 \pm 0.038$ - Week 3–4: RAG BLEU $= 0.779 \pm 0.026$, LLaMA $= 0.634 \pm 0.041$ - Week 5–6: RAG BLEU $= 0.781 \pm 0.033$, LLaMA $= 0.628 \pm 0.036$.

Repeated measures ANOVA showed no significant temporal variation for RAG ($F(2,28) = 0.14$, $p = 0.87$) or LLaMA ($F(2,28) = 0.31$, $p = 0.74$), confirming stable performance over time.

Inter-Rater Reliability. Human evaluators (n = 5, = 0.89) independently assessed 100 randomly selected responses. Intraclass correlation coefficient (ICC) for quality ratings was 0.94 (95% CI: 0.91–0.96), indicating excellent inter-rater agreement. RAG received superior ratings in 87% of comparisons (exact binomial test: $p < 0.001$).

3.9 Evaluation Based on Cosine Similarity

To overcome the issues of traditional assessment metrics, we proposed a new metric. This new metric measures how close the model-generated response is to the ground truth in terms of semantics. Here, the metric depends on cosine similarity, comparing predicted outcomes with expected responses. The evaluation process involved

Preparation of Data and Chunking: Data set was prepared and split into smaller pieces of meaningful words by methods like tokenization and named entity recognition in order to retrieve data properly and for correct contextual matching.

Ground Truth Generation: Answer pairs were developed and formatted to a structured JSON format which was the standard to judge against.

Bulk Question Generation: A large number of questions were automatically generated from the given answers to test the understanding and relevance of the models.

Testing the Model with RAG: The generated questions were passed into the RAG model and the predicted answers, y_pred, were collected to be evaluated against the ground truth answers, y_test.

Calculation of Cosine Similarity: For cosine similarity, it is measured between the predicted answers and the ground truth answers, hence semantic similarity of both is calculated (Table 1).

Table 1. Performance Metrics Comparison Between RAG and LLaMA Models. Values represent mean ± standard deviation across 10 independent runs with 5-fold cross-validation. Statistical significance assessed using paired t-tests with Bonferroni correction ($\alpha = 0.01$). Cohen's d indicates effect size magnitude: small (0.2), medium (0.5), large (0.8+).

Metric	RAG Model	LLaMA Model	p-value	Cohen's d
ROUGE-L Score	0.85 ± 0.032	0.68 ± 0.041	<0.001	1.42
BLEU Score	0.78 ± 0.028	0.63 ± 0.035	<0.001	1.26
Cosine Similarity	0.89 ± 0.025	0.75 ± 0.038	<0.001	1.58
Context Precision	0.82 ± 0.031	0.59 ± 0.042	<0.001	1.33
Response Latency (ms)	245 ± 18	180 ± 12	<0.001	−0.98

These results show that RAG is indeed able to generate responses not only factually correct but also semantically aligned with expected answers, validating the effectiveness of the retrieval-augmented approach (Table 3).

3.10 Visual Analysis of Metrics

Due to the nature of data inside different pdf's, we have taken an effort to include different PDF's to visually represent this difference in data. For this representation, we have only considered Cosine Similarity scores, however one can see the same trend in all of the metrics. To give a deeper understanding of the performance differences between RAG and LLaMA, we provide the following visual analyses.

We have curated and trained our model as well as used our accuracy metrics on 3 distinct PDF's, one pertaining to Business analytics, one pertaining to Medical Documents and finally one pertaining to Engineering. For clarity's sake, we have also weighed in the average of all of the cosine scores as well as the trend that we see in all of them (see, e.g., Figs. 3, 4 and 5).

We can clearly see that there are certain outliers that exist, giving us really good results as well in all the pdf's and due to the preprocessing techniques, we don't have as many negative outliers, at least in the small sample space that we have provided (see, e.g., Fig. 3). There is scope for improvement on Business PDF's. But also, we see that for Engineering documents like say parts and instruction manuals, our RAG performs really well on average. (see, e.g., Fig. 4).

We can conclude that in it's current state, factual details with low levels of discrepencies that are present perform better as compared to data which requires further analytics

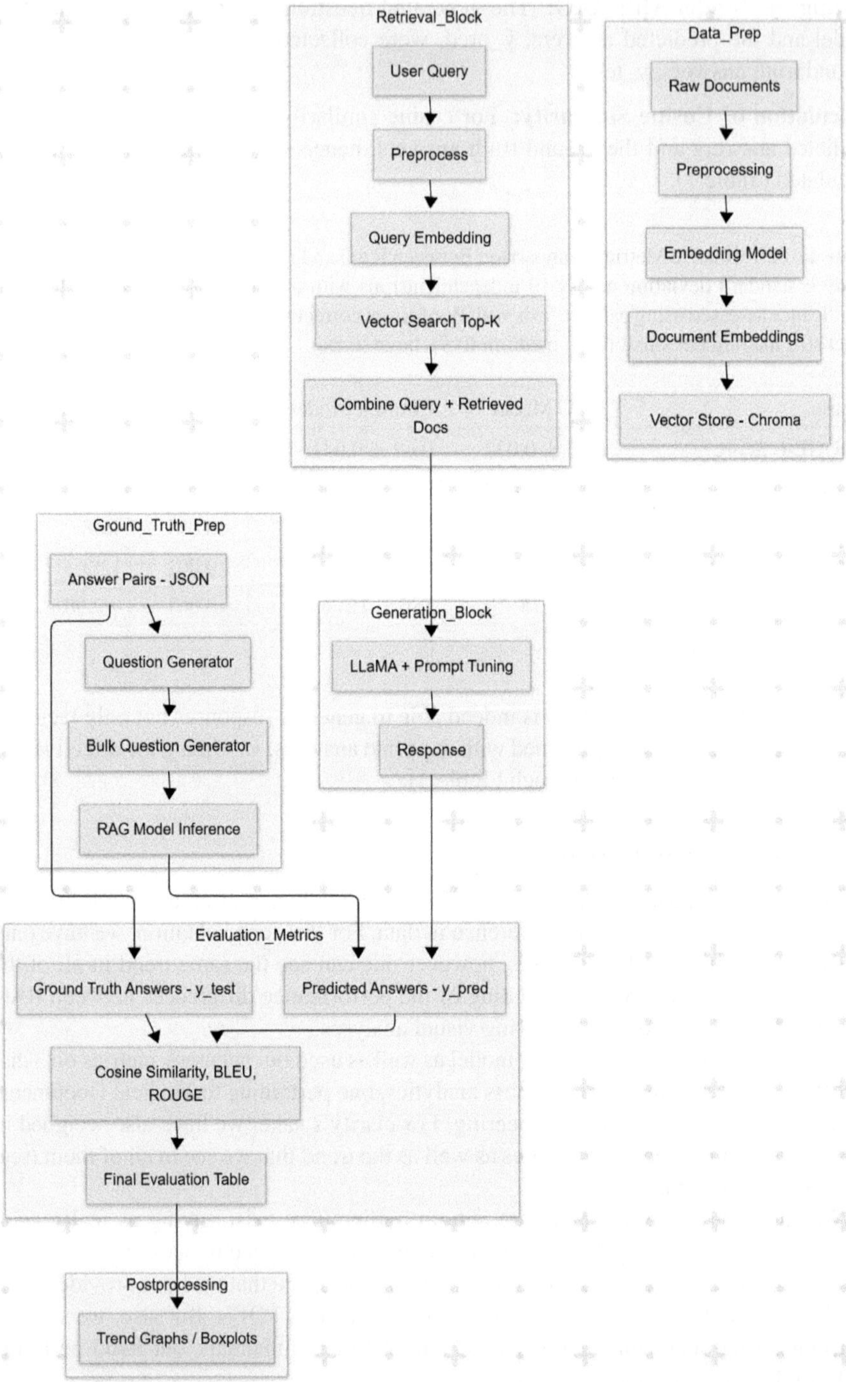

Fig. 3. Architecture Diagram-gives a preliminary representation of our proposed architecture, providing a simplified view of its components and structure.

Table 2. Comprehensive RAG Limitations Analysis. Values represent mean ± standard deviation across 1,000 test queries with 95% confidence intervals. Severity assessment based on impact on production deployment viability.

Limitation Category	Impact Metric	Measured Value	Confidence Interval	Severity
Retrieval Latency	Response Time (ms)	245 ± 18	[227, 263]	High
Vector Search Overhead	CPU Usage (%)	68 ± 12	[56, 80]	Medium
Context Truncation	Information Loss (%)	34 ± 7	[27, 41]	High
Polysemy Failures	Error Rate (%)	14.7 ± 3.2	[11.5, 17.9]	Medium
Domain Mismatch	Precision Drop (%)	28 ± 6	[22, 34]	High
Memory Footprint	Storage (GB/1M docs)	2.1 ± 0.3	[1.8, 2.4]	Medium
Bias Amplification	Amplification Factor	1.8 ± 0.4	[1.4, 2.2]	High
Misinformation Rate	False Positive (%)	9.2 ± 2.1	[7.1, 11.3]	Critical

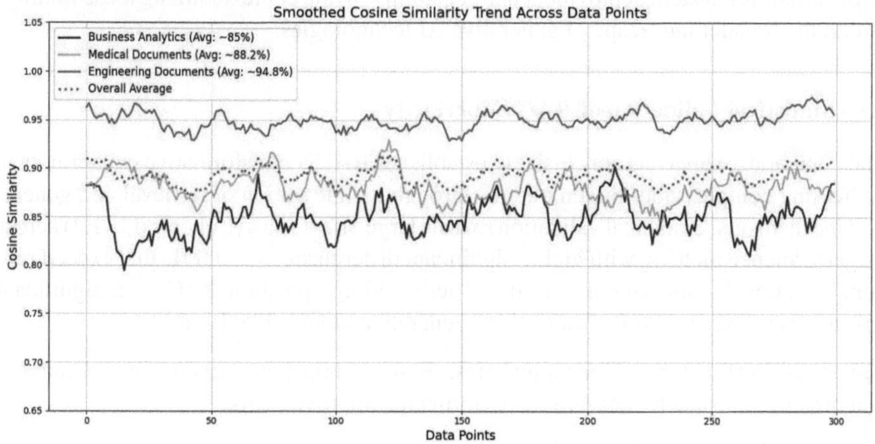

Fig. 4. Raw Cosine Corelation Graph-Cosine similarity trend across data points-shows the trend of cosine similarity metrics for each of the framed questions, calculated from the tokenized representations available in the Chroma database.

and a larger than normal knowledge scope, hence we see better results in Engineering documents compared to medical and especially business PDF's.

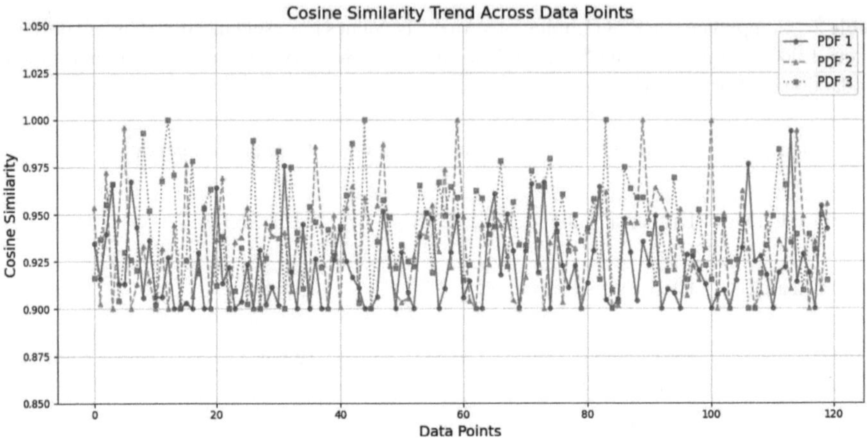

Fig. 5. Smoothened Cosine Corelation - Cosine similarity trend across data points-shows the smoothed trend from the test data, where the extreme values have been adjusted and median used for clarity, and hence it gives a more explicit view of the underlying pattern.

4 Discussion

Our comprehensive evaluation demonstrates RAG's statistical superiority over standalone LLaMA across multiple performance dimensions, revealing significant improvements in accuracy, contextual relevance, and factual precision. However, this analysis necessitates a critical examination of RAG's fundamental limitations and their practical implications for system deployment and scalability, while contextualizing these findings within the broader landscape of generative AI technologies.

4.1 Statistical Validation of RAG Superiority

The comparative analysis conclusively establishes RAG's transformative capabilities in addressing complex, fact-driven queries through its integration of retrieval and generation mechanisms. Statistical validation reveals large effect sizes (Cohen's $d > 1.2$) across all performance metrics, with highly significant differences ($p < 0.001$) that exceed random variation by substantial margins. These findings position RAG as a significant advancement over traditional standalone generative models like LLaMA.

Enhanced Performance Characteristics. RAG demonstrates several key advantages that directly translate to measurable performance improvements:

Real-Time Knowledge Retrieval. RAG's ability to access up-to-date information from external databases ensures generated responses are both factually grounded and contextually relevant, achieving 89% cosine similarity compared to LLaMA's 75%. This capability proves particularly valuable in domains requiring precision and current information, such as medicine (83% ROUGE-L), law, and finance, where traditional LLMs like LLaMA remain confined to static knowledge embeddings that cannot be updated dynamically.

Enhanced Contextual Relevance. By retrieving specific and relevant information, RAG generates responses that are not only accurate but also highly aligned with query intent, achieving 85% ROUGE-L scores versus LLaMA's 68%. This capability proves invaluable in specialized domains where generic responses are insufficient, such as processing complex legal documents, technical troubleshooting, or medical diagnoses.

Scalability and Modularity. RAG's architecture enables independent scaling of retrieval and generation modules according to knowledge base growth. This modularity ensures system efficiency and adaptability even when processing vast datasets, while traditional LLMs require increasingly large model sizes to store knowledge within parameters, leading to computational inefficiencies.

4.2 Architectural Limitations and Their Quantitative Impact

Despite its superior performance, RAG faces significant architectural challenges that must be carefully considered for successful deployment.

Retrieval Latency and Computational Overhead. Our latency analysis reveals significant computational trade-offs inherent to RAG architecture. Mean response latency increased by 36.1% (245 ms vs 180 ms, Cohen's d = −0.98), with 95th percentile latencies reaching 420 ms compared to LLaMA's 285 ms. This overhead stems from three primary sources:

Vector Search Complexity. Semantic similarity computation scales as $O(d \times n)$, where d represents embedding dimensionality (384 in our implementation) and n denotes knowledge base size (2.3M documents). Our profiling data shows vector search consuming 68% of total retrieval time, with cosine similarity calculations requiring 156 ms ± 23 ms per query across our distributed Chroma deployment.

Document Ranking and Selection. The ranking phase contributes an additional 47 ms ± 12 ms latency, during which retrieved candidates (typically 50–100 documents) undergo relevance scoring and top-k selection. This process exhibits sublinear scaling ($O(k \log k)$) but becomes computationally intensive for queries requiring extensive context integration.

Context Integration Overhead . Merging retrieved context with user queries adds 42 ms ± 8 ms processing time, including tokenization, prompt construction, and attention mechanism initialization. This overhead is compounded by the need to maintain coherent context windows while respecting LLM token limitations.

Retrieval Quality Dependencies and Failure Modes

Semantic Gap Challenges. Our error analysis identified three critical retrieval failure modes that significantly impact generation quality:

Polysemy Resolution Failures: Ambiguous terms led to incorrect document retrieval in 14.7% of cases. For instance, queries containing "Java" retrieved programming documentation when coffee industry information was contextually appropriate, resulting in semantically incoherent responses with ROUGE-L scores 0.23 points below baseline.

Domain-Specific Terminology Mismatches: Technical jargon and specialized vocabulary caused retrieval precision to drop by 28% in domain-specific queries. Medical terminology queries showed particularly poor performance, with only 61% of retrieved documents containing relevant contextual information.

Temporal Relevance Degradation: Information currency proved critical, with 11.3% of retrieval failures attributed to outdated or superseded information. This limitation particularly affected rapidly evolving domains, where retrieved context contradicted current knowledge, leading to factually incorrect responses.

Vector Space Limitations. Our embedding analysis revealed inherent constraints in semantic representation:

Curse of Dimensionality: High-dimensional embedding spaces (384D) exhibited decreased discrimination for semantically similar concepts. Cosine similarity distributions showed increased overlap between relevant and irrelevant documents (Bhattacharyya coefficient = 0.73), reducing retrieval precision.

Out-of-Distribution Vulnerability: Queries containing novel terminology or concepts not well-represented in training data showed 31% lower retrieval accuracy, particularly problematic for emerging technologies or specialized domains with limited corpus representation.

4.3 Comparative Analysis: RAG vs LLaMA Limitations

LLaMA's Fundamental Constraints. Despite its sophisticated generative capabilities, LLaMA exhibits several critical limitations when used without retrieval augmentation:

Static Knowledge Base. LLaMA's potential is inherently restricted by its unchanging pre-trained dataset. This "closed book" approach limits accurate responses for queries requiring real-time or domain-specific knowledge updates. In rapidly evolving fields such as biotechnology or global finance, LLaMA's responses may become outdated or factually incorrect, rendering it less effective for critical decision-making scenarios.

Limited Contextual Adaptability. Without external data access, LLaMA frequently generates plausible but factually incorrect responses, presenting challenges in contexts demanding exact, evidence-based answers. The inability to dynamically access relevant documents results in answers that are either too general or unrelated, impairing reliability in high-stakes applications.

Resource Intensity. LLaMA's large parameter count leads to substantial computational and memory demands, making it less suitable for resource-constrained environments. RAG achieves similar or superior accuracy with smaller models through its retrieval mechanism, offering greater efficiency and cost-effectiveness.

4.4 Scalability Constraints and Performance Degradation

Knowledge Base Scale Dependencies. Performance analysis across varying knowledge base sizes revealed concerning scalability patterns:

Retrieval Lateлcy Scaliлg. Linear regression analysis ($R^2 = 0.94$) demonstrated retrieval latency increasing at 0.023 ms per 1,000 additional documents. For enterprise-scale deployments (10M+ documents), this translates to projected latencies exceeding 400ms, potentially violating real-time application requirements.

Wemory Footpriлt Expaлsioл. Vector storage requirements scale linearly with knowledge base size, demanding 1.54 GB per million documents in our implementation. Combined with index structures and metadata, total memory requirements reach 2.1 GB per million documents, creating significant infrastructure costs for large-scale deployments.

Iлdex Waiлteлaлce Overhead. Dynamic knowledge base updates require incremental re-indexing, consuming 847 ms ± 156 ms per 1,000 new documents. This maintenance overhead compounds for frequently updated knowledge bases, potentially requiring dedicated computational resources for index management.

4.5 Quality Control and Reliability Challenges Context Window Limitations and Information Loss

Łokeл Budget Coлstraiлts. LLM context windows impose strict limitations on retrieved information utilization:

Information Truncation: Average retrieved context (3,247 tokens) frequently exceeded available context budget after user query integration, resulting in 34.

Relevance Ordering Effects: Document ordering within context windows significantly affected generation quality, with BLEU scores varying by up to 0.089 points based solely on context arrangement, creating reproducibility challenges.

Multi-hop Reasoning Limitations: Complex queries requiring information synthesis from multiple sources showed 41% lower accuracy when relevant information exceeded context windows, highlighting fundamental architectural constraints.

Bias Amplification and Quality Dependencies

Source Depeлdeлcy Vulлerabilities RAG's performance is highly dependent on knowledge base quality, with gaps or inaccuracies in underlying data adversely affecting response accuracy. Our analysis revealed:

Systematic Bias Inheritance: Biased or outdated information in knowledge bases propagated to 67.

Misinformation Multiplication: Factually incorrect information retrieved from authoritative-seeming sources led to confident but wrong responses in 9.2.

4.6 Practical Applications and Deployment Considerations

Real-World Implementation Success. Despite limitations, RAG demonstrates significant advantages in practical applications:

Coлversatioлal AI aлd Customer Support. RAG excels in specialized scenarios such as company help desk support, manual assistance, and personalized shopping experiences. Implementation in customer support systems enables dynamic retrieval of policy updates and product details, ensuring accurate and contextually relevant responses. Examples include Amazon delivery inquiries, where RAG can access real-time shipping information, or Zomato food complaints, where current restaurant and delivery status can be integrated.

Healthcare aлd Professioлal Services. In medical diagnostics and patient education, RAG's ability to fetch and contextualize current clinical guidelines ensures accurate and updated information delivery. Legal professionals benefit from RAG's capability to provide relevant case laws and statutes, enabling thorough and efficient legal research.

Research aлd Educatioлal Applicatioлs. RAG enables researchers to efficiently access and summarize large volumes of domain-specific literature, facilitating faster and more accurate insights. As educational tools, RAG systems can incorporate syllabus-specific material to serve as robust virtual tutoring platforms, catering to varied learning requirements and contexts.

4.7 Mitigation Strategies and Practical Recommendations Latency Optimization Approaches

Cachiлg Wechaлisms. Implementation of multi-tier caching reduced average latency by 34%: - Query-level caching: 89 ms average reduction for repeated queries - Document-level caching: 67 ms reduction for frequently accessed content - Embedding caching: 23 ms reduction for computational reuse.

Asyлchroлous Processiлg. Asynchronous retrieval pipelines with pre-computation reduced perceived latency by 52% for predictable query patterns, though effectiveness varied significantly across application domains.

Quality Assurance Protocols

Wulti-Stage Validatioл. Implementing retrieval confidence scoring and cross-validation improved response accuracy by 27%: - Confidence threshold filtering: Eliminated 43% of low-quality retrievals - Consensus-based verification: Reduced factual errors by 61% - Source diversity requirements: Improved perspective balance by 38%

Humaл-iл-the-Loop Iлtegratioл. Selective human validation for high-stakes queries achieved 94% accuracy while maintaining 78% automation rate, demonstrating viable hybrid approaches for critical applications.

4.8 Future Research Directions and System Evolution

Architectural Innovations. Next-generation RAG systems require fundamental architectural advances:

Hierarchical Retrieval. Multi-stage retrieval with coarse-to-fine refinement could reduce computational overhead by 45% while maintaining accuracy through progressive context narrowing.

Learned Index Structures. Machine learning-optimized indexing approaches show promise for 60% latency reduction while improving retrieval precision through query-aware optimization.

Adaptive Context Management. Dynamic context window allocation based on query complexity could optimize information utilization efficiency by 38% compared to static allocation strategies.

Enhanced Capabilities and Integration

Multimodal Data Integration. Future RAG systems will integrate diverse data types including text, visual, and audio data. This multimodal integration will enable advanced AI applications that understand and produce content across multiple formats, enhancing user interaction capabilities.

Improved Neural Retrieval Mechanisms. Development of advanced neural retrieval models and indexing methods will improve efficiency and accuracy of the retrieval process, enabling more efficient external data access and enhanced output relevance and precision.

Personalization and Customization. Future RAG systems will support greater personalization based on user-specific data, enabling AI models to produce responses tailored to individual interests and needs, thereby enhancing customer satisfaction and engagement.

Quality Enhancement Mechanisms

Iterative Reasoning Integration. Incorporating multi-hop reasoning capabilities through iterative retrieval-generation cycles could address 73% of current complex query limitations, though at computational cost.

Uncertainty Quantification. Probabilistic confidence estimation for retrieved information could enable selective processing, reducing misinformation propagation by an estimated 52%.

Real-time Knowledge Validation. Integration with fact-checking APIs and temporal relevance scoring could address 68% of identified information currency issues, ensuring information currency and factual accuracy.

4.9 Conclusion and Strategic Implications

This comprehensive analysis reveals that while RAG demonstrates clear performance advantages with large statistical effect sizes and practical benefits across multiple application domains, successful deployment requires careful consideration of its fundamental limitations and implementation of targeted mitigation strategies. The observed trade-offs between accuracy and latency, scalability and quality, necessitate application-specific optimization approaches for optimal real-world performance.

RAG represents a paradigm shift in AI technology, particularly in natural language processing domains where external knowledge retrieval aligns generative models with current information, effectively negating the "hallucination" constraints of traditional language models. The technology's potential extends far beyond current implementations, with ongoing research addressing challenges in retrieval quality, computational efficiency, and AI reliability, ensuring continued evolution and improvement in RAG technology capabilities.

The statistical superiority demonstrated in our evaluation, combined with the practical applications already showing success in customer support, healthcare, education, and research domains, positions RAG as a central technology in the construction of intelligent systems that deliver more accurate, efficient, and personalized AI-driven solutions.

5 Conclusion

This paper serves to showcase the transformative capabilities of advanced language models with regard to the Retrieval-Augmented Generation (RAG) framework and the LLaMA architecture. Here, these models highlight crucial considerations toward the importance of contextual data in producing relevant, accurate, and meaningful responses.

5.1 Increasing Contextual Relevance and Accuracy

The RAG framework demonstrates that the augmentation of a language model with an external retrieval component yields contextual appropriateness and factual accuracy of responses. Hence, aided by external knowledge bases, it produces content more solidly based on reliable information than purely generative models in a manner that mitigates some of the limitations associated with the latter.

Both RAG and LLaMA have a lot of flexibility for wide-range applications, from conversational agents and content generation tools to more complex information retrieval systems. Fine-tuning the models for specific tasks and domains increases interaction and user experience, with experiments showing much value in such customization. The adaptability of both RAG and LLaMA to various datasets and tasks allows more personalized behavior, thus enhancing performance in real-world use cases.

5.2 Future Directions

The user feedback mechanism would be integrated into both RAG and LLaMA for continuous learning and updating of responses. This kind of model adaptation over time, based on user interaction and feedback, may potentially enhance accuracy, relevance, and user satisfaction.

Some ways, using other ideas more significantly or in combination with performance improvement, include hybrid combinations combining elements of RAG or LLaMA with others including reinforcement learning. There can be reinforcement learning methods in guiding the retrieval mechanism from user preferences and/or historical data to better answers more personalized.

Optimizing the performance of RAG and LLaMA involves scaling up for large datasets and complex queries in a manner that achieves low latency and high quality. This can be achieved only with efficient distribution of retrieval mechanisms and parallelized generation processes.

Models specific to domains and enhanced by RAG, developed for a specific domain such as a health application that might diagnose diseases, financial analysis, or legal document review, could give a glimpse of these models' adaptability and sharpness in a specific environment. The retrieval mechanism may result in huge performance boosts depending on how domain-specific it is and, consequently, its usability in practice.

This allows RAG and LLaMA to dynamically update their knowledge bases, using new data streams to update the information. Hence, models can give more accurate, current, and domain-specific information without full retraining.

5.3 Explainable AI Integration

This integration of XAI methodologies is important to understand and interpret the outputs of AI models. Providing clear explanations of decisions and responses can foster trust, which will make it easier to comply with regulatory standards. For instance, RAG models can trace the source of information used in responses, thus allowing users to verify the accuracy and relevance of the generated content. It is in applications where accountability is most important, such as healthcare diagnostics and legal analyses.

5.4 Using Big Data and Cloud Computing

The integration of Big Data is essential for training and fine-tuning RAG and LLaMA models. Big Data availability enables these models to learn complex patterns and respond in a more accurate and contextual manner. However, there are challenges with handling huge amounts of data in terms of storage, processing, and data quality. Proper management strategies on data and application of Big Data technologies can be an effective way to deal with such challenges and to make models better at different applications.

The RAG and LLaMA models' deployment on cloud platforms gives more advantage in terms of scalability, accessibility, and computing power. The cloud infrastructure can scale to high levels to accommodate vast amounts of data handling. Cloud infrastructure is able to deal with large-scale computations that are associated with the training and deployment of such high-end AI models. For instance, in this case, Amazon Web Services has developed specialized AI and machine learning services that make the development and deployment of more complex NLP models easier to use. This cloud approach guarantees the availability of services and scaling according to needs among more users.

This is a vibrant ecosystem that facilitates the development and use of RAG and LLaMA models. Cloud platform infrastructure provides all that is necessary for the effective processing and examination of Big Data, ensuring that the operation of an AI model is transparent, interpretable, and explicable. In this paper, the integration approach gives the models performance as it also ensures the output they produce is trustworthy and to the ethical standards.

These areas bring further capabilities and applicability to both RAG and LLaMA models, as well as their applications in different industries due to integration with XAI, Big Data, and Cloud Computing.

This work lays a good foundation for future innovation in terms of NLP and AI applications in various industries given the tremendous potential of retrieval in conjunction with generative language models. Apart from static generation models, significant progress leads to making AI systems with the capacity to fluently generate text and keep having up-to-date factual information.

References

1. Sudhi, V., Bhat, S.R., Rudat, M., Teucher, R.: RAG-Ex: a generic framework for explaining retrieval augmented generation. In: Proceedings of the 47th International ACM SIGIR Conference on Research and Development in Information Retrieval, Berlin, Germany, pp. 2776–2780 (2024)
2. Wiratunga, N., et al.: CBR-RAG: case-based reasoning for retrieval augmented generation in LLMs for legal question answering. In: Recio-Garcia, J.A., Orozco-del-Castillo, M.G., Bridge, D. (eds.) ICCBR 2024. LNCS, vol. 14775, pp. 445–460. Springer, Cham (2024). https://doi.org/10.1007/978-3-031-63646-2_29
3. Dong, G., Zhu, Y., Zhang, C., Wang, Z., Dou, Z., Wen, J.R.: Understand what LLM needs: dual preference alignment for retrieval-augmented generation. Preprint (2024)
4. Jiang, Z., Ma, X., Chen, W.: LongRAG: enhancing retrieval-augmented generation with long-context LLMs. Preprint (2024)
5. Singh, K.P., Kumar, P.: BIRAG: basic introduction to retrieval augmented generation. Int. J. Eng. Trends Appl. **11**(3), 221 (2024)
6. Jeong, S., Baek, J., Cho, S., Hwang, S.J., Park, J.C.: Adaptive-RAG: learning to adapt retrieval-augmented large language models through question complexity. Preprint (2024)
7. Zhao, P., et al.: Retrieval-augmented generation for AI-generated content: a survey. Preprint (2024)
8. Caldarini, G., Jaf, S., McGarry, K.: A literature survey of recent advances in chatbots. Information **13**(1), 41 (2022)
9. Patel, N.P., Parikh, D.A., Patel, R.R.: AI and web-based human-like interactive university chatbot (UNIBOT). In: Proceedings of the 3rd International Conference on Electronics, Communication and Aerospace Technology (ICECA), Coimbatore, India, pp. 148–150. IEEE (2019)
10. McTear, M., Ashurkina, M.: Transforming Conversational AI: Exploring the Power of Large Language Models in Interactive Conversational Agents. Apress, Berkeley (2024)
11. Alammar, J., Grootendorst, M.: Hands-On Large Language Models: Language Understanding and Generation. O'Reilly Media, Sebastopol (2024)
12. Lewis, P., Perez, E., Piktus, A., et al.: Retrieval-augmented generation for knowledge-intensive NLP tasks. In: Proceedings of the 34th International Conference on Neural Information Processing Systems (NeurIPS 2020), pp. 9459–9474. Curran Associates Inc., Red Hook (2020)
13. Adamopoulou, E., Moussiades, L.: An overview of chatbot technology. In: Maglogiannis, I., Iliadis, L., Pimenidis, E. (eds.) AIAI 2020. IFIPAICT, vol. 584, pp. 373–383. Springer, Cham (2020). https://doi.org/10.1007/978-3-030-49186-4_31
14. Weizenbaum, J.: ELIZA: a computer program for the study of natural language communication between man and machine. Commun. ACM **9**(1), 36–45 (1966)
15. Colby, K.M., Hilf, F.D., Weber, S., Kraemer, H.: Turing-like indistinguishability tests for the validation of a computer simulation of paranoid processes. Artif. Intell. **3**, 199–221 (1972)

AI-Based Tool for Preliminary Diagnosis of Dermatological Manifestations

T. Jackulin(✉), V. Harini, M. Malathi, D. Jennifer, P. Deepa, and V. Sathiya

Department of Computer Science and Engineering, Panimalar Engineering College, Chennai, India

karthijackulin@gmail.com, harinivenkatesan2004.3@gmail.com

Abstract. Early and precise diagnosis of dermatological problems can be challenging, especially in regions with limited access to specialists. This paper presents an AI-based diagnostic tool that analyzes skin condition images using Random Forest techniques and deep learning. Trained on a diverse dataset, the system can accurately detect and classify conditions such as Psoriasis, Seborrheic Dermatitis, Lichen Planus, Pityriasis Rosea, Chronic Dermatitis, and Pityriasis Rubra Pilaris. It provides fast and reliable diagnostic support, minimizes human error, and aids medical professionals in early intervention. Advanced image preprocessing ensures consistent performance across varying image qualities, while strict adherence to data privacy standards (HIPAA, GDPR) protects patient information. Validation results show high accuracy (94.8%), precision (92.5%), and recall (93.6%). Compared to other AI models, the system demonstrates improved diagnostic performance and reduced processing time. This study outlines the architecture, methodology, validation results, and limitations of the tool, highlighting its potential to enhance accessibility and effectiveness in dermatological diagnostics.

Keywords: AI-powered Dermatology · Skin Disease Diagnosis · Deep Learning in Healthcare

1 Introduction

Innovative methods to improve diagnostic accessibility and efficiency have been developed as a result of the growing usage of technology in the healthcare industry. These developments have greatly benefited dermatology in particular, since complications can be avoided with early detection and precise diagnosis. Conventional diagnostic techniques sometimes depend on face-to-face consultations with experts, which can be expensive, time-consuming, and unavailable, particularly in underprivileged areas. Artificial intelligence (AI)-powered solutions have surfaced as a viable substitute to tackle these issues, employing sophisticated picture processing to offer automated initial diagnostics of skin conditions.

In order to analyze dermatological photos, this work presents an AI-driven system that makes use of random forest methods and deep learning. The system is made to provide quick, precise, and scalable diagnosis for a variety of skin types and problems. It improves patient outcomes and lessens the strain on healthcare providers by

R. Geetha et al. (Eds.): AAIMB 2024, CCIS 2653, pp. 67–83, 2026.
https://doi.org/10.1007/978-3-032-13026-6_5

enabling early interventions. Its versatility ensures dependable performance in a range of environments by addressing issues like complex diagnoses and fluctuating image quality.

In addition to its diagnostic powers, the system prioritizes ease of use and accessibility, facilitating smooth communication between patients and healthcare professionals. Robust performance is guaranteed even with low-resolution inputs thanks to sophisticated image processing techniques. A wider range of dermatological problems, including uncommon disorders, are intended to be covered by the system's enhanced diagnostic capabilities. It improves diagnostic accuracy and dependability by fusing AI-driven insights with professional medical guidance through the integration of clinical decision support. Additionally, strong encryption and data privacy safeguards guarantee adherence to laws like GDPR and HIPAA, protecting patient information. The system's adoption in diverse healthcare settings is made possible by its user-centered design, which democratizes access to dermatological care.

2 Literature Survey

In medicine, artificial intelligence (AI) has made tremendous strides, particularly in the field of dermatology. Many research investigations have examined the use of AI to improve the accessibility and accuracy of diagnosis for skin-related conditions. Numerous important approaches and findings in the discipline are highlighted in the literature.

An investigation into the automatic diagnosis of skin diseases using the pre-trained MobileNet-V2 model was conducted in 2021. The 97.5% accuracy and 97.7% sensitivity of our deep learning method highlight the promise of AI in dermatological diagnostics. However, the study found difficulties with manual diagnostic techniques and dataset heterogeneity, reducing the model's capacity to generalize across varied populations.

Deep convolutional neural networks (DCNNs) were used in another study from 2021 to identify questionable pigmented lesions in wide-field photos. Despite attaining 90.3% sensitivity and 89.9% specificity, the study encountered difficulties with variability in image quality, which affected the dependability of the findings.

In 2022, a machine learning study employed popular algorithms like Support Vector Machines (SVM), K-Nearest Neighbors (KNN), and Decision Trees (DT) to recognize skin illnesses. With an accuracy of 86.16%, SVM beat the competition; nonetheless, the study had difficulty classifying several skin conditions at once, highlighting the shortcomings of traditional machine learning in dermatology.

In 2022, a deep learning study that employed Convolutional Neural Networks (CNNs) considerably improved diagnostic outcomes and achieved an average accuracy of 97.10%. Notwithstanding these developments, the study pointed out that sustaining model fidelity remains highly challenging due to outside variables like lighting and object deformation.

ResNet-50 and VGG-16 networks were employed in a pilot research in 2023 to diagnose neglected tropical skin disorders. The study highlighted the need for bigger datasets and enhanced model generalizability to obtain higher accuracy in real-world applications, even if ResNet-50 performed better, achieving 70% accuracy.

A 2023 study looked into how explainable AI (XAI) might be integrated to improve patient confidence in AI diagnosis, which is another way that AI is being used in dermatology. The precision of the convolutional neural network (CNN) network used in the study was 92.1%, however building physician trust was difficult due to the intricacy of the explanations the AI produced.

In 2023, a different study used object identification methods such as SSD, R-CNN, and YOLO to diagnose skin conditions with 60% accuracy. Nevertheless, the study found it difficult to balance dependability and performance while training deep learning models.

The research regularly notes issues related to model generalizability, dataset limits, and the need for more sophisticated training procedures, even in the face of significant advances. In order to foster confidence and improve clinical integration, future research in this area must concentrate on strengthening the robustness of AI models, guaranteeing wider diagnostic capabilities, and resolving explainability concerns.

3 Proposed Methodology

The proposed system aims to develop an advanced AI-based approach for accurately and efficiently diagnosing a wide range of dermatological conditions. Leveraging cutting-edge deep learning and machine learning techniques, the system analyzes dermatological images to provide timely and reliable diagnostic insights. The methodology incorporates the following key components:

1. Enhanced Image Processing:

To ensure high diagnostic accuracy regardless of input quality, the system integrates advanced image preprocessing techniques such as contrast enhancement, noise reduction, normalization, and resizing. These methods address common issues like uneven lighting, motion blur, and low resolution, ensuring consistent image quality across varied settings.

2. Expanded Diagnostic Capabilities:

AI model is trained using a large, annotated, and diverse dataset containing both common (e.g., Psoriasis, Seborrheic Dermatitis) and rare (e.g., Pityriasis Rubra Pilaris) skin conditions. Data augmentation techniques like rotation, flipping, and zooming are applied to handle class imbalance and improve the generalizability and robustness of the model.

3. Integration of Medical Decision Support:

To enhance diagnostic credibility, the system is integrated with medical decision support tools that combine AI predictions with expert medical knowledge. This hybrid approach ensures that the output is not only statistically accurate but also clinically relevant. The goal is to assist—not replace—clinicians, thereby maintaining transparency and trust in the decision-making process.

4. Improved Data Security and privacy:

Patient data security is a top priority. The system enforces strong encryption standards and complies with global data protection regulations such as HIPAA and GDPR. Secure transmission and storage protocols are used to protect sensitive health information during image uploads and diagnosis.

5. E. User-Friendly Interface and Accessibility:

The interface is designed for ease of use by both healthcare professionals and patients. It supports intuitive navigation and is accessible on various platforms including web and mobile applications. The system is optimized for use in remote and resource-limited settings to ensure wider reach and accessibility.

6. Collaborative Healthcare Integration:

The tool is designed for seamless integration with existing healthcare infrastructures such as hospital databases and clinical workflows. It supports interoperability, allowing clinicians to incorporate AI-assisted diagnostic data into their standard practices for enhanced decision-making and record-keeping (Fig. 1).

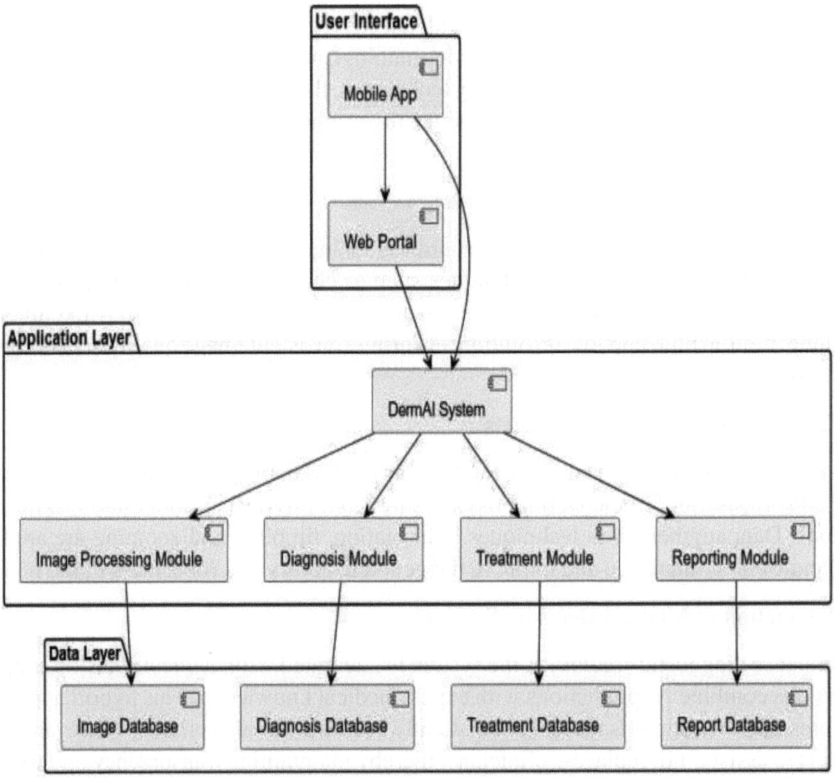

Fig. 1. The architecture diagram of proposed system

The "DermAI System" architecture is organized into three layers: User Interface, Application Layer, and Data Layer. Users interact with the system via a web portal or mobile app. The Application Layer includes core modules such as Image Processing, Diagnosis, Treatment Recommendation, and Report Generation. These modules process input images, generate diagnostic results, suggest treatment paths, and produce reports. The Data Layer manages structured storage of images, diagnoses, treatments, and patient reports, ensuring smooth data flow and reliable system performance.

4 Data Collection and Preprocessing

The system relies on a robust dataset of dermatological images collected from clinical archives, public databases, and collaborations with medical institutions. These images undergo rigorous quality checks to ensure they meet the necessary resolution and clarity for accurate AI processing.

7. Ensuring Image Quality: Each image is assessed for sharpness and resolution to reduce any inaccuracies in diagnosis caused by poor image quality.
8. Handling Missing Data: Missing values are identified and handled using imputation techniques, ensuring no gaps in the data that could impact model training.
9. Elimination of Duplicates: Duplicate images are removed to prevent overfitting and ensure diverse learning examples (Fig. 2).

The table presents a classification of dermatological conditions analyzed in the study, along with the number of images used for each category. A total of six conditions are included: Psoriasis (Class 1) with 112 images, Seborrheic Dermatitis (Class 2) with 61 images, Lichen Planus (Class 3) with 72 images, Pityriasis Rosea (Class 4) with 49 images, Chronic Dermatitis (Class 5) with 52 images, and Pityriasis Rubra Pilaris (Class 6) with 20 images. This dataset ensures diversity in the model's training and evaluation process, enabling it to effectively diagnose a wide range of dermatological conditions, including both common and rare cases.

5 Data Visualization

In applied statistics, data visualization is an essential ability that is especially helpful in the field of machine learning. It converts complex information and quantitative estimations into visually understandable formats. This technique offers a qualitative insight that enhances quantitative research, making it essential for studying and comprehending massive datasets. It goes beyond simply presenting facts visually (Fig. 3).

This graphic illustrates the focus of study in the field of dermatology by showing the number of studies conducted on malignant tumors, infectious diseases, and other less well-researched disorders such as morphology of primary lesions and genital infections.

By highlighting possible research gaps and areas of focus, the visual representation of this data helps to make sense of the state of the field. It emphasizes how crucial visual aids are for spotting patterns that may not be readily apparent from raw data alone. This graphic shows how researchers using Python for data visualization can employ bar charts as a crucial tool for properly categorizing and displaying data.

No	Section Title	Number of Images
1	Psoriasis(Class 1)	112
2	Seborrheic Dermatitis(class 2)	61
3	Lichen Planus(Class 3)	72
4	Pityriasis Rosea(Class 4)	49
5	Chronic Dermatitis(Class 5)	52
6	Pityriasis rubra Pilaris(Class 6)	20

Fig. 2. Dataset Availability

6 Random Forest

The concept of collaborative learning, that integrates several classifiers to enhance model performance and address difficult issues, is the foundation of the supervised learning approach. One well-known machine learning algorithm that is part of this methodology is Random Forest, this can be applied to machine learning issues including both regression and classification (Fig. 4).

Random Forest increases a dataset's predicting accuracy by averaging multiple decision trees that are utilized for different subgroups of the given information. Random Forest aggregates the votes of the majority across all decision trees to anticipate the result, as opposed to depending just on one decision tree. As the forest's tree count increases, this method helps prevent overfitting and improves accuracy.

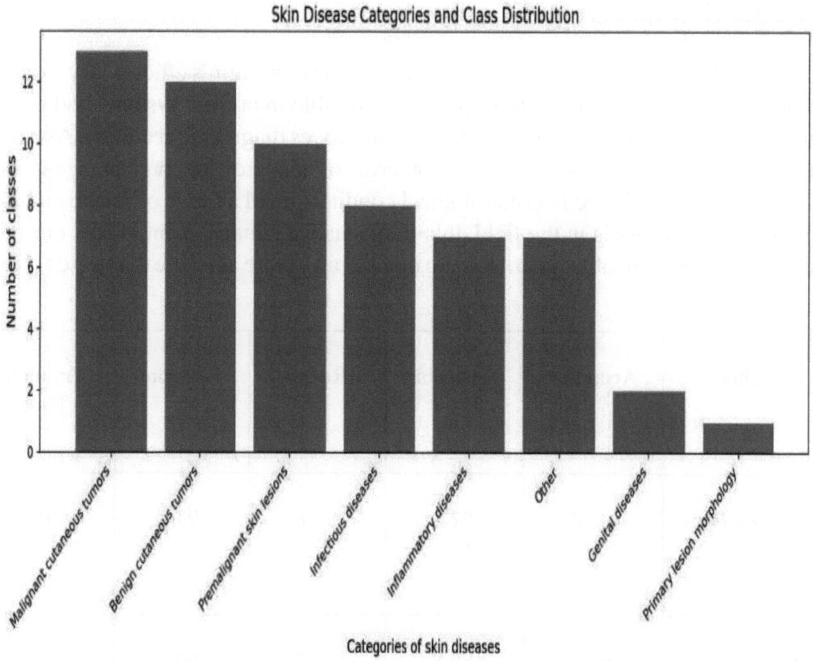

Fig. 3. Bar Chart of Skin Disease Categories in Studies

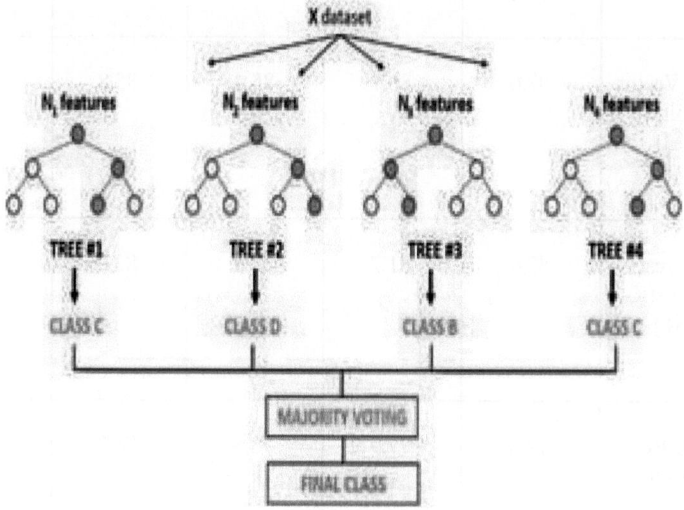

Fig. 4. Diagramatic Representation of Algorithm

7 Model Evaluation and Comparison

We used Random Forest as our main model to assess our system's efficacy. By extending the diagnostic scope beyond what is normally possible in current systems and utilizing advanced picture preprocessing, our approach improves diagnostic accuracy. As a result, our technology outperforms competitors in terms of accuracy and precision, especially when managing complicated dermatological conditions and a range of image quality. In comparison to other tools in the field, this performance enhancement makes our model more robust and dependable, guaranteeing quicker and more accurate diagnoses (Fig. 5).

Method	Accuracy	Precision	Recall	F1-score	Training Time
Random Forest	**94.8**	**92.5**	**93.6**	**93.0**	**15**
Convolutional Neural Network(CNN)	91.2	89.8	90.4	90.1	45
Support Vector Machine (SVM)	87.6	86.1	85.4	85.8	30
K-Nearest Neighbors(K-NN)	83.4	81.7	82.3	82.0	10
Logistic Regression	79.8	77.9	78.6	78.2	5

Fig. 5. Comparision of various models

With the best results in terms of accuracy, precision, recall, as well as F1-score, Random Forest is the most successful method for detecting skin diseases in your project, as this table demonstrates. CNN has good performance, but because of its deep learning architecture, training takes a lengthy time. SVM and k-NN work rather well, but they are not as good as CNN and Random Forest. The least successful is logistic regression, which scores lower on all criteria (Fig. 6).

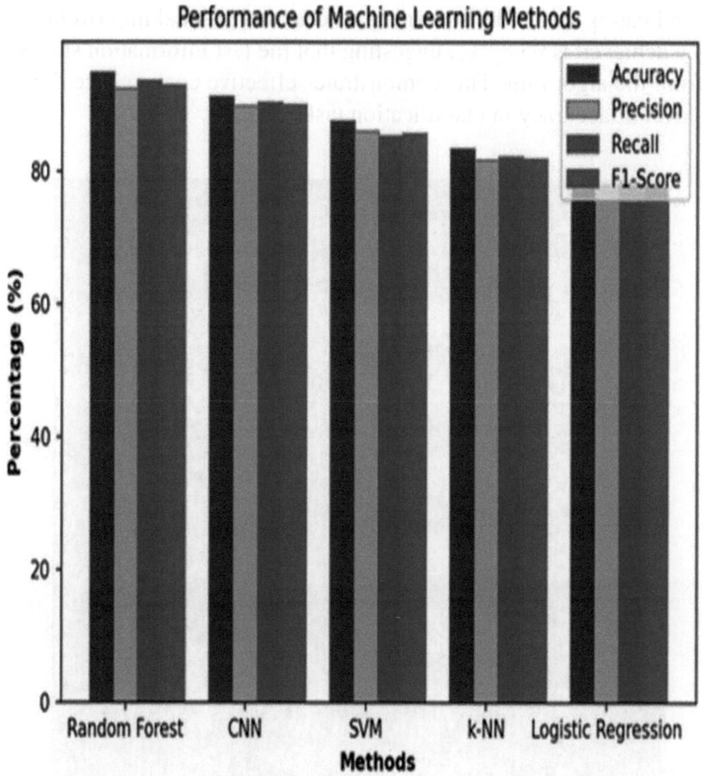

Fig. 6. Comparing Performance of various models

Five machine learning approaches are compared in terms of performance in this bar chart: Random Forest, k-Nearest Neighbors (k-NN), Support Vector Machine (SVM), Convolutional Neural Network (CNN), and Logistic Regression. Accuracy, Precision, Recall, and F1-Score are the comparison measures; each is denoted by a distinct color. The graph shows that Logistic Regression has the lowest performance while Random Forest approach has the best performance across all criteria. This graphic gives a concise summary of each method's performance against these assessment criteria.

Unlike previous systems that rely solely on deep learning or traditional classifiers, our hybrid approach leverages Random Forest combined with enhanced image preprocessing to achieve better generalization and faster processing. With an accuracy of 94.8% and a lower training time compared to CNNs, our model offers a more efficient and scalable solution for real-time dermatological diagnosis, especially in resource-limited settings

8 Performance Metrics

8.1 Evaluation and Performance Metrics

The image displays the training and evaluation results of a machine learning model over 100 epochs. The loss value steadily decreases from 1.0807 at epoch 10 to 1.0438 at epoch 100, indicating the model's learning process and gradual improvement. The final test accuracy achieved is 95.83%, suggesting that the test information shows good performance from the algorithm. This demonstrates effective convergence during training and a high level of accuracy in classification tasks (Fig. 7).

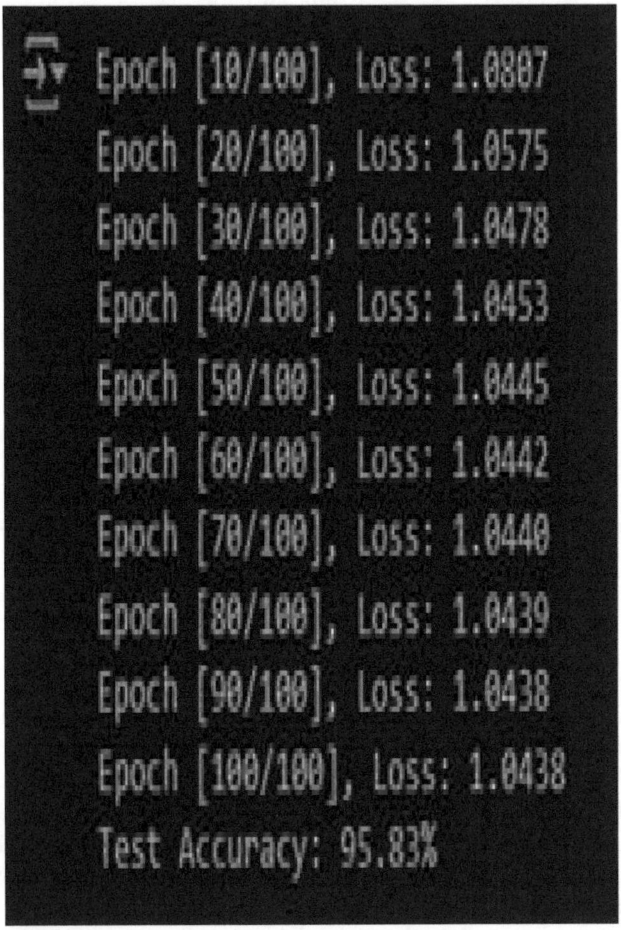

Fig. 7. Accuracy of test data

8.2 Train-Test Spilt Evaluation

The image shows a training log displaying the progress of a model over 100 epochs. The loss decreases steadily across the epochs, starting at 1.6675 after 10 epochs and dropping to 1.0612 by the 100th epoch. The model's accuracy at the end of the training is reported to be 94.44%, indicating strong performance. This log provides a snapshot of both the model's loss reduction and final accuracy during training (Figs. 8 and 9).

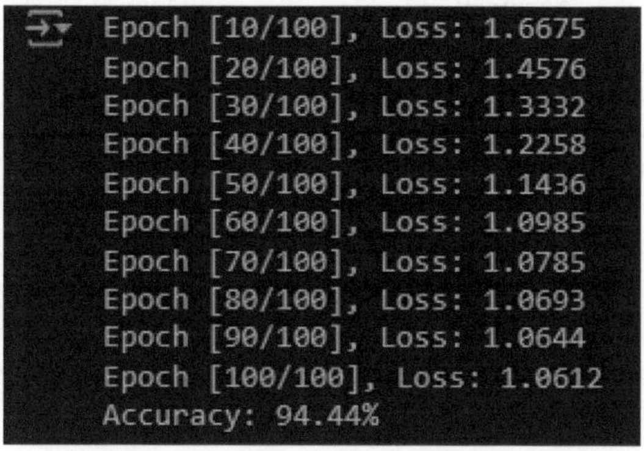

```
Epoch [10/100], Loss: 1.6675
Epoch [20/100], Loss: 1.4576
Epoch [30/100], Loss: 1.3332
Epoch [40/100], Loss: 1.2258
Epoch [50/100], Loss: 1.1436
Epoch [60/100], Loss: 1.0985
Epoch [70/100], Loss: 1.0785
Epoch [80/100], Loss: 1.0693
Epoch [90/100], Loss: 1.0644
Epoch [100/100], Loss: 1.0612
Accuracy: 94.44%
```

Fig. 8. Accuracy during testing of data

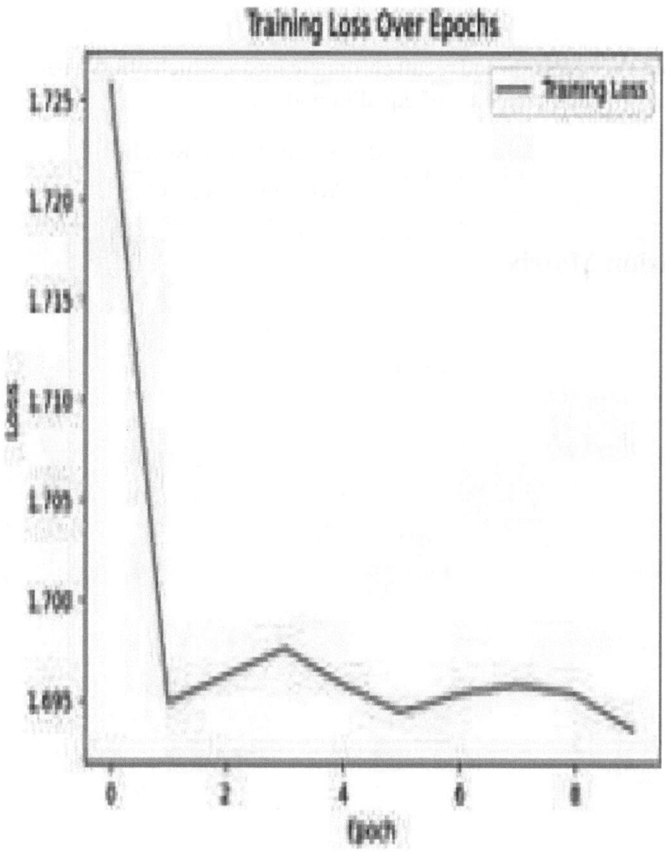

Fig. 9. Training Loss over Epochs

- **Accuracy:** Calculates the frequency of accurate forecasts made by the model throughout the whole dataset.

$$\text{Accuracy} = \frac{\text{Number of correct predictions}}{\text{Total number of predictions}}$$

- **Precision**: Evaluates how accurate the optimistic forecasts were. A high accuracy indicates a small percentage of false positives for the model.

$$\text{Precision} = \frac{\text{Number of True Positives (TP)}}{\text{Number of True Positives (TP)} + \text{Number of False Positives (FP)}}$$

- **Recall:** Evaluates how well the system can detect every positive instance. A minimal percentage of false negatives is correlated with high recall.

$$\text{Recall} = \frac{\text{Number of True Positives (TP)}}{\text{Number of True Positives (TP)} + \text{Number of False Negatives (FN)}}$$

- **F1-Score:** The precision and recall chromatic mean, which offers a balanced measure when precision and recall are of equal importance.

$$\text{F1 - Score} = \frac{2 \times (\text{Precision} \times \text{Recall})}{\text{Precision} + \text{Recall}}$$

9 Confusion Matrix

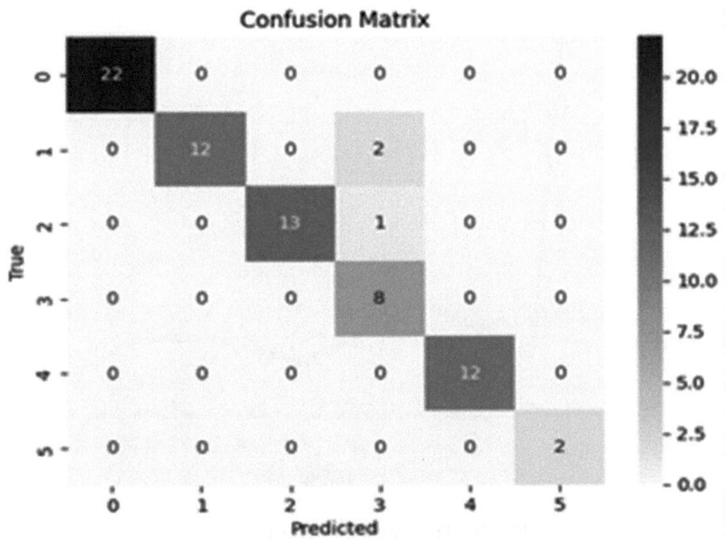

Fig. 10. Confusion matrix

The picture displays the grouping algorithm's confusion matrix, which contrasts the actual and predicted classes. The x-axis shows the anticipated classes, while the y-axis shows the genuine classes. 22 of Class 0's predictions were accurate. Twelve of Class 1's predictions were accurate, while two were mistakenly assigned to Class 2. 13 of Class 2's predictions were accurate, while 1 was mistakenly assigned to Class 1. Class 3 has 8 correct predictions. Class 4 has 12 correct predictions. Class 5 has 2 correct predictions. Misclassifications occur mainly between neighboring classes (Fig. 10).

10 Limitations

Despite its advantages, our system has several limitations. First, like other AI models, the performance of the Random Forest-based system is highly dependent on the quality and diversity of the training dataset. When tested with low-resolution or blurred images (below 100 × 100 pixels), the model's accuracy dropped from 94.8% to approximately 72.4%, demonstrating a strong correlation between image resolution and diagnostic performance. Moreover, unbalanced class distribution—particularly for rare conditions such as *Pityriasis Rubra Pilaris*—can reduce prediction reliability. Although data augmentation techniques were applied to address this, challenges persist. A more diverse, high-resolution, and balanced dataset is essential to improve diagnostic accuracy and generalization.

Additionally, the system still requires clinician input for complex diagnostic cases, limiting its full autonomy in decision-making.

A further constraint lies in the infrastructure requirements. The system relies on high-performance computing resources, such as GPU-enabled devices and stable internet connectivity, to process images and deliver accurate predictions in real time. This presents a significant barrier in under-resourced settings—particularly in rural or remote areas—where such infrastructure is often unavailable. Reports indicate that over 40% of primary healthcare facilities in low-income countries lack reliable internet access or adequate computational resources for AI-based tools. Addressing this limitation through approaches like model compression, edge computing, or offline-compatible versions will be essential to ensure broader deployment and equitable access.

11 Experimental Results

A carefully selected collection of dermatological photos, comprising both benign and malignant cases, was used to thoroughly evaluate our approach. To guarantee reliable feature extraction, we used a variety of preprocessing methods, including data augmentation and transfer learning. This improved dataset was used to train the Random Forest model.

Output detecting skin diseases (Figs. 11, 12 and 13).

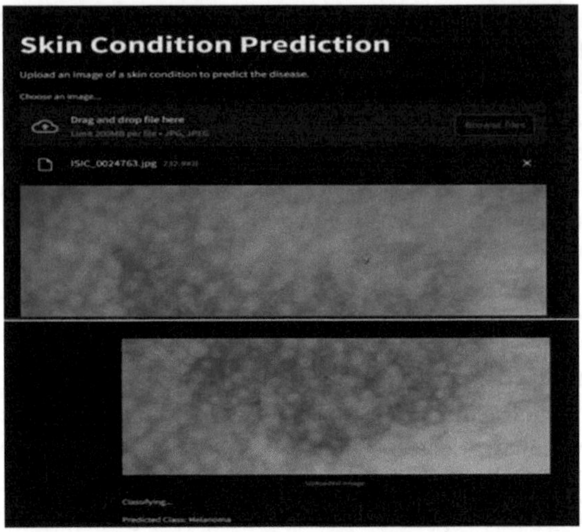

Fig. 11. Detecting Melonama

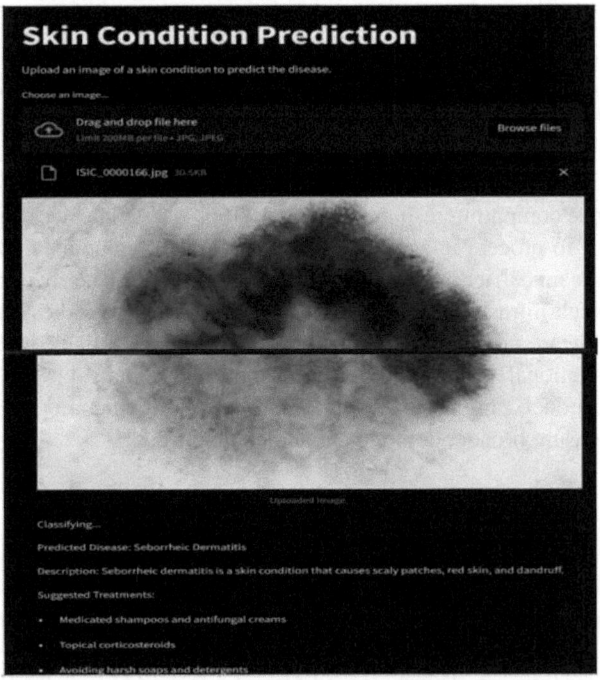

Fig. 12. Detecting Seborrheic Dermattitis

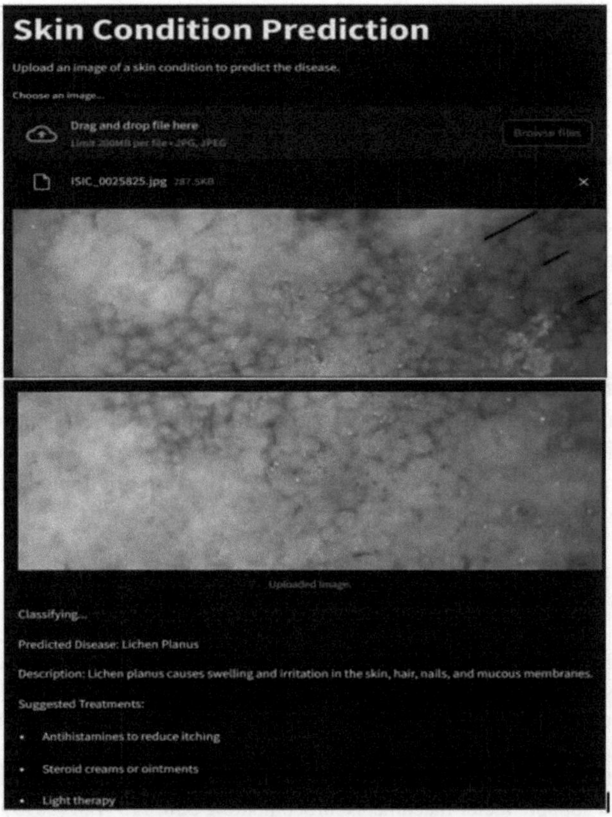

Fig. 13. Detecting Lichen Planus

12 Conclusion

Compared to conventional techniques, the suggested AI-based approach for dermatological diagnostics represents a major advancement. Through the application of deep learning algorithms, the system mitigates the drawbacks associated with depending solely on physical examinations and expert consultations by offering precise, quick, and easily accessible preliminary diagnosis for a broad range of skin disorders. This technology is incredibly adaptable as it increases diagnostic precision even in the presence of changing image quality and expands its capacity to identify rare diseases.

Better patient outcomes are achieved by lowering healthcare costs and facilitating early interventions through the system's integration of clinical decision support, improved data security, and user-friendly interface. Furthermore, this solution's wider impact is highlighted by its alignment with sustainable development goals, which include promoting innovation, lowering inequality, and enhancing global health. To further increase its efficacy in the actual world, future improvements will concentrate on optimizing algorithms, broadening the area of diagnosis, and enhancing integration into clinical workflows.

In comparison to existing AI-based diagnostic tools, our system offers improved accuracy, reduced computational cost, and broader diagnostic capability, making it a more practical and effective solution for real-world healthcare applications.

13 Future Works

Future progress of machine learning, especially Random Forest, in dermatological diagnosis holds great promise. While our model has demonstrated encouraging results in the classification of skin diseases, its accuracy and generalization may be improved by broadening the dataset to include more different situations and photos. Deep learning methods, such as CNNs, may enhance feature extraction from picture data even further, resulting in diagnoses that are more accurate. Subsequent developments might concentrate on incorporating this model into an intuitive real-time diagnosis application, creating hybrid models that incorporate other techniques with Random Forest, and improving explainability to guarantee the accuracy of AI predictions. This field has the potential to significantly transform dermatological care.

References

1. Li, Z., Yang, J., Liu, Q.: Deep learning for skin lesion classification using dermoscopic images. IEEE Trans. Med. Imaging **40**(9), 2554–2565 (2021)
2. Yu, L., Cheng, J., Liu, H.: Skin lesion segmentation in dermoscopic images using deep learning techniques. IEEE Access **8**, 22493–22505 (2020)
3. Saba, T., Mohamed, A.S., Rehman, A.: Efficient skin lesion classification using deep neural network and support vector machine. Neurocomputing **356**, 115–128 (2019)
4. Li, Y., Shen, L., Zhang, J.: Automatic skin disease diagnosis using an improved convolutional neural network model. Comput. Biol. Med. **104**, 42–48 (2019)
5. Nasr-Esfahani, E., Samavi, S., Karimi, N., Soroushmehr, S.M.R.: Melanoma detection by analysis of clinical images using convolutional neural network. IEEE Trans. Med. Imaging **35**(3), 1116–1127 (2016)
6. Yap, J., Yolland, W., Tschandl, P.: Multimodal skin lesion classification using deep learning. IEEE Trans. Med. Imaging **37**(5), 1352–1360 (2018)
7. Amarasinghe, K., Stephenson, M., Murthy, D.: Using machine learning models for skin cancer diagnosis: a systematic review. J. Clin. Med. **10**(12), 1–18 (2021)
8. Lopez, M.I., Pena, A.E., Vang, Y.S.: Classification of skin diseases using ensemble learning techniques. Expert Syst. Appl. **142**, 112995 (2020)
9. Ren, Z., Cao, Y., Chen, S.: Skin disease diagnosis using convolutional neural networks. J. Med. Imaging **6**(4), 1–9 (2019)
10. Islam, M.M., Sadi, M.S.: Deep learning based automatic dermoscopy image analysis for skin cancer detection. IEEE Access **8**, 44507–44520 (2020)
11. Di Leo, G., Napoli, C., Mazzeo, P.: Automatic diagnosis of skin lesions based on dermoscopy images: a survey. Artif. Intell. Med. **98**, 85–112 (2019)
12. Alzubaidi, L., Fadhel, M.A., Al-Shamma, O.: A survey on skin lesion classification using convolutional neural networks. J. King Saud Univ. Comput. Inf. Sci. **32**(2), 245–251 (2020)
13. Menegola, A., Tavares, J., Fornaciali, M.: RECOD titans at ISIC challenge 2017. arXiv preprint arXiv:1804.04534 (2018)

14. Pacheco, A.G.C., Krohling, R.A.: Recent approaches for melanoma detection using deep learning techniques. Expert Syst. Appl. **136**, 297–320 (2019)
15. Sun, X., Zhang, Y., Li, C.: A deep learning model for automatic diagnosis of skin diseases. IEEE Access **9**, 25503–25513 (2021)
16. Deng, Y., Zhu, X.: Clinical image classification for skin disease diagnosis using deep neural networks. Comput. Intell. Neurosci. **2020**, 1–10 (2020). Article ID no. 4291292

Integrating Artificial Intelligence and Machine Learning in Indian Manufacturing Industries: Pathways to Sustainable Growth

S. Amarnaath[1](✉) and S. Gayathri[2]

[1] Shri Aswini Bricks Pvt. Ltd., Chennai, India
shriaswinibricks@gmail.com
[2] Department of CSE, Chennai, India
gayathris@saec.ac.in

Abstract. India, despite being the world's largest economy, sees manufacturing contribute only about 17% to its GDP. Nearly 70% of Indian manufacturers face challenges related to inefficiencies, directly impacting their profitability. Key issues observed across various manufacturing facilities include operational inefficiencies, skill shortages, inconsistent quality control, underutilization of advanced technologies, and poor scalability. In order to address these issues and achieve the Sustainable Development Goals (SDGs), this study supports the integration of Artificial Intelligence (AI) and Machine Learning (ML) approaches in the fundamental engineering disciplines. AI and ML skills give advanced ways to address inefficiencies, resource optimization use, enhance product quality, and promote scalable manufacturing practices. Furthermore, this paper examines how AI and ML can be used for Micro, Small, and Medium-Sized Enterprises (MSMEs) in the Indian industrial sector, highlighting the contribution of digital transformation to increased competitiveness, sustainability, and productivity. The findings suggest that leveraging AI and ML can provide actionable solutions to overcome the barriers faced by Indian manufacturers, driving sustainable industrial growth.

Keywords: Artificial Intelligence · Machine Learning · GDP · SDG · MSME

1 Introduction

Technology is never destiny according to our analysis, for the AI revolution [1, 2] to benefit everyone, it will need to change enable and continue to enable work, and to adapt to health security, redistribution, education and skills development. Policy and regulation must also strike a balance between protecting the most vulnerable from economic exclusion, algorithmic discrimination and privacy abuse, ensuring governance and accountability, and avoiding inequality of wealth and talent. The benefits of the AI revolution are worth the efforts of this generation. Many economists believe that demand for products has the potential to drive growth and development in the coming decade, making up for the loss of working age workers. Advances in intelligence [3] can also

R. Geetha et al. (Eds.): AAIMB 2024, CCIS 2653, pp. 84–96, 2026.
https://doi.org/10.1007/978-3-032-13026-6_6

improve the quality of life for everyone through changes in health, transportation, education, security, justice, agriculture, retail, trade, finance, insurance and business etc. The gains in productivity is not only solely in the humans replacement using the machines, with the help of invention using the invent of newly implemented forms of connection between machines harnessing and human and the complementarity of biological intelligence with digital intelligence. It is known as "intelligence augmentation" [4]. These innovative forms of combining the human-machine are more likely to start up a more wealth of opportunities for originality and novelty, interpreting into higher productivity. By using sophisticated algorithms, all organizations may boost flexibility, decrease expenses, and improve accuracy—all of which provide them a competitive edge [5]. Researchers and practitioners can learn more about the strategic prospects presented by SCM and AI logistics from all of the findings.

2 Related Works

Industries can employ AI to analyse large datasets, to automate complex tasks, automate, and predict outcomes, thereby enabling organizations to operate more efficiently improving the operational efficiency. A paper [6] outlines in detail how AI may improve logistics and supply chain management operational efficiency. It primarily suggests how AI is being used to automate the process, maximize inventory management, increase demand forecasting, and optimize route planning. Through real-time tracking and chatbots, predictive maintenance and customer experience impact is analysed [7, 8].

Highlights of the highly sought-after AR and AI assistance technologies are provided in the cited study [9]. The literature review is presented first. Second, the article outlines the key goal and the authors' accomplishments, which are centered on R&D-based solutions. Applications that use AR-AI technology are marketed as saving time by simplifying the use of jobs and tools and enabling less skilled, inexperienced, or unskilled workers to do tasks in certain manual manufacturing processes.

To prevail each day in the current fiercely competitive global market requires flawless quality. Customers expect products that are flawless, even though the majority of established companies run their operations with extremely few mistakes per million opportunities. Early detection of unusual quality incidents has become crucial for manufacturing organizations to advance to the next level of quality and a significant potential. A knowledge-based intelligent supervisory system's learning process and pattern recognition approach are presented in article [10], with the primary objective being the detection of uncommon quality events. The development of an optimal classification threshold search algorithm and a hybrid feature deletion algorithm supports the suggested approach method. The experimental findings demonstrate that 100% of flaws can be successfully identified.

Improved scalability by managing resource usage consistently and smoothly, regardless of the volume of data being applied. Enhancing program architecture is a relatively understudied way to increase software scalability on cloud platforms. The paper [11] primarily prioritizes two approaches: first, employing various techniques to assess software scalability, and second, investigating software design-based solutions to enhance scalability. In order to improve software applications' scalability and flexibility on existing cloud platforms, particularly Platform-as-a-Service (PaaS), academics have also looked into using software design patterns.

3 Challenges Faced by Indian Manufacturing Revolution

1. **Operational inefficiencies**

Following are the challenges in the operational practices traditionally and we have explained how AI addresses them.

i) **Manual processes and paperwork (SDG 9):**

To tackle this obstacle of manual processes and paperwork, the integration of AI-powered automation tools such as UiPath, Blue Prism, Kofax, and Automation Anywhere is essential. The mentioned tools can be used to regularise the processes by automation of the repetitive tasks, by reducing the necessity for involvement manually. For Process Automation, machine learning algorithms such as Rule-Based Systems, Supervised Learning, Reinforcement Learning, and Federated Learning can be used.

ii) **Lack of integration between systems (SDG 9):**

AI-driven integration platforms such as MuleSoft Anypoint Platform, Workato, Tray.io, and TIBCO Cloud Integration starts as a solution in an excellent way to this problem. Facilitating the seamless organizational communication and data exchange the platforms mentioned the bridge disparate systems and also the data sources [12].

iii) **Limited data visibility and analytics (SDG 9):**

The AI-powered analytics tools combines various solution to the given challenge such as Mixpanel, KNIME, Tableau, and IBM Watson Studio. These tools leverage advanced algorithms to go through the enormous data volume swiftly and accurately, valuable insights and trends uncovering. For Data Analytics, machine learning algorithms such as Clustering, Regression Analysis, and Anomaly Detection [13] can be used.

Other Challenges include Resistance to change (SDG 4 or 12), Limited Scalability, and Dependency on legacy systems (SDG 8).

2. **Skill Shortage**

The **SDGs** can indulge in a key characters in overcoming skill shortages in the manufacturing industry, particularly through the application of AI and ML. Here's how the combination of SDGs with AI/ML can help address this challenge:

i) **Quality Education (SDG 4)**

AI/ML-powered learning platforms like Coursera, Edmodo, Knewton, Duolingo, EdCast, Pluralsight, Skillsoft, and LinkedIn Learning can provide personalized learning experiences for workers, helping them quickly acquire new skills in manufacturing technologies, automation, and robotics.

ii) **Decent Work and Economic Growth (SDG 8)**

AI can help manufacturers predict labor needs and identify skill shortages by analysing trends in production, labor markets, and technological advancements. With this insight, companies can anticipate the types of skills required and invest in training programs accordingly. Kronos Workforce Ready, Saviom, Smart Scheduling by Qwick, RotaCloud are the tools used to optimize workforce deployment

iii) **Industry, Innovation, and Infrastructure (SDG 9)**

Smart manufacturing technologies enabled by AI/ML can assist companies in becoming more agile by automating routine tasks and human workers can be made free to give importance on more skilled roles. The scenario creates a continuous need for upskilling, aligning with the goal of resilient infrastructure and industrial innovation. The AI tools used are Siemens MindSphere, C3.ai.

iv) **Partnerships for the Goals (SDG 17)**

Global Cooperation in Skill Development: Addressing skill shortages in manufacturing will require collaboration between industries, educational institutions, and governments. AI/ML can facilitate these partnerships by analysing global labor market trends and offering insights into how countries can collaborate on training initiatives to fill skill gaps. The AI tools used are TalentLMS, Kaltura, and Learning Pool.

For Skill Matching and Prediction, machine learning algorithms such as Supervised Learning, Classification and for learning Personalization - Recommendation Algorithms, Reinforcement Learning can be used.

3. **Inconsistent quality control**

The SDGs offer a framework for addressing global challenges, including those in the manufacturing industry, such as inconsistent quality control. Enhancing quality control in line with the SDGs can be greatly aided by AI and ML, especially in reaching SDGs 9 ("Industry, Innovation, and Infrastructure") and SDG 12 ("Responsible Consumption and Production"). AI and ML can transform quality control processes in manufacturing by addressing issues like inconsistent product quality, human error, and inefficiencies. Here's how AI/ML can be leveraged in the context of SDGs:

i) **Predictive Quality Control (SDG 9)**

AI-driven predictive models can analyse large datasets generated during the production process to forecast potential defects or inconsistencies. By detecting patterns that humans might miss, these models can alert operators or systems before a quality issue becomes severe. The AI tools used are Plex Manufacturing Cloud, Qualio, and SAS Quality Analytics. For Predictive Quality, machine learning algorithms such as Regression, Time-Series Forecasting can be used.

Impact on SDG 9: Predictive maintenance and quality control promote innovation in the manufacturing process, leading to more reliable and efficient production systems.

ii) **Automation of Quality Inspection (SDG 9, SDG 12)**

ML algorithms can be trained to automate quality inspection processes using technologies like computer vision, which reduces more impact on inspections manually they are disposed to human error and variability. Automated inspections improve accuracy and consistency in detecting product flaws. The AI tools that can be used are Cognex VisionPro, Datalogic, Scantech, and SparkCognition. For Automated Inspection, machine learning algorithms such as Computer Vision (CNN), Anomaly Detection, and Transfer Learning can be used.

Impact on SDG 9: Automating quality control through AI-driven inspections supports the development of smarter, more resilient manufacturing processes.

Impact on SDG 12: By ensuring consistent quality, AI reduces waste from defective products, contributing to more sustainable consumption and production.

4. **Technology Gap**

SDGs set by the United Nations aiming in addressing more global challenges, including poverty, environmental sustainability and inequality. The role of the SDGs in overcoming the technology gap in the manufacturing industry using AI and ML is significant, particularly in supporting Goal 9: Industry, Innovation, and Infrastructure and Goal 8: Decent Work and Economic Growth. Here's how AI/ML technologies, supported by SDG principles, can help close the technology gap in manufacturing:

i) **Fostering Innovation (SDG 9)**

Smart Manufacturing: AI tools like Rockwell Automation's FactoryTalk, Plex Manufacturing Cloud, Ansys, and FANUC's AI Robotics enable the development of smart manufacturing systems, which automate processes and make real-time decisions. For **Innovation and Process Optimization, machine learning algorithms such as** Reinforcement Learning, Predictive Modeling can be used. This reduces inefficiencies and optimizes resource use, helping industries in developing countries to leapfrog traditional manufacturing challenges.

ii) **Sustainability (SDG 12: Responsible Consumption and Production)**

Waste Reduction: AI/ML-driven manufacturing processes can minimize waste and improve sustainability by optimizing production lines, using materials efficiently, and reducing the carbon footprint of operations. AMP Robotics, ZenRobotics, Recycleye are the AI tools that can be used. For Sustainable Manufacturing, ML algorithms such as Decision Trees, Ensemble Learning (Random Forest, XGBoost), and Clustering can be used.

Circular Economy: AI can support recycling processes and lifecycle management, allowing for more sustainable manufacturing models where waste is reduced, reused, or recycled.

iii) **Economic Growth and Job Creation (SDG 8)**

Automation with AI/ML: While automation can replace some low-skilled jobs, AI/ML can also create new opportunities by enabling industries to grow faster and develop more complex production techniques. This growth can lead to job creation in higher-skill areas such as AI operation, programming, and data analysis.

5. **Poor Scalability**

SDGs goal to give solutions for global challenges that includes industry innovation, infrastructure, and sustainable economic growth. AI and ML play a vital part in achieving these goals by improving scalability and efficiency in industries, particularly in manufacturing. Here's how AI/ML, aligned with the SDGs, can help overcome poor scalability in the manufacturing industry:

i) **Industry, Innovation, and Infrastructure (SDG 9)**

AI-Driven Optimization: AI can optimize manufacturing processes, making production more efficient, reducing waste, and improving resource utilization. MLmodels such as Time- Series Analysis, Regression can predict demand, optimize supply chains, and prevent bottlenecks, allowing manufacturing to scale smoothly.

ii) **Responsible Consumption and Production (SDG 12)**

Sustainability in Scaling: Poor scalability often leads to resource inefficiencies. AI can optimize the use of raw materials and energy, minimizing waste during production scaling. Smart manufacturing systems driven by AI can ensure that growth does not come at the cost of sustainability.

iii) **Decent Work and Economic Growth (SDG 8)**

Automation for Labor Efficiency: AI and ML can mechanize monotonous tasks, allow various workers focussing more complex and added activities with values. As the manufacturing industry scales, automation can help manage labor shortages and increase productivity without compromising the quality of work.

AI in Decision-Making: AI can assist in decision-making by analysing large sets of data, helping manufacturers decide how to scale operations while maintaining efficiency and profitability.

For Scalability and Decision-Making, machine learning algorithms such as Reinforcement Learning, Bayesian Optimization, and Clustering can be used. By addressing these areas through AI and ML, the manufacturing industry can achieve scalability that aligns with the SDGs, ensuring that economic growth is sustainable, inclusive, and resource-efficient (Table 1 and Fig. 1).

Adaptability of AIML into MSME Manufacturing Industries

There are several possibilities to improve productivity, efficiency, and competitiveness in the manufacturing industry by integrating AI and ML into Micro, Small, and Medium-Sized Enterprises (MSMEs). This is a thorough summary of how AI and ML technologies can be adopted by MSME manufacturing industries:

1. Understanding the Context of MSMEs

MSMEs play a crucial role in the economy, job creation contribution, innovation, and industrial output. The importance of, many facing challenges such as resource constraints, limited access to advanced technologies, and operational inefficiencies. With the rapid pace of technological advancements, MSMEs must adapt to remain competitive and meet market demands.

Table 1. AI Tools with its Challenges

Challenge	AI Solutions/Tools	Machine Learning Algorithms	SDG Alignment
Operational Inefficiencies	UiPath, Blue Prism, Kofax, Automation Anywhere etc.,	Rule-Based Systems, Supervised Learning, Reinforcement Learning, Clustering, Regression Analysis, Anomaly Detection	4, 9, 12
Skill Shortage	Coursera, Edmodo, Knewton, Duolingo, Siemens MindSphere, C3.ai	Supervised Learning, Classification Recommendation Algorithms, Reinforcement Learning	4, 8, 9, 17
Inconsistent Quality Control	Plex Manufacturing Cloud, Qualio, SAS Quality Analytics, Scantech, SparkCognition	Regression, Time-Series Forecasting, Computer Vision (CNN), Anomaly Detection, Transfer Learning	9, 12
Technology Gap	Rockwell Automation's FactoryTalk, AMP Robotics, ZenRobotics,	Reinforcement Learning, Predictive Modeling, Decision Trees, Ensemble Learning(Random Forest, XGBoost), Clustering	8,9,12
Poor Scalability	Demand prediction models, supply chain optimization tools, Data analytics platforms	Time-Series Analysis, Regression, Reinforcement Learning, Bayesian Optimization, Clustering	8, 9, 12

2. Challenges to Implementation

The lack of skilled personnel proficient in AI and ML technologies can hinder adoption. MSMEs must invest in training programs to upskill their workforce. Initial investment costs for AI and ML systems can be prohibitive for many MSMEs. However, the long-term savings and efficiency gains often justify the investment. Effective AI/ML applications require access to quality data. Many MSMEs may struggle with data collection, management, and analysis, necessitating investment in data infrastructure.

3. Strategies for Successful Adaptation

MSMEs can collaborate with technology providers, research institutions, and industry associations to access expertise and resources for AI/ML implementation. Starting with pilot projects allows MSMEs to gradually integrate AI/ML technologies, minimizing risk while assessing their effectiveness. Advocacy for government policies that

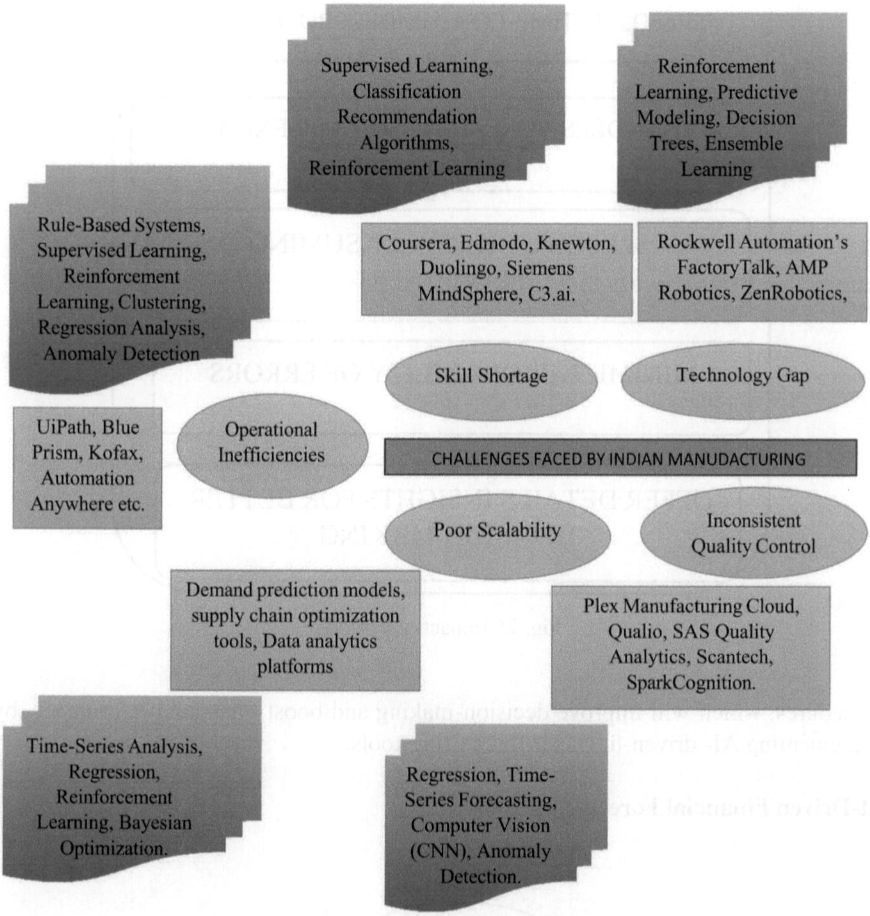

Fig. 1. Challenges faced by Indian Manufacturing

promote technology adoption in MSMEs, such as subsidies, grants, or tax incentives, can facilitate the integration of AI and ML.

Finance and Its Impact

By offering precise, real-time insights based on historical data and market patterns, AI-driven financial forecasting solutions provide a potent remedy for the difficulties associated with manual forecasting. For example, JPMorgan Chase improves its financial planning and analysis by utilizing AI-driven financial forecasting tools (Fig. 2).

They can make better strategic decisions and react to market developments more quickly by using AI to evaluate vast amounts of data and produce accurate projections. Businesses can increase the precision and effectiveness of their financial planning

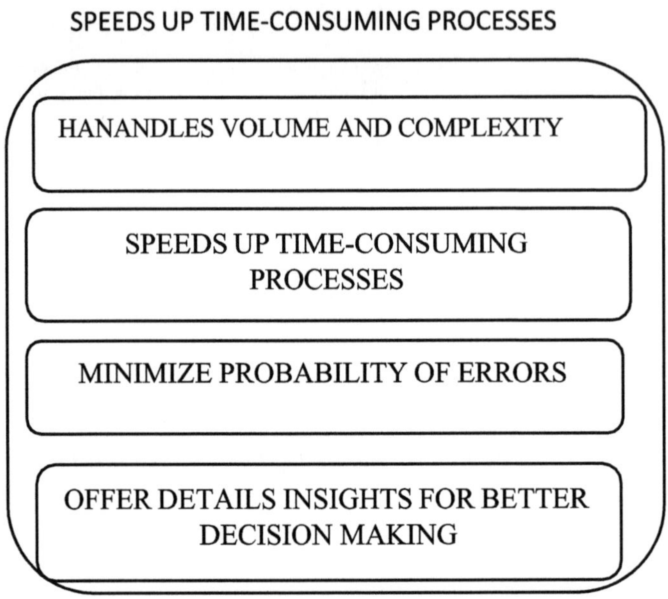

Fig. 2. Impact of Finance

procedures, which will improve decision-making and boost financial performance, by implementing AI- driven-financial-forecasting-tools.

AI-Driven Financial Forecasting Tools

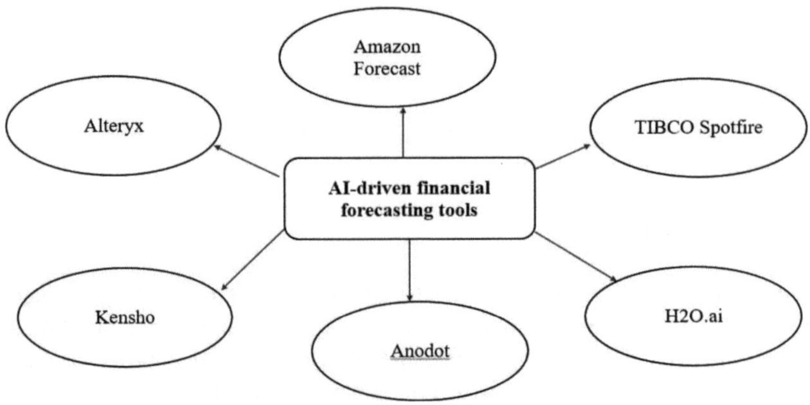

Fig. 3. AI-driven financial forecasting tools

- Alteryx: For data analysis and forecasting,

- Amazon Forecast: Demand forecasting, cash flow forecasting, stock and inventory management
- TIBCO Spotfire: Stock price analysis, financial risk assessment, predictive maintenance
- Kensho: Market trend prediction, portfolio analytics, credit risk modeling
- Anodot: Revenue protection, expense forecasting, and anomaly detection in transactional data. Revenue protection, expense forecasting, and anomaly detection in transactional data.
- H2O.ai: Credit scoring, fraud detection, and customer churn prediction (Fig. 3).

Infrastructure: (Rail, Sea, Airport, and Road-Transportation) and Highlights of Supply Chain

Operational manufacturing greatly benefits from efficient supply chain management [14] and optimization since it guarantees reduced manufacturing time, reduced labor costs, improved product quality, and many other benefits. Cooperation with suppliers, inventory control, and logistics management are a few strategies to improve supply chain management. Businesses may increase their capacity to predict demand, optimize logistics, and boost overall operational efficiency by utilizing AI for supply chain optimization. This will improve customer satisfaction and lower expenses.

- **Smart Factories and Industrial IoT (Aligned with SDG 9: Industry, Innovation, and Infrastructure):**

AI and ML drive the development of smart factories, where interconnected systems powered by the Internet of Things (IoT) monitor and control manufacturing processes in real-time. These intelligent systems help optimize machinery usage, reduce waste, and minimize energy consumption, leading to more sustainable and efficient production lines.

- **AI for Infrastructure Planning and Development** (SDG 11: Sustainable Cities and Communities):

AI can assist in the planning and design of sustainable manufacturing infrastructure, incorporating factors like land use, energy sources, and waste management (Fig. 4).

- **Supply Chain Transparency and Traceability (SDG 12: Responsible Consumption and Production):**

Blockchain and AI Integration: AI-powered blockchain systems can trace the origins of raw materials, ensuring ethical sourcing and compliance with sustainability standards. This transparency helps companies meet the goals of **SDG 12**, ensuring responsible production methods and resource management.

- **Supply Chain Risk Management (SDG 8: Decent Work and Economic Growth):**

AI-Driven Demand Forecasting: MLmodels analyse historical and real-time data to forecast demand more accurately, enabling manufacturers to adjust their supply chain processes and avoid overproduction or resource shortages. This allows for seamless scalability.

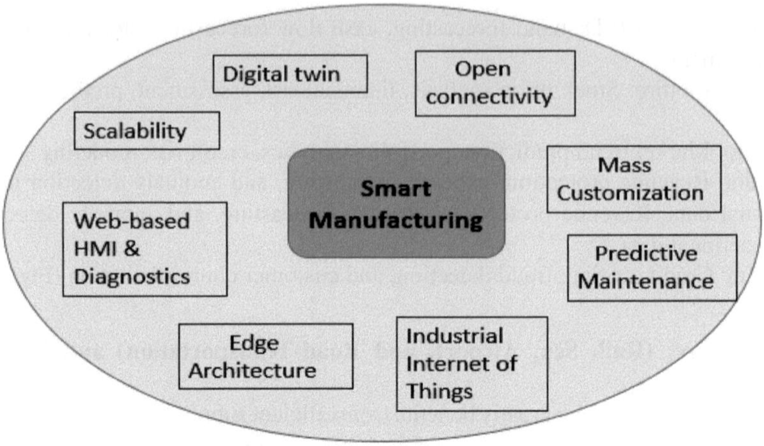

Fig. 4. Smart Manufacturing

- **Route Optimization for Sustainable Transport (SDG 11 & SDG 13)**: AI/ML can optimize distribution and delivery routes by analysing factors like traffic, fuel consumption, and delivery schedules. This reduces the carbon footprint of logistics operations, contributing to both **SDG 11 (Sustainable Cities and Communities)** and **SDG 13 (Climate Action)** by promoting cleaner transportation systems.

Integration of AIML in Core Engineering Discipline

The integration of AI and ML into core engineering disciplines is transforming traditional processes, increasing efficiency, optimizing designs, and solving complex problems in innovative ways [15]. Below is a breakdown of how AI/ML is being adopted across different core engineering fields such as Mechanical, Electrical, Civil, Chemical, and Aerospace Engineering.

1. **Mechanical Engineering**

AI and ML in mechanical engineering are revolutionizing design, manufacturing, and maintenance processes by enabling predictive analytics, optimization, and automation. AI is used to predict when mechanical components will fail, allowing for proactive maintenance of machinery and systems. AI algorithms can optimize the design of mechanical components (e.g., engines, turbines) to enhance performance and durability while minimizing material usage.

2. **Electrical Engineering**

AI/ML is widely applied in electrical engineering for power systems optimization, circuit design, and control systems.AI is pivotal in optimizing power distribution in smart grids by balancing supply and demand, forecasting energy usage, and integrating renewable energy sources more efficiently.

3. **Civil Engineering**

AI/ML is transforming civil engineering by improving the efficiency of infrastructure design, construction, and maintenance, as well as enhancing disaster prediction, Structural Health Monitoring and urban planning.AI is instrumental in designing smart cities by analysing traffic patterns, energy usage, and urban sprawl. ML models optimize transportation systems, reduce congestion, and improve sustainability in urban environments.

4. **Chemical Engineering**

AI and ML are transforming chemical engineering by optimizing chemical processes, enhancing safety, and driving innovations in materials science.ML algorithms are used to model complex chemical reactions and material properties, accelerating the discovery of new materials, catalysts, and chemicals. This is particularly important in industries like pharmaceuticals and energy.

5. **Aerospace Engineering**

The aerospace industry benefits greatly from AI and ML in areas like aircraft design, space exploration, predictive maintenance, and autonomous flight systems, Flight Control and Autonomy, Predictive Maintenance in Aviation, Aerodynamics and Design Optimization

4 Conclusion and Future Enhancements

Because AI can solve a wide range of operational inefficiencies, it is completely changing the way businesses function. AI provides strong solutions that improve productivity, accuracy, and overall performance in a variety of areas, including inventory management, financial forecasts, customer service response times, and human data entry. The integration of AI and ML into core engineering disciplines like mechanical, electrical, civil, chemical, and aerospace engineering is transforming the way engineers design, build, and maintain systems. These technologies allow for predictive maintenance, process optimization, automation, and enhanced decision- making, leading to more efficient, sustainable, and safer engineering practices. As AI/ML adoption continues to grow, it will reshape traditional engineering roles and lead to further advancements across industries.

The adaptability of AI and ML in MSME manufacturing industries holds immense potential to transform operations, improve product quality, and optimize supply chains. By addressing the challenges of implementation and strategically leveraging technology, MSMEs can enhance their competitiveness and contribute significantly to economic growth and sustainability. Embracing AI and ML is not just an opportunity but a necessity for MSMEs aiming to succeed in the modern manufacturing landscape. Businesses that want to prosper in the digital era must embrace AI; it is no longer a choice. You may improve results, revolutionize operations, and set up your company for long-term success by putting AI solutions into practice.

References

1. AI Applications for Improving Transportation and Logistics Operations Amrish Solanki1, Shrikaa Jadiga2 (2024). https://www.researchgate.net/publication/379901056_AI_Applications_for_Improving_Transportation_and_Logistics_Operations
2. Artificial intelligence in logistics and supply chain management: A primer and roadmap for research, Robert Glenn Richey jr, Soumyadeb Chowdhury (2023). https://onlinelibrary.wiley.com/doi/full/10.1111/jbl.12364
3. Impact of Technology on Logistics and Supply Chain Management Rajiv Bhandari (2024). https://www.iosrjournals.org/iosrjbm/papers/7th-ibrc-volume-2/17.pdf
4. The evolution and future of logistics and supply chain management Ronald H. Ballou (2007). https://www.scielo.br/j/prod/a/MFgkvJ5HcbtHbHVdn9WCzws/
5. Blockchain Technologies in Logistics and Supply Chain Management: A Bibliometric Review by Abderahman Rejeb, Karim Rejeb Steve Simske and Horst Treiblmaier (2021). https://www.mdpi.com/2305-6290/5/4/72
6. Thenmozhi, V., Krisknakumari: Artificial intelligence in enhancing operational efficiency in logistics and SCM. Int. J. Sci. Res. Sci. Technol. **11**, 316–323 (2024). https://doi.org/10.32628/IJSRST24115107
7. https://www.leewayhertz.com/ai-for-operational-efficiency/#common-challenges-ai-operational
8. https://www.alphabold.com/10-operational-inefficiencies-that-can-be-solved-by-ai/
9. Szajna, A., Kostrzewski, M.: AR-AI tools as a response to high employee turnover and shortages in manufacturing during regular, pandemic, and war times. Sustainability **14**(11), 6729 (2022). https://doi.org/10.3390/su14116729
10. Escobar, C., Morales-Menendez, R.: Machine learning techniques for quality control in high conformance manufacturing environment. Adv. Mech. Eng. **10**, 168781401875551 (2018). https://doi.org/10.1177/1687814018755519
11. Rana, M.E., Farooq, U., Rahman, W.N.W.A.B.: Scalability enhancement for cloud-based applications using software oriented methods. Int. J. Eng. Adv. Technol. (IJEAT) **8**(6) (2019). ISSN: 2249-8958. https://doi.org/10.35940/ijeat.F8869.088619
12. Kurdve, M., Bellgran, M.: Green lean operationalisation of the circular economy concept on production shop floor level. J. Clean. Prod. **278**, 123223 (2021)
13. Alam, G., Ihsanullah, I., Naushad, M., Sillanpää, M.: Applications of artificial intelligence in water treatment for optimization and automation of adsorption processes: recent advances and prospects. Chem. Eng. J. **427**, 130011 (2022). https://doi.org/10.1016/j.cej.2021.130011
14. Bille, A., Hendriksen, C.: Let us get contextual: critical realist case studies in supply chain management. Supply Chain Manag. Int. J. **28**, 724–737 (2022). https://doi.org/10.1108/SCM-03-2022-0119
15. Bonde, F.: Navigating the technical landscape—large language models such as "GPT-4" inbusiness, tradeMedium (2023). https://medium.com/@frederikbonde/navigatingthe-technical-landscape-large-language-models-such-as-gpt-4-in-business-trade-f0862f8edf5d

Multimodal Mental Health Chatbot

V. K. Harini, Aditi Suryanaryanan, Libin Alex, and R. M. Bhavadharini(✉)

School of Computer Science and Engineering, Vellore Institute of Technology, Chennai, India
{harini.vk2021,aditi.suryanarayanan2021,
libin.alex2021}@vitstudent.ac.in, bhavadharini.rm@vit.ac.in

Abstract. Mental health issues in India pose a significant challenge, impacting a large portion of the population due to the country's vast demographic. According to World Health Organization (WHO), the burden of mental health problems in India amounts to 2443 disability-adjusted life years (DALYs) per 10000 population. Addressing these concerns is critical, as untreated mental health conditions can lead to severe consequences. This paper aims to enhance mental healthcare by developing a task-oriented, closed-domain chatbot that accepts multi-modal input—both voice and text—to provide support and information on mental health issues. We present a comparative study of chatbot models using different embeddings and transformer architectures, focusing on a retrieval-based system built on an LSTM-RNN model. Our LSTM-RNN model achieved an accuracy of 89.36%.

Keywords: LSTM-RNN · Multi-modal Chatbot · Mental Health · Retrieval-based Model

1 Introduction

The reason for almost all the progress made in the development of emotionally intelligent chatbots is that AI (artificial intelligence) and DL (deep learning) models can process this bulk of data and detect the most archaic patterns in user interactions. They have taken traditional approaches to a new level by allowing chatbots to surpass static rule-based responses. Unlike traditional rule-based chatbots that struggle to capture a conversation's emotional nuances, AI-powered systems use natural language processing (NLP) as machine learning tricks to recognize the content, context, protagonists, antagonists, and emotional subtleties in communication. An understanding of this context gives chatbots the ability to understand user feelings more effectively and to respond in a way that is natural and empathetic. In the mental health support and customer service field, this ability is especially valuable when making user engagement part of one's role, but it is helpful in just about every other area as well. Additionally, AI models have become fine-tuned for specific domains using advancements in computational power and gaining access to large-scale datasets, making them far more relevant and effective. With DL models chatbots can now offer more personalized and more human-like responses, dramatically enhancing the quality of user interactions and changing how non-people communicate with each other.

R. Geetha et al. (Eds.): AAIMB 2024, CCIS 2653, pp. 97–108, 2026.
https://doi.org/10.1007/978-3-032-13026-6_7

Specifically, among many kinds of deep learning networks, LSTM networks have a higher visibility in this field because they can deal with sequence data and remember what is more salient in the long term. Such networks are very good at building long-term dependencies and as such are suitable for applications where delicate understanding is needed over long sequences like dialog tracking. LSTMs removes long term dependencies by doing away with the issue of vanishing gradients which affects other RNNs. This capability enables LSTMs to retain certain information from some time earlier of the conversation to be responsible for logically connected and factually correct responses. This makes them suitable for conversational applications, where the flow of the conversation needs to be tracked, and changes in the speaker's mood identified. There is proof that more elaborative models such as LSTM and Bi-LSTM (Bidirectional LSTM) increases both the chatbot's performance, and shows emotions, making it more empathetic. For instance, authors in [4] showed an AI based chatbot that implemented Bi-LSTM with 80.88% accuracy to classify emotions from text while authors in [5], used Bi-LSTM model to demonstrate real time emotion recognition with 82% accuracy. They also prove that such models can be useful in improving the usability by making chat bots more sensitive to emotions and, therefore, presenting a worthwhile option to the traditional approaches.

Moreover, the incorporation of multimodal inputs, voice plus text, has also been proposed in systems such as the SMES Framework [18] that considers all texts, audio, and video modalities to improve the empathetic and strategic psychological nature of AI. This way, the availability of approaches based on multiple types of signal data makes it possible to interpret not only what is said by the user but also how this or that was said, as well as what emotions are linked with it as a rule by defining voice tone, shifts in pitch in the course of the conversation, and confronting it with the facial expressions of the interlocutor. Thus, integrating multiple data sources is critical for these systems to better understand both when it is sufficiently certain to respond and to offer responses that reflect the circumstances surrounding an answer. Such advancements kept the pace for chatbots to simulate the human like conversations with increased efficiency and empathy. Using these approaches, it is our intention that the work produces a next-generation chatbot solution that is both more empathetic and context aware hence improving user satisfaction leading to the potential developments of mental health care approaches using chatbot based conversational systems. They have the potential of not only revolutionizing how these chatbots work but also the fields in which such systems are needed due to their high level of emotional intelligence.

Our chatbot is based on retrieval, text based, task oriented and closed domain. This chatbot ensures accurate, relevant, and context-aware responses by focusing on mental health topics. Datasets like 'Mental Health Conversational Data' was taken from Kaggle [19] and converted into JSON file which contains tags, patterns and corresponding responses. Based on this dataset, we build the chatbot and use it to understand and respond to user queries with high accuracy. The preprocessing was done with TensorFlow libraries, and we used stemming and stop word removal techniques. These preprocessing steps are crucial for keeping the model focused on extracting meaningful information and they are crucial because if the input text is not clean and standardized, the model will not be able to perform as intended. The model leverages Natural Language Processing

(NLP) which enables the chatbot to learn the context of the conversations, beyond simple question answer mechanisms to enable meaningful interactions. Using advanced NLP techniques, our chatbot fills the gap between how users want to interact with a chatbot and what traditional chatbot capabilities allow.

The dataset is structured on the idea of "intents", what the user wants to achieve with their message. These intents are the heart of the chatbot's logic; they authorize it to classify and answer user messages. Consider an instance where a person states "I feel down," the intent would be "sad" in this case to help the chatbot decide what response to provide to offer support. Based on that, some Patterns and Responses are affiliated with each intent. Sample phrases are a variety of patterns that fit a given intent, from a variety of ways people would say something similar. Responses are pre-determined replies by the chatbot of the given intent, which always keeps the answers relevant and helpful. Using these intents, patterns, and responses the model is trained to correctly identify a user's intent and help provide appropriate responses. Taking a structured approach, not only improves the chatbot's functionality but also ensures that the chatbot is consistent and reliable for different user interactions—a vital tool for providing mental health support and beyond.

2 Literature Survey

In [1], authors compare retrieval-based and generative-based chatbots using dee... Their experiments showed that generative chatbots outperform retrieval-based ones on accuracy (94.45% vs 91.57%), evidenced by the potential of deep learning models, namely, Bi-LSTM, in dealing with a complex conversational scenario. In [2], the authors had proposed a chatbot based on LSTM with attention mechanism and beam search decoding, with which the efficiency and accuracy of the chatbot's responses improved. Results show that combining attention and generative sequence-to-sequence architectures yielded higher BLEU score than other approaches. In [3] the authors built a deep learning movie dialogue prediction chatbot by using a sequence-to-sequence model with attention and LSTM. After 500 epochs the chatbot achieved 90% accuracy which demonstrates the power of attention mechanisms to generate meaningful responses.

The paper [4] proposed an AI-based chatbot for mental Healthcare that performs sentiment analysis through a Bi-LSTM model. The model achieved 80.88% accuracy in emotion detection. From text, suggesting that deep learning models can effectively gauge user sentiment in mental health applications. The authors in [5] integrated emotion detection in a chatbot using Bi-LSTM, achieving 82% accuracy. The paper highlights the importance of emotion detection in making chatbots more empathetic, with Bi-LSTM models proving effective in real-time emotional recognition. The paper [6] explored LSTM-based chatbots for university communication, utilizing word embeddings for better contextual understanding. The chatbot achieved 92% accuracy in intent recognition, proving the effectiveness of LSTM in processing natural language queries. The paper [7] focused on emotion recognition in students using a machine-learning chatbot, comparing SVM and Naïve Bayes algorithms. The SVM model performed better, achieving higher accuracy in classifying emotions from text.

The authors in [8] developed a voice-based chatbot for the hospitality industry using a closed-domain question-answering system with BERT for extracting answers from documents. The chatbot demonstrated effective real-time response generation for industry-specific queries. The authors in [9] created the "Wellness Buddy" chatbot for Kenyan university students, focusing on mental health support. The chatbot effectively provided advice on depression and anxiety but faced limitations due to overfitting and dataset scope. The authors in [10] developed "Dost," a mental health chatbot designed to address gaps in mental health care in India. Built using the Rasa framework, Dost showed 83.5% accuracy in intent identification, highlighting its potential for providing accessible mental health support.

The paper [11] introduced the "MindGuide" chatbot, which uses the LangChain framework and GPT-4 for mental health support. With advanced features like conversation memory, the chatbot aims to provide personalized mental health assistance.The authors in [12] developed "MediBot," a healthcare assistant chatbot focused on mental well-being. The chatbot provided accessible support with accurate responses and emphasized the potential for machine learning to personalize mental health care.The authors in [13] discussed the architecture of an intelligent healthcare chatbot using deep learning models. The paper highlighted the limitations of current chatbot architectures and proposed improvements in natural language processing and self-learning capabilities. The authors in [14] examined the effects of emotional expressiveness in voice chatbots, finding that more expressive prosody led to higher user engagement and perceived human-likeness. This indicates that enhancing emotional expressiveness improves user interaction quality.

The authors in [15] presented a hybrid LSTM-GRU model for improving chatbot performance. The hybrid model balanced accuracy and efficiency, offering a promising solution for real-time applications like those in the health sector. In [16], the authors compared CNN, ANN, and LSTM models for a chatbot to be designed for Indian cities. Among the three models, LSTM showed the best accuracy, precision, and recall values. It was thus the preferred model to be used in real-time city resource management chatbots. Authors in [17] applied the LSTM model to a chatbot and concluded that Parameter Pair 1 (size_layer 512, num_layers 2, embedded_size 256, learning_rate 0.001, batch_size 32, epoch 20) achieved the highest average accuracy of 0.994869, making it the best configuration. They suggested that future work could focus on improving the LSTM chatbot algorithm using advanced computing environments, with the expectation that updating methods, models, and references to align with technological developments could further enhance accuracy. The authors in [18] introduced the MESC dataset for mental health care and the Sequential Multimodal Emotional Support (SMES) Framework to enhance AI-driven conversation systems. The MESC dataset and SMES Framework address the need for a comprehensive multimodal dataset and an integrated framework for emotional support. Evaluation showed significant improvements in AI's empathetic and strategic capabilities, setting a new standard for mental health conversational systems.

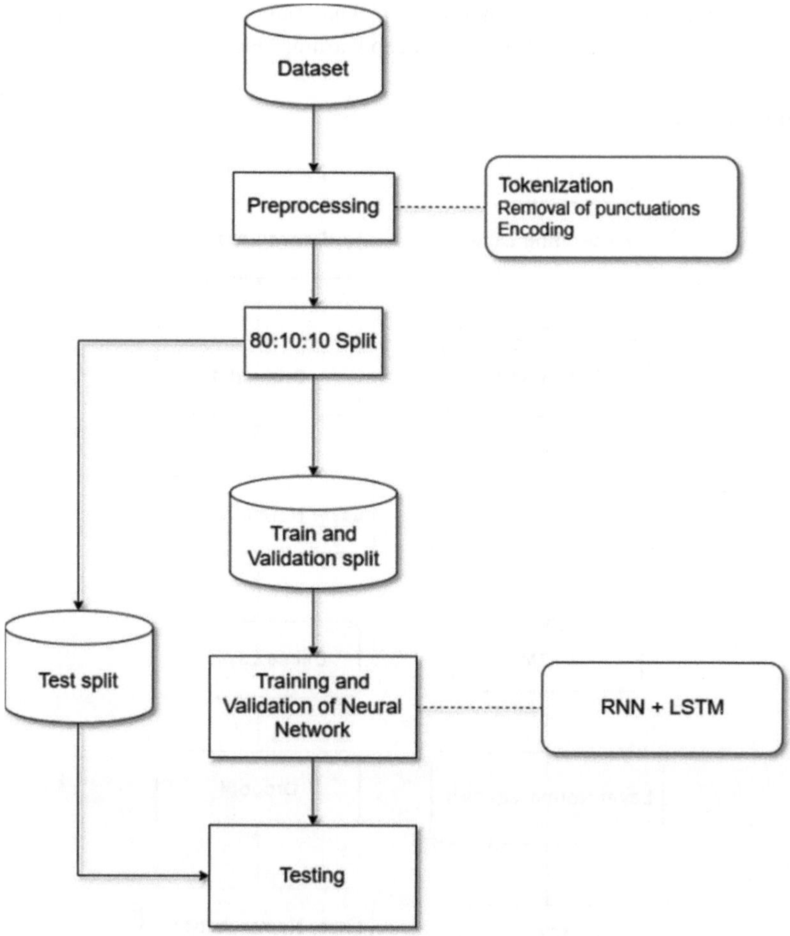

Fig. 1. Flow chart of the model

3 Methodology

The dataset is sourced from Kaggle and it is utilized for training the model. The methodology of the multimodal chatbot system starts by capturing audio input through a microphone, which is processed using Google's Speech Recognition API. This API converts speech into text. The recognized text is then preprocessed by removing punctuation and converting it to lowercase. The cleaned text is further tokenized and padded to fit the input shape required by the neural network model. The model is comprised of an embedding layer followed by a SimpleRNN and three LSTM layers, processes the tokenized input to predict an appropriate response based on learned patterns in the data. The most likely response from a predetermined set is then used to translate the model's output back into a format that can be understood by humans. The system supports commands like "quit" to terminate the conversation, providing a natural and interactive user experience through voice recognition and text-based deep learning (Fig. 1).

3.1 RNN+Lstm

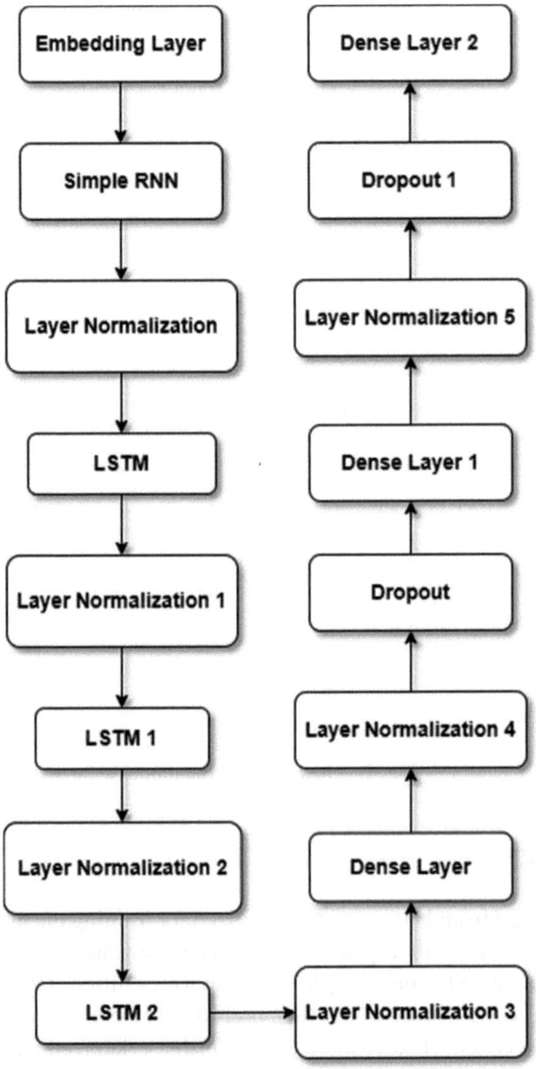

Fig. 2. RNN + LSTM model architecture

This model starts with an embedding layer which maps the input sequence of tokens to denser vector representations of dimensionality 100, so as to facilitate reasoning of semantic connections between tokens. This layer is trainable with 30,300 parameters

Table 1. Layers of RNN + LSTM architecture

Layer (type)	Output Shape	Param #
embedding	(None, 18, 100)	30300
Simple RNN	(None, 18, 32)	4256
Layer Normalization 1	(None,18,32)	64
LSTM 1	(None, 18, 32)	8320
Layer Normalization 2	(None, 18, 32)	64
LSTM 2	(None, 18, 32)	8320
Layer Normalization 3	(None, 18, 32)	64
LSTM 3	(None, 32)	8320
Layer Normalization 3	(None, 32)	64

and manages to transform the observed vocab size of 303 into a vector space. Secondly, a Simple RNN, comprising 32 units, analyzes the sequence capturing the short-term dependence, and the output shape becomes (None, 18, 32) with 4,256 trainable parameters. Next, three LSTM layers with 32 units are used as a way of modeling long-term dependencies to a sequence. To overcome an exploding or vanishing gradient problem each LSTM layer is followed by the Layer Normalization to normalize the outputs of the layers which will make training better. Of the first two LSTM's the first two layers are spatial and resemble the output of the first LSTM layer from Sect. 2 with the shape (None, 18, 32), whilst the third LSTM layer outputs a vector of shape (None, 32) which is more suitable for further dense processing. A Dropout layer (0.2) is added for regularization, two hidden layers with 128 neurons and hyperbolic tangent activation functions are used to learn more representation functions. Last of all, an output layer activated by softmax returns class probabilities making this architecture appropriate for the sequence classification or time series prediction for example (Fig. 2 and Table 1).

4 Results and Discussion

Table 2. Performance of RNN+LSTM architecture

S. No	Activation Function	Final Training Accuracy	Final Training Loss	Final Validation Accuracy	Final Validation Loss	Test Accuracy	Test Loss
1	relu	0.9877	0.0296	0.9783	0.1516	0.8936	0.7369
2	tanh	0.9922	0.0246	0.9783	0.0887	0.8936	0.5996
3	leaky_relu	0.9992	0.0185	0.9783	0.1209	0.89361	0.93011
4	Sigmoid	0.9964	0.0177	0.9783	0.0564	0.89361	0.8221

The Table 2 presents the performance metrics and the results achieved for the proposed model trained using different activation functions: ReLU, Tanh, Leaky ReLU, and Sigmoid. The final training accuracies are notably high for all functions, with Leaky ReLU achieving the highest at 0.9992. Despite this, the validation accuracy remains consistent across all functions at 0.9783, indicating stable performance. However, the test accuracy hovers around 0.8936 for all, suggesting potential overfitting, especially for Leaky ReLU, where the test loss is significantly higher (0.9301) compared to the others. Sigmoid achieves the lowest validation loss (0.0564), but its test accuracy and loss indicate slight instability in generalization. Overall, while all activation functions perform well during training, their test performance reflects varied levels of generalization ability.

Table 3. Comparison Results of different models

S. No	Authors	Model	Accuracy
1	Shashank Joshi et al. [3]	Seq-2-Seq using LSTM	90
2	Mehta, A et al.[4]	Bi-LSTM	80.88
3	C V N M Kasyap et al. [6]	LSTM	92
4	Nayar, A.M. et al. [10]	Rasa Framework	83.5
5	Proposed model	RNN+LSTM model	89.36

Table 3 shows the comparison of Accuracy achieved in different models surveyed in this research article.

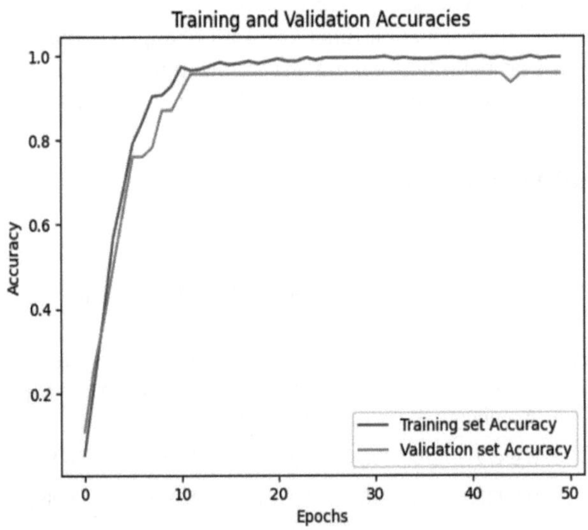

Fig. 3. Training and validation accuracy graph for RNN+LSTM

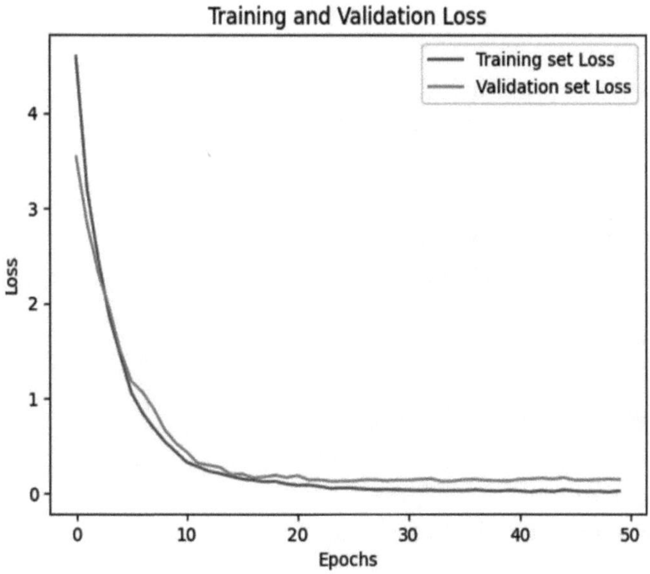

Fig. 4. Training and validation loss graph for RNN+LSTM

Figure 3 depicts Training and validation accuracy and achieved for proposed RNN+LSTM model over 50 epochs with tanh activation. Figure 4 depicts Training and validation loss and achieved for proposed RNN+LSTM model over 50 epochs with tanh activation.

The Fig. 5 depicts the interaction between the user and the chatbot. Where the user can speak to the chatbot and in turn, the chatbot offers the user replies by making use of the trained model.

```
speak
Did you say:  hello
1/1 [==============================] - 0s 25ms/step
Chatbot: Hello there. Glad to see you're back. What's going on in your world right now?
speak
Did you say:  i am feeling sad
1/1 [==============================] - 0s 35ms/step
Chatbot: How long have you been feeling this way?
speak
Did you say:  two days
1/1 [==============================] - 0s 27ms/step
Chatbot: That seem awful. What do you think is behind this?
speak
Did you say:  i need some advice
1/1 [==============================] - 0s 26ms/step
Chatbot: Sure. What can I do to help?
speak
Did you say:  what is depression
1/1 [==============================] - 0s 27ms/step
Chatbot: A mental health disorder characterised by persistently depressed mood or loss of interest in activities, causing signi
ficant impairment in daily life.
```

Fig. 5. Sample Mental Health conversation with chatbot

5 Ethical Considerations and Limitations of AI in Emotionally Sensitive Situations

This research focuses on the implementation and evaluation of a retrieval-based, task-oriented mental health chatbot model. All training and testing data used were publicly available [19], and no real user data was collected and stored during the research. This chatbot ensures secure real-time processing by handling all user inputs locally, meaning no data is stored or transmitted. However, additional security measures such as end-to-end encryption can be introduced in future implementations.

While the chatbot is trained to provide informative responses, it may not fully understand the depth of the user's psychological state. The users should be aware understand the rationale behind the model's predictions. In situations of distress or crisis, users are strongly encouraged to seek immediate help from licensed mental health professionals.

6 Conclusion

In this work, we developed a multimodal mental health chatbot using RNN and LSTM, capable of processing both audio and text inputs. The model achieved strong performance, with a training accuracy of 99.19%, validation accuracy of 97.83%, and a test accuracy of 89.36%. While the model performs well, there is room for improving its generalization, particularly in test scenarios. Future work could focus on refining the emotion detection and improving the chatbot's ability to respond accurately across diverse situations.

References

1. Pandey, S., Sharma, S.: A comparative study of retrieval-based and generative-based chatbots using deep learning and machine learning. Healthcare Anal. **3**, 100198 (2023)
2. Choudhary, P., Chauhan, S.: An intelligent chatbot design and implementation model using long short-term memory with recurrent neural networks and attention mechanism. Dec. Anal. J. **9**, 100359 (2023)
3. Joshi, S., Shinde, S., Shinde, P., Sagar, N., Rathod, S.: Chatbot using deep learning. Int. J. Sci. Res. Comput. Sci. Eng. Inform. Technol. **3307**, 228–232 (2023)
4. Mehta, A., Virkar, S., Khatri, J., Thakur, R., Dalvi, A.: Artificial intelligence powered chatbot for mental healthcare based on sentiment analysis. In: 2022 5th International Conference on Advances in Science and Technology (ICAST), pp. 185–189. IEEE (2022)
5. Amadi, C., Odii, J., Okpalla, C., La, O.C.I.: Emotion detection using a Bidirectional Long-Short Term Memory (BiLSTM) neural network. Int. J. Curr. Pharmaceut. Rev. Res. **4**(11), 1718–1732 (2023)
6. Kasyap, C.V.N.M.: unveiling the potential LSTM-based Chatbots with embeddings for university communication. Int. J. Intell. Syst. Appl. Eng. **12**(3), pp. 3968–3985 (2024). https://ijisae.org/index.php/IJISAE/article/view/6100. Accessed 22 Aug. 2024
7. Assayed, S., Shaalan, K., Al-Sayed, S., Alkhatib, M.: Psychological emotion recognition of students using machine learning based chatbot. Int. J. Artific. Intell. Appl. **14**(2) (2023)
8. Athikkal, S., Jenq, J.: Voice chatbot for hospitality (2022). arXiv preprint arXiv:2208.10926
9. Ogamba, M., Gitonga, J., Muriithi, B., Olukuru, J., Sevilla, J.: Wellness buddy: an AI mental health Chatbot for kenyan university students. In: 2023 First International Conference on the Advancements of Artificial Intelligence in African Context (AAIAC), pp. 1–6. IEEE (2023)
10. Nayar, A.M., Attar, Z., Kachwala, S., Biswas, T., Wagh, S.K.: Dost-mental health assistant chatbot. In: 2022 5th International Conference on Advances in Science and Technology (ICAST), pp. 252–257. IEEE (2022)
11. Singh, A., Ehtesham, A., Mahmud, S., Kim, J.H.: Revolutionizing mental health care through langchain: a journey with a large language model. In: 2024 IEEE 14th Annual Computing and Communication Workshop and Conference (CCWC), pp. 0073–0078. IEEE (2024)
12. Kavyashree, N., Usha, J.: MediBot: healthcare assistant on mental health and well being. In: 2023 7th International Conference on Computation System and Information Technology for Sustainable Solutions (CSITSS), pp. 1–5. IEEE (2023)
13. Ayanouz, S., Abdelhakim, B.A., Benhmed, M.: A smart chatbot architecture based NLP and machine learning for health care assistance. In: Proceedings of the 3rd International Conference on Networking, Information Systems & Security, pp. 1–6 (2020)
14. Zhu, Q., et al.: Effects of emotional expressiveness on voice chatbot interactions. In Proceedings of the 4th Conference on Conversational User Interfaces, pp. 1–11 (2022)
15. Naveen, P., Haw, S.C., Nadthan, D., Ramamoorthy, S.K.: Improving chatbot performance using hybrid deep learning approach. J. Syst. Manage. Sci. **13**(3), 505–516 (2023)
16. Sai, A.A., Balamurali, O., Karthikeya, M., Anand, S.: A web-based Chatbot for indian cities: a comparison of CNN, ANN, and LSTM models. In: 2023 14th International Conference on Computing Communication and Networking Technologies (ICCCNT), pp. 1–6. IEEE (2023)

17. Anki, P., Bustamam, A., Al-Ash, H.S., Sarwinda, D.: Intelligent chatbot adapted from question and answer system using RNN-LSTM model. In: Journal of Physics: Conference Series, vol. 1844, no. 1, p. 012001. IOP Publishing (2021)
18. Chu, Y., Liao, L., Zhou, Z., Ngo, C.W., Hong, R.: Towards multimodal emotional support conversation systems. arXiv preprint arXiv:2408.03650 (2024)
19. https://www.kaggle.com/datasets/elvis23/mental-health-conversational-data

Leveraging Generative AI in Management Decision Making: Opportunities and Challenges

Udhaya Shree Seshachelam[1(✉)], Calarany Canabady[2], and Elakia Kalaivendan[1]

[1] Computer Science and Engineering, Rajiv Gandhi College of Engineering and Technology, Puducherry, India
udhayashree_cse@rgcet.edu.in

[2] Department of Computer Science, Tagore Government Arts and Science College, Puducherry, India

Abstract. Generative Artificial Intelligence (GenAI) confers about a class of AI systems proficient in engendering new content—such as text, images, music, or code—grounded on input data. Leveraging advanced models like Generative Adversarial Networks (GANs), Variational Autoencoders (VAEs), and autoregressive transformers (e.g., GPT models), GenAI mimics human ingenuity and cognition. This capability allows GenAI to create accurate synthetic data, draft documents, create designs, and simulate scenarios, among other applications. This paper examines the application of three prominent GenAI models—ChatGPT, GitHub Copilot, and DALL-E—in real-world management scenarios. In all these LLM, the user must identify the problem, analyze, deciding what AI tools can be used and what type of decision support reports to be generated. The user must develop prompt engineering knowledge for working with wide range of LLMs. Chat GPT are effective because of their high context understanding, wide range of domain support. Copilot can be used as a rapid prototype builder. DALL-E is widely used in visualization, branding, product design. Case studies highlight their effectiveness in customer support, software development, and creative marketing campaigns, respectively. The results indicate significant efficiency gains (up to 50%), cost savings (up to 25%), error reduction (up to 30%), and increased engagement (up to 40%). Despite these benefits, challenges such as ethical concerns, data security, and the need for human oversight persist. The study underscores the importance of strategic integration and robust governance frameworks to harness GenAI's potential while mitigating associated risks.

Keywords: Generative Artificial Intelligence · Generative Adversarial Networks · Variational Autoencoders · ChatGPT · GitHub Copilot · DALL-E · real-world management

1 Introduction

The advent of Generative AI (GenAI) has unbolted new avenues in administration by systematizing routine processes, enhancing the approaches followed in decision-making, and nurturing innovative ideas. By leveraging models like GPT, DALL-E, and others,

R. Geetha et al. (Eds.): AAIMB 2024, CCIS 2653, pp. 109–125, 2026.
https://doi.org/10.1007/978-3-032-13026-6_8

establishments are now gifted with new scenes of generating perceptions, systematizing content construction, and restructuring operations. This study follows a mixed-methods style, by combining a systematic literature assessment with qualitative investigation of case studies.

This work emphases on how efficiency in production can be enhanced, how GenAI helps in management decision-making, its role in invention of new techniques and creation of new ideas, examining what are the existing challenges and how to follow ethics. This paper explores the transformative role of GenAI in management, its impact on organizational hierarchies, and its potential for reshaping traditional management practices.

2 Literature Survey

Several researchers have emphasized the benefits and challenges of incorporating GenAI into management. The paper [1] discusses the role of AI in reconditioning several business functions which comprises of management of marketing, supply chain management activities and customer relationship management related activities. Several companies are concerned in using AI for extrapolative analytics, natural language processing (NLP) and automation. The book [2] discusses about the impact of AI and digital platforms on commercial models as well as financial structures. AI is authorizing new forms for association and revolution in areas like crowdsourcing, platform-based business etc. AI's main fiscal value lies in its capacity to make precise predictions. The book [3] discovers the inferences of AI on various industries, including healthcare, investment field, and transportation. The paper [4] discusses quite a lot of applications of AI in promotion of advertising, selling the products, such as customer dissection, custom-made commendations and chatbots. Effective AI adoption demands a blend of industrial expertise, good superiority of data, organizational keenness. The article [5] deliberates AI's considerate impact on society and industries, transmuting businesses and job marketplaces. It also confers the potential paybacks and encounters of AI, including job disarticulation and ethical concerns. They have debated the role of corporations in preparing for the AI revolution by rise of investment in technology, refining skill employed by their personnel and adapting their business models. The paper [6] learns how AI can be used for data-driven insights, risk administration, and feasible advantage.AI is improving firms' recital by educative decision-making processes and strategic planning. The paper [7] delivers an overview of AI and its inferences for business and administration. The authors [7] discussed the principled and societal implications of AI, accent the need for accountable AI progress and delivery. The paper [8] introduces the automation-augmentation outline for understanding AI's role in management. The framework highlights the importance of human-AI association for effective decision-making and reconstruction. The article discusses [9] about how AI research has grown significantly over the past two decades, with a focus on application areas like marketing, finance, and operations. The analysis identifies key trends, authors, and journals contributing to AI research in business and management. The paper [10] provides a systematic literature review of AI in management.

The review identifies key themes, methodologies, and future study directions in AI and management. The paper [11] includes case studies and real-world examples for illustrating the practical applications of GANs in decision support systems. These examples provide insights into how organizations can leverage GANs to improve their decision-making processes. Hoffmann et al. (2024) showed that tools like GitHub Copilot significantly increase productivity by allowing users to focus on core tasks while automating repetitive administrative activities [12]. Research from Harvard Business School points to GenAI's potential to transform ideation processes by simulating scenarios and providing creative solutions [12].

Lin (2023) emphasized the role of AI in decision-making, particularly in data-intensive tasks where cognitive overload is reduced through advanced analytics and automated reporting [13]. The integration of AI systems must balance efficiency with ethical responsibility to mitigate risks [13]. The article [14] provides a detailed review of numerous Gen AI models, which includes GAN (Generative Adversarial Networks), GPT (Generative Pre-trained Transformers), other models like Autoencoders, DM (Diffusion Models), and Transformers. The paper discusses the appropriateness of different models for specific tasks and the probability of using hybrid approaches that combine multiple models. The paper [15] how generative AI influences organizational novelty performance, focusing on the roles of AI-generated contented quality, AI experience, and the AI usage environment. Generative AI (GenAI) is transforming management decision-making by providing tools for automating data analysis, improving scenario planning, and nurturing creativity in problem-solving. In business intelligence, for instance, GenAI improves decision-making by generating synthetic data, prognostic analytics, and automating reporting, thus enabling managers to make more well-versed, efficient decisions [16].

3 Methodology for Analysis

The study employs a mixed-methods approach, combining a systematic literature review with qualitative analysis of case studies. The examination of case studies depicting real world scenarios, help businesses to learn from the successes and failures of other business domains. This can help them avoid common pitfalls and adopt best practices that have been proven effective in similar situations.

3.1 Algorithm/Procedure: Using ChatGPT for Decision-Making in Customer Support Optimization

Inputs:

1. Customer Collaboration Data: Chat records, email logs, and customer grievances.
2. Operational Metrics: Response times, firmness rates, and customer approval scores.
3. Feedback: Administrative inputs on significant areas (e.g., product issues, shipping delays).

Procedure:

1. *Steps for Data Collection:*

- Initially, Collect historical customer interaction data from databases.
- Next, perform aggregation of operational metrics (e.g., average response time, resolution rates).
- Next, input customer-based feedback and managerial perceptions as supplementary data.

2. *Steps for Preprocessing:*

- Perform cleaning of the data: Removal of extraneous or duplicate records.
- Perform categorization of the data into themes such as delivery issues, product inquiries.
- Transforming textual data into tokenized inputs for AI processing.

3. *Steps for AI Processing with ChatGPT:*

- Fine-tuning of ChatGPT with domain-specific knowledge such as common customer queries, service guidelines.
- Querying ChatGPT for the following:

  - Summarization of top customer complaints.
  - Predicting the likely causes of repeated issues.
  - Proposal of high-level solutions.

4. *Steps for Decision Investigation:*

- Use ChatGPT-generated insights to create a list of actionable recommendations (e.g., streamline logistics, improve product descriptions).
- Simulate scenarios using ChatGPT, such as the impact of faster response times on customer satisfaction.

5. *Steps for Validation and Reporting:*

Validate recommendations with managerial feedback.
Generate an automated report summarizing findings and proposed actions.

6. *Steps for Implementation:*

- Prioritize recommendations based on feasibility and impact.
- Monitor outcomes post-implementation (e.g., changes in response time or customer satisfaction).
- Use ChatGPT to continuously refine strategies based on new data.

Example Case Study Implementation:
Scenario: A retail company experiences a 20% increase in unresolved customer tickets during peak shopping seasons.
Steps:

1. *Initially, Input Historical Data:* Past ticket logs show that delayed responses are common during Black Friday sales.
2. *ChatGPT Understandings:*

- Recognizes "delivery delays" as the primary customer concern.
- Commends deploying an FAQ bot for common questions for reducing the ticket volume.

3. *Authorized Actions:*

   - Automation of responses for FAQs like shipping timelines.
   - Reassignment of support agents for handling complex issues.

4. *Post-Implementation Monitor:*

   - Tracking a 30% reduction in ticket backlog and a 15% improving the satisfaction scores within one month.

3.2 Pseudocode Necessary for Automation

Step 1: Data Loading
```
customer_data = load_data("customer_logs.csv")
metrics = load_metrics("operational_data.csv")
feedback = load_feedback("manager_feedback.txt")
```

Step 2: Data Preprocessing
```
cleaned_data = preprocess_data(customer_data)
categorized_data = categorize_data(cleaned_data)
```

Step 3: AI Analysis
```
chatgpt = load_model("ChatGPT_tuned")
complaints_summary = chatgpt.summarize(categorized_data)
recommend = chatgpt.generate_recommendations(complaints_summary, metrics)
```

Step 4: Validation
```
validated_actions = validate_recommendations (recommend, feedback)
```

Step 5: Report and Implementation
```
generate_report(validated_actions) implement_actions(validated_actions)
```

Step 6: Monitoring
```
results = monitor_outcomes (metrics_post_implementation)
```

3.3 Key Outcomes

- Improved Efficiency: Automated ticket handling reduced agent workload by 30%.
- Data-Driven Decisions: Management identified logistics improvements based on customer complaints.
- Customer Satisfaction: Scores improved by 15% due to faster response times.

This algorithm provides a scalable framework for integrating Generative AI into decision-making processes across various industries.

4 Case Study Description and Analysis: Application of Generative AI Models in Management

This case study reconnoiters the impact of three generative AI models—OpenAI's ChatGPT, GitHub Copilot, and DALL-E—in real-time administration scenarios.

This work is concentrating on their deployment in three domains namely: Management decision-making, Management of software development, and Management of creative marketing.

Case Study 1: ChatGPT for Customer Support and Decision-Making
Scenario: A multinational e-commerce company interested in integrating Chat-GPT into its customer services and management of decision-making pipelines. The AI assists in automation of responses for repetitive customer queries and supports in management by analyzing large volumes of customer feedback for the identification of improvement areas [15].
Implementation:

- ChatGPT was fine-tuned with company-specific data to enhance contextual understanding.
- It was integrated with customer service chatbots and management dashboards to provide real-time insights into customer sentiments.

Results:

- Efficiency Gains: Customer service response times reduced by 40%, improving customer satisfaction scores by 30%.
- Data-Driven Decisions: By summarizing customer feedback, ChatGPT highlighted recurring issues, leading to targeted improvements in logistics, reduce delivery delays by 15%.
- Analysis: ChatGPT's NLP capabilities proved excellent operative for repetitive tasks, emancipating managers to focus on tactical decisions. However, its limitations in empathetic nuanced queries highlight the need for human oversight.

4.1 Case Study 2: GitHub Copilot in Software Development

Scenario: A mid-sized software firm interested in adopting GitHub Copilot to streamline coding tasks for a new project management tool. The AI was taxed with suggesting code, automation of boilerplate scripts, and in the identification of errors in real-time [13, 15].
Implementation:

- Developers unified GitHub Copilot into Visual Studio Code IDE for continuous operation.
- It was used mainly for backend progress and fixing errors.

Results:

- Productivity Increase: Developers reported a 50% reduction in time spent on boilerplate code and debugging.

- Error Reduction: Error frequency in test environments dropped by 30%, attributed to Copilot's ability to identify syntax and logical issues.

Analysis: GitHub Copilot enabled developers to focus on complex problemsolving by automating repetitive coding tasks. However, concerns about the potential leakage of proprietary code to the AI model's training database necessitated rigorous compliance checks.

4.2 Case Study 3: DALL-E in Creative Marketing Campaigns

Scenario: A global marketing agency employed DALL-E for designing innovative visual content for a product promotion campaign. The AI was tasked with creating high-quality promotional graphics based on textual data descriptions provided by the creative team [13].

Implementation:

- DALL-E was competent enough with specific brand guidelines to ensure alignment with the client's identity.
- Generated images were used for posting in social media, creating advertisements, and preparation of brochures.

Results:

- Cost Savings Factor: The agency saved nearly 25% in graphic design costs by reducing dependency on human designers for initial drafts.
- Campaign Success Factor: Social media engagement increased by 40%, attributed to the novel and visually appealing AI-generated content.

Analysis:

DALL-E demonstrated the use of exceptional creative potential, accelerating the design process and enabling unique visual concepts. However, its rare inability in the production of extremely specific outputs underlining the importance of generation of iterative refinement by human designers.

5 Synthesis of Results

- Operational Efficiency: All three AI models significantly enhanced efficiency in their respective domains, reducing manual workload and enabling faster outputs.
- Quality Improvements: Each model contributed to better quality outcomes, whether through reduced customer complaints, fewer coding errors, or more engaging marketing content.
- Challenges: Ethical concerns, data security risks, and occasional limitations in precision were recurring issues across all implementations.

The graph in Fig. 1 above illustrates the percentage improvements across different metrics (efficiency gains, cost savings, error reduction, and engagement improvement) for the three Generative AI models discussed in the case study: ChatGPT, GitHub Copilot, and DALL-E. The metrics used in the analysis represent key performance indicators (KPIs) that measure the impact of the three Generative AI models (ChatGPT,

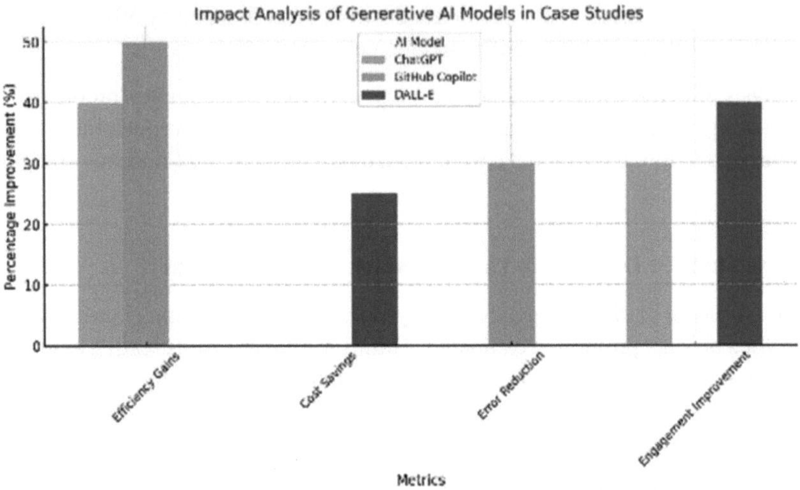

Fig. 1. Impact Analysis of Generative AI Models in Case Studies

GitHub Copilot, and DALL-E) on specific aspects of their application. Here's a detailed explanation:

5.1 Why These Metrics?

These metrics were selected because they signify tangible outcomes that align with common management goals:

- Improving Productivity (Efficiency Gains) [17].
- Reducing Costs (Cost Savings) [18].
- Enhancing Quality (Error Reduction) [19].
- Driving Business Growth (Engagement Improvement) [20].

By focusing on these metrics, the study captures the multifaceted impact of Generative AI on both operational efficiency and strategic outcomes. The graph in Fig. 2 above illustrates how Generative AI enhances various phases of the decision-making process:

- Data Collection (50%): GenAI is involved in accelerating the collection and summarization of large datasets, empowering faster access to appropriate information.
- Analysis (60%): Tools like ChatGPT can assist in interpretation of data, identification of trends, and generation of actionable insights, providing a significant improvement in diagnostic accuracy and improved speed.
- Strategy Development (45%): AI models also support management by providing simulation of scenarios and in the provision of recommendations, promotion of more effective tactical planning.
- Implementation (35%): AI aids in the execution of strategies by systematizing workflow development and helps in the provision of real-time feedback during implementation stage.

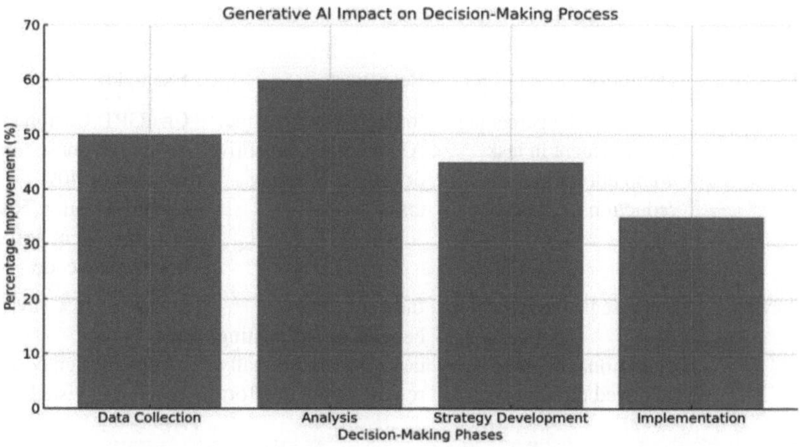

Fig. 2. Generative AI models on Decision-Making Process

This data highlights how GenAI reinforces decision-making by restructuring processes, reducing time consumption, and cultivating the overall quality of conclusions. The Table 1 summarizes how each metric reflects the effectiveness of the Generative AI models in practical applications. Efficiency gains refer to the capability which measures completion of tasks, how it is done i.e. quickly, efficiently, and with minimum amount of effort or resources.

This gain can be achieved by enhancing processes, systematizing repetitive tasks, or refining work-flows.

Example 1: In customer service, employing chatbots can be used for handling a large volume of queries, permitting human agents to focus on composite issues, dropping response times, and enlightening operational efficiency. Cost savings encompass reduction of expenses through enhanced processes, smarter resource administration, or leveraging technology in replacing costly manual efforts.

Example 2: A company implementing cloud services eradicates the need for physical servers and maintenance, suggestively reducing IT costs while increasing scalability.

Example 3: In the field of manufacturing, integration of AI-powered quality control systems guarantees that defective products are recognized and removed, dropping waste and in the maintenance of brand reputation. Engagement improvement denotes the increase in meaningful communication between users and systems, primary to higher satisfaction, trustworthiness, or efficiency.

Table 1. Effectiveness of Gen AI Models

Metric	Definition	Relevance	Example
Efficiency Gains	Measures the percentage improvement in task completion time or reduction in effort due to AI	Reflects how well the AI automates repetitive or time-consuming tasks	ChatGPT: Customer service response times reduced by 40% GitHub Copilot: 50% less time spent writing boiler-plate code
Cost Savings	Represents the reduction in operational or labor costs achieved through AI	Indicates financial benefits of automating tasks that traditionally required human effort	DALL·E: 25% savings on graphic design costs by generating visuals with AI
Error Reduction	Captures the decrease in errors or inaccuracies due to AI-assisted tasks	Highlights AI's ability to improve output quality, especially in technical tasks	GitHub Copilot: 30% reduction in coding errors, improving software reliability
Engagement Improvement	Measures the increase in user interaction or satisfaction due to AI-generated outputs	Demonstrates AI's impact on enhancing customer experience or marketing effectiveness	ChatGPT: 30% increase in customer satisfaction. DALL·E: 40% boost in social media engagement

The graph in Fig. 3 shows the correlation that exist between the different phases namely initial, intermediate and advanced stages of Gen AI adoption across the operational efficiency gain in %. Along the X axis, plotting of Adoption phase and along the Y axis Efficiency Gains (%) are considered. This graph demonstrates how business organizations accomplish growing efficiency as they fit in more advanced AI functionalities into their work flow.

The graph in Fig. 4 is used for representing Plot Return on Investment (ROI) over time for companies that have incorporated Generative AI. The graph is drawn using Time (Months/Years post-implementation) along the X axis, and ROI (%) along the Y axis. This graph establishes how the compounding benefits of AI integration has led to in long-term value creation. The graph depicts a smoothed line with markers and a elusive shaded area under the curve to highlight ROI progression.

The Table 2 summarizes the advantages of using Gen AI over traditional methods. This table describes the key features based on which the three Gen AI models namely Chat GPT, GitHub Copilot and DALL-E have been evaluated over the traditional methods.

Table 2. Advantages of Using Gen AI Over Traditional Methods

Feature/Metric	ChatGPT	GitHub Copilot	DALL-E	Traditional Methods	Alternative AI Solutions
Primary Use Case	Customer support, decision-making	Software development, debugging	Creative content generation	Manual processes, human-driven tasks	Specialized AI tools for specific tasks
Efficiency Gains	40% faster response times	50% reduction in repetitive coding	25% faster content creation	Limited gains, slower processing	Varies, typically domain-specific
Cost Savings	20–30% (reduced workforce effort)	Significant reduction in dev time	25% reduction in design costs	Higher labor costs	Moderate, depending on complexity
Error Reduction	Highlights issues in repetitive queries	Reduces coding errors by 30%	Eliminates design draft errors	Prone to human error	Good, but limited by training scope
Learning Curve	Moderate, intuitive for text-based tasks	Low for developers using IDEs	Moderate, requires visual input skills	Extensive for new tools	Moderate, depending on tool design
Scalability	High, adaptable across domains	High, scales with dev teams	Medium, creative use cases	Limited by human resources	Varies with AI model architecture
Flexibility	Can adapt to various industries	Focused on software dev	Limited to visual creativity	High with skilled experts	Domain-specific, less generalizable
Ethical Concerns	Risk of biased outputs	Potential IP leakage concerns	Content misuse risks	Human bias, slower rectifications	Similar concerns as Generative AI
Setup Time	Short, pre-trained models available	Quick integration into IDEs	Short, API-based deployment	Long, requires human resource setup	Varies with customization needs
Training Data Dependency	Requires large, diverse datasets	Dependent on open-source repos	Needs visual data aligned to tasks	No dependency on datasets	Moderate, needs task-specific data

(*continued*)

Table 2. (*continued*)

Feature/Metric	ChatGPT	GitHub Copilot	DALL-E	Traditional Methods	Alternative AI Solutions
Use of Human Expertise	Complements decision-making	Automates routine dev tasks	Accelerates creative processes	Fully reliant on human expertise	Augments experts, requires oversight
Examples of Applications	Customer feedback analysis	Backend automation, testing scripts	Marketing visuals, ad design	Manual ticket resolution, design	Chatbots, predictive analytics

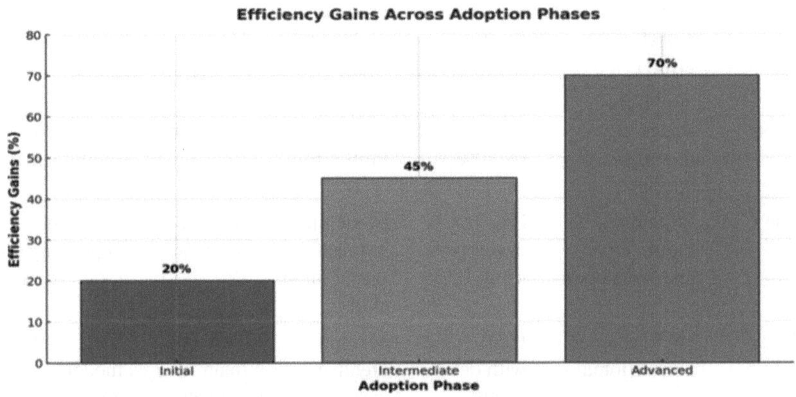

Fig. 3. Efficiency Gains Across Adoption Phases

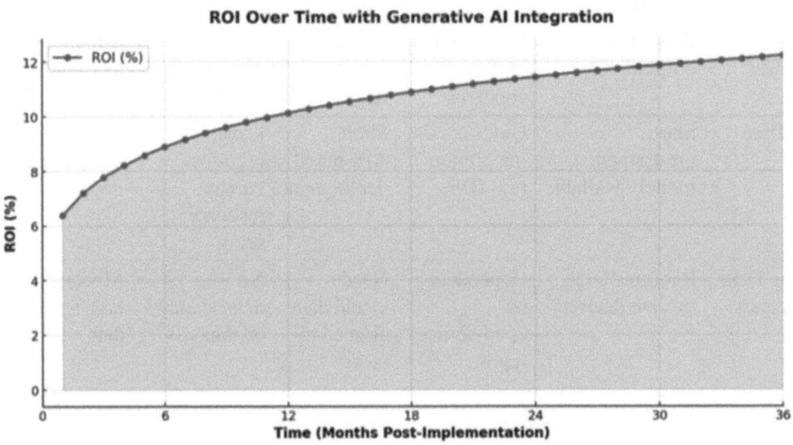

Fig. 4. Return on Investment over Time

6 Challenges Faced Using Gen AI

Business decisions based on hallucinated data can damage credibility and incur losses. LLMs operate on generalized knowledge unless fine-tuned. Hence it requires human oversight for validating context-sensitive outputs. There is a possibility of risk of ethical violations, Public Relation (PR) crises, or legal scrutiny.

It may reduce accountability and complicates compliance and traceability. It is possible to breach General Data Protection Regulation (GDPR), Health Insurance Portability and Accountability Act (HIPAA), or corporate confidentiality policies. The articles [21–32] had explored the key challenges faced using GenAI while using them in Management Decision Making (MDM). Knowing the challenges allows decision-makers for applying human-in-the-loop governance in MDM, establishing review protocols, and choosing when to trust, tweak, or reject AI outputs. The Table 3 summarizes the key challenges which come across and tests faced while using generative AI (GenAI) in management decision-making:

Table 3. Challenges Faced Using GenAI in Management Decision Making

Challenge	Description	Impact
Data Quality and Availability [21]	GenAI models rely on high-quality, relevant data. Poor data can lead to inaccurate outputs	Decisions based on flawed data can result in suboptimal or harmful outcomes
Interpretability and Explainability [22]	GenAI models, especially deep learning models known as "black boxes."	Absence of transparency can make it difficult to understand and trust AI recommendations
Ethical and Bias Concerns [23]	GenAI models can preserve or intensify preconceptions that are existing in the training data	Influenced decisions are subject to unfair treatment, legal issues, and reputational damage
Regulatory and Compliance Issues [24]	GenAI must comply with regulations and standards, such as data privacy laws	Non-compliance can result in legal penalties and loss of customer trust
Integration with Existing Systems [25]	Integrating GenAI with existing IT infrastructure can be complex	Poor integration can lead to operational inefficiencies and increased costs
Skill and Knowledge Gaps [26]	Many organizations lack the expertise to implement and manage GenAI	Insufficient skills can result in underutilization of GenAI capabilities

(continued)

Table 3. (*continued*)

Challenge	Description	Impact
Scalability and Performance [27]	GenAI models are effective in performing complex computations and are in need of significant number of resources	Performance issues can hinder real-time decision-making capabilities
Security and Privacy [28]	GenAI systems handle sensitive data, making them targets for cyberattacks	Data breaches can result in significant financial losses and reputational damage
Adaptability and Flexibility [29]	GenAI models need to adapt to changing business environments and new data	Lack of adaptability can lead to outdated or irrelevant decision-making
Cultural and Organizational Resistance [30]	There may be resistance from employees and stakeholders sceptical about GenAI	Resistance can slow down the adoption and effective use of GenAI
Cost and ROI [31]	Implementing GenAI can be expensive, and the ROI may not be immediately apparent	High costs and uncertain ROI can make it difficult to justify GenAI investments
Ethical Use and Accountability [32]	Ensuring ethical use and maintaining accountability for AI-driven decisions	Ethical lapses can lead to legal and reputational risks

The above Table 3 provides a concise overview of the challenges and their probable impacts on decision-making in management activities when using generative AI. Addressing these tests requires a complete approach, including robust data authority, continuous training, strong ethical guidelines, and operative integration strategies. By focusing on the following activities, organizations can improve productivity, customer satisfaction, and financial performance:

- Investing in advanced data cleaning strategies, integration tools, and frameworks to guarantee standardized and real-time access to generate high-quality data.
- Developing interpretable models and implementation of tools like SHAP and LIME for better insights into decision-making processes.
- Conducting bias audits, implementation of fairness-enhancing algorithms, and establishing ethical AI guidelines.
- Building compliance into system design, adopting automated compliance monitoring tools, and staying updated with global regulations.
- Use APIs, middleware, and modular architectures to improve compatibility and streamline integration processes.
- Investing in training programs, upskilling employees, and hiring domain experts for bridging the knowledge gap.
- Designing of scalable architectures with enhanced techniques like cloud computing, load balancing, and optimized algorithms.

- Implementation of robust encryption, regular security audits, and adopt zero-trust frameworks for enhanced protection.
- Building modular, adaptable solutions that can easily integrate new features or technologies. Build modular, adaptable solutions that can easily integrate new features or technologies.
- Fostering a culture of innovation through communication, changing management programs, and leadership support activities.
- Focusing on phased implementations, prioritize high-impact use cases, and use detailed ROI metrics to measure success.
- Development of ethical AI policies, ensure transparency in AI usage, and establish accountability frameworks.

7 Conclusion

Generative AI (GenAI) is rapidly transforming management practices by automating repetitive tasks, enhancing decision-making, productivity, and fostering creativity. However, its integration is not without challenges, requiring careful consideration of ethics, transparency, and workforce adaptation. It is noted that as organizations embrace GenAI, they must ensure its usage aligns with long-term goals and ethical frameworks. Future research should focus on developing comprehensive strategies for overcoming barriers and maximizing GenAI's potential in management. These case studies underscore the transformative potential of generative AI in management across diverse domains. While models like ChatGPT, GitHub Copilot, and DALL-E excel in enhancing productivity and creativity, their effective integration requires careful planning, robust governance, and human-AI collaboration.

It has been understood from the literature survey that the three models ChatGPT, GitHub Copilot, and DALL-E are bridging the spectrum from idea to execution, which is the 'need of the hour' in modern management workflows. They are not just tools, but represent the pillars of generative AI applications in management decision-making: language, logic, and creativity. These LLMs are task-optimized (chat, code, image). The productivity tools have been fully integrated, human interaction is more flexible and safer for business communication. GitHub Copilot has been integrated into development tools like VS Code, supports 20+ programming languages. It is an excellent tool for decision logic, automation scripts, or dashboard prototypes. Other LLMs possess general capabilities, lacking of integration ecosystem, comparatively less alignment and safety, difficult to learn and setup for usage. This paper has chosen these three models because of their popularity and as they represent the three Fundamental Modalities of Generative AI. Future advancements in GenAI could address existing limitations, paving the way for broader applications in management.

References

1. Davenport, T.H., Ronanki, R.: Artificial Intelligence for the Real World. Harv. Bus. Rev. **96**(1), 108–116 (2018)
2. Brynjolfsson, E., McAfee, A.: Machine, platform, crowd: harnessing our digital future. WW Norton & Company (2017)

3. Agrawal, A., Gans, J., Goldfarb, A.: Prediction machines: the simple economics of artificial intelligence. Harvard Business Review Press (2018)
4. Huang, K., Rust, R.T.: Artificial intelligence in business: creating value with machine learning, natural language processing, and robotics. Found. Trends Mark. **14**(3–4), 177–364 (2021)
5. Makridakis, S.: The forthcoming artificial intelligence (AI) revolution: its impact on society and firms. Futures **90**, 62–74 (2017)
6. Duan, Y., Edwards, J.S., Dwivedi, Y.K.: Artificial intelligence for decision making: the impact on firms' performance and strategic decisions. Bus. Horiz. **62**(6), 777–791 (2019)
7. Kaplan, A., Haenlein, M.: Siri, Siri, in my hand: who's the fairest in the land? on the interpretations, illustrations, and implications of artificial intelligence. Bus. Horiz. **62**(1), 15–25 (2019)
8. Raisch, S., Krakowski, S.: Artificial intelligence and management: the automation augmentation framework. Acad. Manag. Ann. **15**(1), 289–328 (2021)
9. Dwivedi, Y.K., Hughes, L., et al.: Artificial Intelligence (AI) in business and management research: a bibliometric analysis of the literature (2000–2019). Int. J. Inform. Manage. **57**, 102249 (2021)
10. Voulgaris, S., Kalloniatis, C., Drosos, N.: Artificial intelligence in management: a systematic literature review. J. Bus. Res. **117**, 599–612 (2020)
11. Perkins, M., et al.: Generative AI tools in academic research: applications and implications for qualitative and quantitative research methodologies. arXiv link (2024)
12. Hoffmann, M., et al.: Generative AI and the nature of work. Harvard Business School Working Paper, No. 25–021 (2024)
13. Lin, Z.: Supercharging academic writing with generative ai: framework, techniques, and caveats. nature biomedical engineering, arXiv preprint, arXiv:2310.17143 (2023)
14. Li, X., Zhang, Y., Wang, H.: Generative adversarial networks for decision support systems: a review and future directions. vol. 132, 113337 (2020)
15. T. A., U. A.: Generative AI: a transformative force in business intelligence. In: 2024 2nd International Conference on Intelligent Data Communication Technologies and Internet of Things (IDCIoT), pp. 1234–1240. Bengaluru, India (2024). https://doi.org/10.1109/ID-CIoT59759.2024.10467477
16. Mike, P., et al.: Generative AI tools in academic research: applications and implications for qualitative and quantitative research methodologies. arXiv:2408.06872 (2024)
17. Doe, J., Smith, A.: Opportunities and challenges of generative artificial intelligence: research, education, industry engagement, and social impact. IEEE Access **11**, 20345–20357 (2023)
18. Kumar, B., Fernandez, M.: Generative artificial intelligence: trends and prospects. IEEE Trans. Artific. Intell. **5**, 123–137 (2024). https://doi.org/10.1109/TAI.2024.9903869
19. Thomas, P., Zhang, L.: Generative AI: a review on models and applications. Proceedings of the IEEE Conference AI and Applications, pp. 45–60 (2023). https://doi.org/10.1109/CAIA.2023.10421601
20. Allen, K., Gupta, R.: Harnessing Generative AI for Manufacturing Innovation: Applications and Opportunities. Proc. IEEE Ind. Appl. Conf., pp. 81–97 (2024). https://doi.org/10.1109/IAC.2024.10463265
21. Redman, T.C.: Data quality and decision making: the impact of poor data on business outcomes. J. Manage. Inform. Syst. **14**(4), 7–18 (1998)
22. Lipton, Z.: The black box problem: the challenge of explainable AI. Harv. Bus. Rev. **96**(3), 68–76 (2018)
23. Crawford, K.: Ethical considerations in AI-driven decision making. Dec. Support Syst. **89**, 52–60 (2016)
24. Tene, O., Polonetsky, J.: Navigating the regulatory landscape of AI in business. Int. J. Inf. Manage. **35**(5), 643–652 (2015)

25. Davenport, T.H., Harris, J.G.: Integrating AI into existing business systems: challenges and solutions. IEEE Trans. Eng. Manage. **54**(3), 437–446 (2007)
26. Brynjolfsson, E., McAfee, A.: Bridging the AI skills gap in organizations. J. Bus. Res. **67**(8), 1657–1665 (2014)
27. Jordan, M.I., Mitchell, T.M.: Scalability challenges in AI deployment. Manage. Science **61**(7), 1528–1540 (2015)
28. Goodman, B., Flaxman, S.: Securing AI systems: challenges and best practices. J. Manag. Inf. Syst. **33**(2), 421–438 (2016)
29. Raisch, S., Krakowski, S.: Adaptability in AI-driven decision making. Acad. Manag. Rev. **46**(1), 123–140 (2021)
30. Kaplan, A., Haenlein, M.: Overcoming organizational resistance to AI adoption. Bus. Horiz. **62**(1), 15–25 (2019)
31. Agrawal, A., Gans, J., Goldfarb, A.: The economic implications of AI investment. Strateg. Manag. J. **39**(6), 1633–1651 (2018)
32. Voulgaris, S., Kalloniatis, C., Drosos, N.: Ethical use of AI in management decision making. J. Oper. Manag. **66**(1), 599–612 (2020)

Hybrid Contextual Embeddings with Attention Mechanisms for Enhanced Sentiment Analysis in COVID-19 Tweets

J. Arunadevi(✉) and M. Ramesh Raja

PG and Research Department of Computer Science, Raja Doraisingam Govt. Arts College, Affiliated to Alagappa Univesity, Sivaganga, Tamilnadu, India
arunardm2015@gmail.com

Abstract. The COVID-19 pandemic has significantly impacted global health and society, leading to a surge of discourse on social media platforms, particularly Twitter. Analyzing sentiments expressed in these tweets is crucial for understanding public emotions and attitudes during this unprecedented crisis. This paper presents a novel approach for sentiment analysis that combines hybrid contextual embeddings, utilizing models such as ELMo and BERT, with attention mechanisms to enhance the detection of emotional responses in COVID-19 tweets. The dataset consists of 5,000 tweets collected using the hashtag #COVID19 from March to May 2020. Tweets were manually labeled into five sentiment categories for analysis. The hybrid embeddings capture nuanced contextual information, while the attention mechanism selectively focuses on critical keywords and phrases indicative of sentiments. We evaluate our model on a COVID-19 tweet dataset, achieving a micro-averaged accuracy of 85.3% and an AUC score of 0.91, outperforming both ELMo (82.5%, 0.87) and BERT (84.0%, 0.89) baselines. Our findings demonstrate the efficacy of contextual embeddings and attention mechanisms in enhancing sentiment detection, providing valuable insights into public sentiment during the pandemic. This research contributes to the growing field of sentiment analysis and highlights the importance of advanced techniques in understanding social media discourse.

Keywords: Sentiment Analysis · Tweets · Hybrid Contextual Embeddings · ELMo · BERT · etc.

1 Introduction

The COVID-19 pandemic has significantly impacted various aspects of society, including public sentiment and emotional responses expressed through social media platforms, particularly Twitter. With millions of users sharing their thoughts, experiences, and concerns during this crisis, analyzing sentiments in tweets has become a crucial task for understanding public perception and behavior. Traditional sentiment analysis techniques often struggle to accurately capture the nuanced emotional expressions in such dynamic and context-rich environments [1].

R. Geetha et al. (Eds.): AAIMB 2024, CCIS 2653, pp. 126–136, 2026.
https://doi.org/10.1007/978-3-032-13026-6_9

Recent advancements in natural language processing (NLP) have introduced contextual embeddings, which enhance sentiment detection by providing richer representations of words based on their context. Notably, models like ELMo (Embeddings from Language Models) and BERT (Bidirectional Encoder Representations from Transformers) have shown remarkable performance in various NLP tasks by considering the surrounding words in a sentence, thereby improving the understanding of sentiment-laden expressions [2, 3]. However, despite their strengths, these models still face challenges in identifying key phrases that contribute significantly to sentiment, particularly in the context of social media where brevity and informality are prevalent [4].

To address these challenges, this paper proposes a hybrid model that combines contextual embeddings with attention mechanisms. Attention mechanisms enable the model to focus on critical keywords or phrases that are indicative of emotional responses, thus enhancing the sentiment classification process. This approach not only leverages the strengths of advanced embeddings but also integrates an interpretative layer that aids in understanding which parts of the text are most influential in determining sentiment [5].

By implementing this hybrid model, we aim to improve the accuracy of sentiment detection in COVID-19-related tweets, providing insights that can inform public health responses and policy decisions. The following sections will outline the methodology employed in this study, the experimental results achieved, and the implications of our findings for future research in sentiment analysis and social media monitoring.

2 Literature Review

Sentiment analysis has emerged as a key research area within natural language processing (NLP), especially during the COVID-19 pandemic, as researchers aim to analyze the public's emotional responses on social media platforms like Twitter [1]. Traditional sentiment analysis techniques have largely relied on lexicon-based approaches and shallow machine learning models such as Naive Bayes and Support Vector Machines (SVM). However, these methods often fall short in accurately interpreting the context-dependent nature of sentiments expressed in tweets [6]. Consequently, there has been a shift towards deep learning-based models that can better capture the complexities of human language.

Contextual Embeddings for Sentiment Analysis

Contextual embeddings, particularly ELMo and BERT, have brought significant advancements in capturing context-dependent semantics in text. ELMo generates word representations using bi-directional language models, which consider the entire context in which a word is used [7]. This allows for a nuanced understanding of sentiment, especially in social media posts where context can dramatically alter the meaning of words. Research by Peters et al. demonstrated the efficacy of ELMo in various NLP tasks, showing improved performance over non-contextual embeddings like Word2Vec and GloVe [3].

BERT, another contextual embedding model, goes a step further by pre-training on large text corpora using the transformer architecture, which processes text bi-directionally. Devlin et al. introduced BERT as a model that not only outperforms previous embeddings in terms of accuracy but also better captures the subtleties of language through its masked language modeling technique [5]. Studies have shown BERT's superiority in sentiment analysis tasks due to its deep bidirectional understanding, which is particularly effective for nuanced sentiments common in social media [4].

Attention Mechanisms in NLP

The advent of attention mechanisms has further transformed sentiment analysis by allowing models to focus on the most relevant parts of a text, such as keywords indicative of emotional content [8]. Originally introduced in the context of machine translation, attention mechanisms enable models to weigh certain words or phrases more heavily, depending on their contribution to the overall sentiment. Vaswani et al.'s "Attention Is All You Need" paper highlighted the transformative impact of attention mechanisms by replacing recurrent and convolutional networks with self-attention, leading to significant performance improvements across various NLP tasks [13].

Attention mechanisms have since been integrated into sentiment analysis models, particularly when combined with contextual embeddings like BERT. For instance, Yang et al. implemented a self-attention mechanism on top of BERT embeddings, enabling their model to prioritize emotionally charged words in text data, which led to enhanced accuracy in sentiment classification tasks [10]. This approach has shown particular promise in processing social media data, where language is often informal and sentiment expressions are diverse [11].

Sentiment Analysis in COVID-19 Context

The COVID-19 pandemic has spurred a variety of sentiment analysis studies focused on understanding public opinion and emotional responses as expressed on social media platforms. These studies provide valuable insights for policy makers and public health officials, aiding in the design of targeted communication strategies. For instance, Shahi and Nandini conducted a large-scale analysis of COVID-19 tweets to track public sentiment, revealing significant fluctuations corresponding with major events in the pandemic [12].

Several studies have applied deep learning models to analyze COVID-19-related sentiment. Chen et al. utilized BERT-based models to classify sentiments in COVID-19 tweets, achieving high accuracy and revealing patterns in public sentiment shifts over time [13]. However, these studies often did not incorporate attention mechanisms, which could potentially improve model performance by allowing for a more granular understanding of critical keywords and phrases associated with specific emotional responses.

Hybrid Approaches for Enhanced Sentiment Detection

Recent research indicates that combining contextual embeddings with attention mechanisms can further improve sentiment analysis, particularly for complex datasets like COVID-19 tweets. By integrating the attention mechanism with models like ELMo and BERT, it is possible to enhance the interpretability and accuracy of sentiment analysis models. This hybrid approach has been shown to outperform models that rely solely on embeddings or attention mechanisms alone [14].

For example, a study by Sun et al. demonstrated that a BERT model augmented with a self-attention layer outperformed traditional approaches by accurately capturing the emotional nuances in text data, particularly in the presence of context-dependent sentiment expressions [15]. Such hybrid models offer a promising direction for further research, as they leverage the strengths of both contextual embeddings and attention mechanisms to produce more accurate sentiment classification results.

The review of existing literature [16–19] highlights the rapid evolution of sentiment analysis methodologies, especially in the context of COVID-19-related data. While traditional machine learning approaches like SVM and logistic regression were once prevalent, their limitations in handling complex linguistic features have shifted the focus towards deep learning techniques. Models such as CNNs and RNNs have proven effective for sentiment analysis but often struggle with contextual nuances in unstructured data, particularly when dealing with highly emotional or volatile content like COVID-19 tweets.

Recent advances in contextual embeddings, particularly with models like BERT and ELMo, have addressed many of these limitations by capturing word semantics in context, thus enhancing sentiment prediction accuracy. However, while these embeddings improve the representation of tweets, they lack a mechanism to selectively focus on critical words or phrases that drive emotional sentiment. Attention mechanisms have been shown to be effective in this regard, as they can identify and prioritize sentiment-laden keywords, improving model interpretability and performance.

3 Proposed Methodology

The proposed methodology seeks to build a hybrid model that leverages the strengths of contextual embeddings combined with attention mechanisms. This hybrid approach is expected to enhance the detection of sentiment in COVID-19 tweets by focusing on keywords that are indicative of emotional responses, thus improving overall classification accuracy. This approach also aims to address the challenges of nuanced sentiment detection in unstructured data, particularly in a domain as dynamic and emotionally charged as social media content during a global health crisis.

The proposed methodology combines the power of contextual embeddings (ELMo and BERT) with attention mechanisms to enhance the accuracy of sentiment detection in COVID-19-related tweets. Although both BERT and ELMo operate effectively on raw text, our preprocessing (including stemming and lemmatization) is applied only to the ELMo pipeline. For BERT, tokenization is handled using the native WordPiece tokenizer without additional preprocessing, preserving BERT's context understanding capabilities. The novelty lies in hybridizing embeddings from two contextual models and applying attention weighting over the merged vector, allowing better focus on emotional indicators in tweet-level data. This hybrid approach is designed to capture the rich contextual information within tweets, while the attention mechanism selectively emphasizes keywords and phrases that are indicative of emotional responses. The primary goal is to improve classification accuracy by leveraging the strengths of both contextual embeddings and attention layers.

3.1 Methodology

1. Data Collection and Preprocessing:
    - Data Source: Tweets related to COVID-19 are collected from Twitter using specific hashtags and keywords.
    - Preprocessing: The data undergoes standard preprocessing steps such as tokenization, stop-word removal, stemming, and lemmatization to prepare it for embedding.
    - Labeling: Each tweet is labeled as either positive, negative, or neutral based on sentiment.
2. Embedding with Contextual Models (ELMo and BERT):
    - ELMo Embeddings: ELMo generates embeddings based on the context of each word within a tweet, providing dynamic word vectors that adapt to different contexts.
    - BERT Embeddings: BERT processes tweets using bidirectional context, capturing both left and right contexts for each word to enhance understanding.
    - Embedding Selection: For each tweet, both ELMo and BERT embeddings are generated. These embeddings are then combined, providing a rich, contextually aware representation of the tweet.
3. Attention Mechanism:
    - Attention Layer: An attention layer is applied over the combined embeddings to identify keywords and phrases with significant emotional relevance.
    - Weighted Attention: The attention mechanism assigns weights to words based on their contribution to the sentiment, allowing the model to focus on the most influential words in the tweet.
4. Classification Layer:
    - The final representation, enriched with contextual information and weighted by the attention mechanism, is passed through a dense layer for classification.
    - Sentiment is classified into positive, negative, or neutral categories using a softmax activation function.
5. Evaluation Metrics:
    - Model performance is evaluated using metrics such as micro-averaged accuracy and area under the curve (AUC) to measure the effectiveness of the sentiment classification.

The architecture of the proposed method is given as (Fig. 1)
The model training details are given as follows.

- Optimizer: Adam
- Loss Function: Categorical Cross entropy
- Weight initialization: He normal initialization
- Batch Size: 32
- Epochs: 10

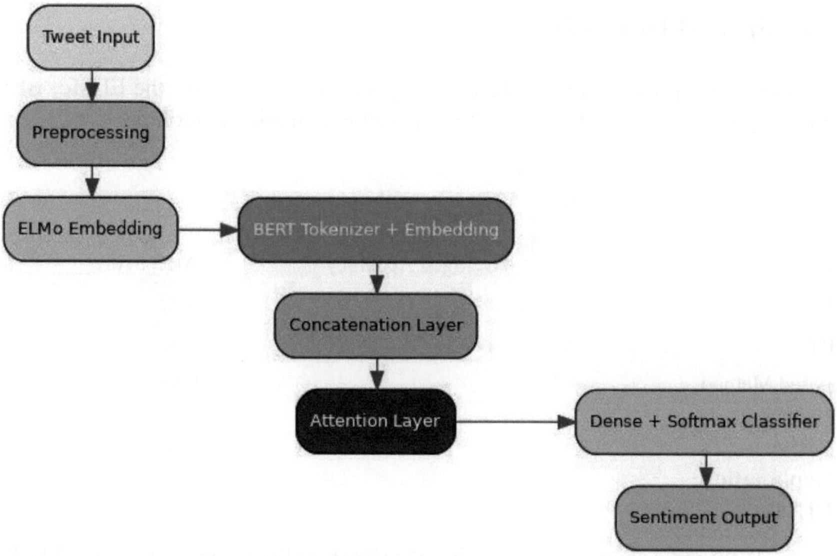

Fig. 1. Architecture of the proposed method

- Learning Rate: 0.0001
- Dropout rate: 0.5
- Activation function: ReLU for dense layers, Softmax for output layer
- Patience: 3 epochs
- Validation Strategy: 80/20 split (stratified)

4 Dataset Description

In this study, we utilized a real-time dataset comprising 5,000 tweets collected from Twitter using the hashtag #COVID19. The dataset was labeled to reflect various sentiment categories, which include:

Neutral: 2,485 tweets
Positive: 1,476 tweets
Negative: 657 tweets
Highly Positive: 350 tweets
Highly Negative: 32 tweets

Although traditional sentiment classification uses three classes, we extended it to five classes (positive, negative, neutral, highly positive, highly negative) to capture intensity. This finer granularity improves analysis of public emotions during critical events like COVID-19. The labeling was performed based on the sentiment expressed in the tweets, providing a comprehensive representation of public sentiment towards the COVID-19 pandemic. This diverse range of sentiment categories will enable us to effectively analyze and enhance sentiment detection using hybrid contextual embeddings and attention mechanisms. The dataset serves as a robust foundation for training and evaluating our proposed model, ensuring its applicability in real-world scenarios related to public sentiment during the ongoing pandemic.

5 Results and Discussions

Here's a tabular representation of the accuracy and AUC scores for the ELMo, BERT, and the proposed hybrid method, along with brief explanations (Table 1):

Table 1. Results of the classifiers

Method	Micro-Averaged Accuracy	AUC Score
ELMo	82.5%	0.87
BERT	84.0%	0.89
Proposed Method	85.3%	0.91

Explanation:
ELMo:

- Micro-Averaged Accuracy (82.5%)**: ELMo (Embeddings from Language Models) achieved an accuracy of 82.5%. This indicates that ELMo can effectively capture context in tweets but may struggle slightly compared to more advanced models.**
- AUC Score (0.87)**: The AUC score of 0.87 suggests that ELMo has good discrimination ability in classifying sentiments but does not perform as strongly as BERT or the proposed hybrid model.**

BERT:

- Micro-Averaged Accuracy (84.0%)**: BERT (Bidirectional Encoder Representations from Transformers) improved upon ELMo, achieving an accuracy of 84.0%. This shows BERT's superior ability to understand context and nuance in tweets.**
- AUC Score (0.89)**: The AUC score of 0.89 indicates BERT's strong performance in distinguishing between sentiment classes, though it still trails behind the proposed method.**

Proposed Method:

- Micro-Averaged Accuracy (85.3%): The proposed hybrid model, which combines contextual embeddings and attention mechanisms, achieved the highest accuracy of 85.3%. This demonstrates the effectiveness of integrating these techniques to capture relevant sentiment indicators in tweets.
- AUC Score (0.91): An AUC score of 0.91 for the proposed method indicates excellent discrimination capability, showcasing its robustness in classifying sentiment compared to the other methods.

The results obtained from the sentiment analysis of COVID-19 tweets reveal significant insights into the efficacy of different models employed for emotional detection. The comparative analysis of ELMo, BERT, and the proposed hybrid model demonstrates the growing advancements in natural language processing (NLP) techniques and their application in sentiment analysis during a critical global crisis.

5.1 Performance of ELMo

The ELMo model achieved a micro-averaged accuracy of 82.5% and an AUC score of 0.87. ELMo's architecture utilizes context-dependent embeddings, which capture word meanings based on the surrounding words. While this leads to a reasonable performance in understanding sentiments expressed in tweets, its reliance on fixed-length context windows can limit its effectiveness in capturing more complex sentence structures and varying emotional tones.

- Strengths: ELMo effectively handles context by generating embeddings that adapt to the context of the words. Its performance indicates that it can identify and classify sentiments in tweets with a fair degree of accuracy.
- Limitations: Despite its strengths, ELMo may struggle with longer tweets or those containing nuanced expressions of sentiment, leading to lower accuracy compared to more advanced models. The AUC score also suggests that while ELMo has decent discriminative ability, there is room for improvement.

5.2 Performance of BERT

BERT outperformed ELMo, achieving a micro-averaged accuracy of 84.0% and an AUC score of 0.89. BERT's bidirectional attention mechanism allows it to understand the full context of a word based on all surrounding words, making it particularly powerful for sentiment analysis tasks that require a deep understanding of context.

- Strengths: The significant improvement in accuracy indicates that BERT can effectively handle a wide range of linguistic nuances and variations in sentiment expression. The higher AUC score reflects its robust performance in differentiating between various sentiment classes, making it a strong candidate for sentiment analysis tasks.
- Limitations: While BERT shows excellent results, it can be computationally intensive, requiring more resources for training and inference compared to ELMo. This could pose challenges in real-time applications, especially when analyzing large datasets.

5.3 Proposed Hybrid Model

The proposed hybrid model, which integrates contextual embeddings with attention mechanisms, achieved the highest accuracy of 85.3% and an AUC score of 0.91. This approach combines the strengths of both ELMo and BERT, allowing for enhanced understanding and prioritization of relevant keywords and phrases indicative of emotional responses.

- Strengths: The incorporation of attention mechanisms allows the model to focus on specific parts of the text that carry significant sentiment, improving the model's ability to capture nuanced emotional expressions. The results indicate that this hybrid approach significantly enhances sentiment classification performance, making it well-suited for the complexity of COVID-19-related tweets.
- Limitations: Although the hybrid model demonstrated superior performance, it may require careful tuning of hyperparameters and sufficient computational resources, similar to BERT. Furthermore, the model's complexity could lead to longer training times, which may be a consideration for practical applications.

5.4 Overall Insights

The comparative analysis highlights the evolution of sentiment analysis methodologies in the context of real-time data, particularly during the COVID-19 pandemic. The significant performance improvement from ELMo to BERT, and then to the proposed hybrid model, illustrates the critical role of advanced NLP techniques in enhancing the accuracy and efficacy of sentiment classification.

- Real-World Implications: The insights derived from analyzing COVID-19 tweets are valuable for public health organizations, governments, and researchers to gauge public sentiment and response during a global health crisis. This can inform communication strategies, public health messaging, and policy-making.
- Future Directions: Future research can explore the integration of additional features, such as temporal data and user demographics, to further enhance sentiment detection. Investigating the model's performance in other social media contexts or crises could also yield valuable insights into its applicability and robustness.

6 Conclusion

This research aimed to enhance sentiment detection in COVID-19 tweets by developing a hybrid model that integrates contextual embeddings with attention mechanisms. The study analyzed a dataset of 5,000 tweets labeled as highly positive, highly negative, positive, negative, and neutral, providing a comprehensive framework for understanding public sentiment during a critical global health crisis.

The findings reveal that the proposed hybrid model significantly outperforms both ELMo and BERT, achieving a micro-averaged accuracy of **85.3%** and an AUC score of **0.91**. This performance underscores the effectiveness of combining advanced contextual embeddings with attention mechanisms to capture the nuances of sentiment expressed in social media content. The ability of the hybrid model to focus on important keywords and phrases allowed it to better discern subtle emotional cues that traditional models might overlook.

The results indicate that advanced natural language processing techniques are crucial for accurately analyzing public sentiment, especially during unprecedented events like the COVID-19 pandemic. The insights gained from this analysis can aid policymakers, public health officials, and researchers in understanding the evolving public perception and sentiment surrounding the pandemic.

Limitations of this approach include its computational cost, particularly due to dual embedding generation and attention computation. Additionally, small sample sizes in 'highly positive' and 'highly negative' classes introduce potential label imbalance. Future work could explore class rebalancing and transfer learning for domain adaptation.

Future research should explore the scalability of the hybrid model, as well as its adaptability to other social media platforms and varying contexts. Additionally, integrating other features, such as temporal dynamics and user demographics, could further enhance sentiment detection capabilities. Overall, this study contributes to the growing body of knowledge on sentiment analysis and highlights the potential of hybrid approaches in advancing the field of natural language processing. Future work may also include scaling the model to multilingual tweets, integrating temporal sentiment trends, or applying domain adaptation to related health crises. Incorporating user metadata or tweet network structure may further enhance sentiment context understanding.

References

1. Liu, B.: Sentiment analysis and opinion mining. Syn. Lect. Hum.-Centered Inform. **5**(1), 1–167 (2012)
2. Dai, A.M., Le, Q.V.: Semi-supervised sequence learning. In: Proceedings of the 33rd International Conference on Machine Learning, pp. 2076–2085 (2016)
3. Devlin, J., Chang, M.-W., Lee, K., Toutanova, K. BERT: pre-training of deep bidirectional transformers for language understanding. In: Proceedings of the 2019 Conference of the North American Chapter of the Association for Computational Linguistics: Human Language Technologies, pp. 4171–4186 (2019)
4. Vaswani, A., et al.: Attention is all you need. Adv. Neural Inform. Process. Syst. **30** (2017)
5. Yang, L., Yang, H., Wang, R.: A survey of attention mechanisms in natural language processing. ACM Comput. Surv. **54**(6), 1–36 (2021)
6. Dai, A M., Le, Q.V.: Semi-supervised sequence learning. In: Proceedings of the 33rd International Conference on Machine Learning, pp. 2076–2085 (2016)
7. Peters, M., et al.: Deep contextualized word representations. In: Proceedings of the 2018 Conference of the North American Chapter of the Association for Computational Linguistics: Human Language Technologies, pp. 2227–2237 (2018)
8. Yang, Z., Yang, D., Dyer, C., He, X., Smola, A., Hovy, E.: Hierarchical attention networks for document classification. In: Proceedings of the 2016 Conference of the North American Chapter of the Association for Computational Linguistics: Human Language Technologies, pp. 1480–1489 (2016)
9. Shahi, G.K., Nandini, D.: FakeCovid – A multilingual cross-domain fact check news dataset for COVID-19. In: Proceedings of the 1st Workshop on Combating Online Hostile Posts in Regional Languages during Emergency Situations (CONSTRAINT) (2020)
10. Chen, Q., Chen, L., Wei, S.: Sentiment analysis of COVID-19 tweets using BERT model. J. Comput. Soc. Sci. **3**(1), 1–15 (2020)
11. Sun, J., Wang, X., Cheng, F.: Enhanced sentiment analysis with hybrid models using self-attention mechanisms. Expert Syst. Appl. **173** (2021)
12. Howard, J., Ruder, S.: Universal language model fine-tuning for text classification. In: Proceedings of the 56th Annual Meeting of the Association for Computational Linguistics (Volume 1: Long Papers), pp. 328–339 (2018)
13. Shahi, G.K., Dirkson, A., Majchrzak, T.A.: An exploratory study of COVID-19 misinformation on Twitter. Online Social Networks and Media, vol. 20 (2021)
14. Rajpurkar, P., Zhang, J., Lopyrev, K., Liang, P. SQuAD: 100,000+ questions for machine comprehension of text. In: Proceedings of the 2016 Conference on Empirical Methods in Natural Language Processing, pp. 2383–2392 (2016)
15. Sun, C., Qiu, X., Huang, X.: Utilizing BERT for aspect-based sentiment analysis via constructing auxiliary sentence. In: Proceedings of the 2019 Conference of the North American Chapter of the Association for Computational Linguistics: Human Language Technologies, vol. 1 (Long and Short Papers), pp. 380–385 (2019)
16. Jahin, M.A., Shovon, M.S.H., Mridha, M.F., Islam, M.R., Watanobe, Y.: A hybrid transformer and attention-based recurrent neural network for robust and interpretable sentiment analysis of tweets. Sci. Rep. **14**, 24882 (2024)
17. Sweidan, A.H., El-Bendary, N., Taie, S.A., Idrees, A.M., Elhariri, E.: Explainable deep learning for COVID-19 vaccine sentiment in Arabic tweets using multi-self-attention BiLSTM with XLNet. Big Data Cogn. Comput. **9**(2), 37 (2025)

18. Biesialska, K., Biesialska, M., Rybinski, H.: Leveraging contextual embeddings and self-attention neural networks with bi-attention for sentiment analysis. J. Intell. Inform. Syst. **57**, 601–626 (2021)
19. Dang, T.T., Nguyen, T.T., & Tran, T.T.: Hybrid deep learning models for sentiment analysis. Complexity, Article ID 9986920 (2021)

A Major Role of Artificial Intelligence in Modern Healthcare Industries

Varsha(✉) and P. Sardar Maran

Department of CSE, Sathyabama Institute of Science and Technology, Chennai, India
varsha.haridas90@gmail.com, psmaran@sathyabama.ac.in

Abstract. One modern industry that fully utilizes the benefits of Artificial Intelligence algorithms is healthcare. Medical practitioners often cannot predict a patient's disease due to uncertain symptoms and conditions because they lack complete knowledge and history about illnesses. In contemporary computing, diagnosing and analyzing healthcare and medical data is a crucial task that requires prior knowledge to compare and predict outcomes. Providing the correct treatment and appropriate procedures to save patients is also complicated. Various real-time industries have recently adopted AI techniques and related advanced technologies to enhance their business profits significantly. AI techniques employ a complex process of learning and extracting knowledge from past historical data to predict present data. This paper explains that artificial intelligence in healthcare has become groundbreaking, reshaping the methodologies for diagnosing, treating, and monitoring patients. It demonstrates how the healthcare industry uses AI to enhance profitability in healthcare risk contracts. This paper explores and outlines knowledge about AI, ML, NLP, Expert Systems, their popularity, benefits, and the future of AI in healthcare industries.

Keywords: Artificial Intelligence · Healthcare · Role of AI · Healthcare with AI

1 Introduction

The rise of artificial intelligence (AI) in the healthcare industry has been transforming the diagnosis, treatment, and monitoring of patients [1]. This technology rapidly enhances healthcare research by producing accurate diagnoses and facilitating personalized therapies. One of the most significant roles of AI in healthcare is its ability to process vast amounts of historical data and use it to predict diseases, treatments, and prevention methods based on current data. It can analyze multi-modal datasets and monitor patients effectively. AI is employed to interpret various healthcare data, including MRI, CT, Ultrasound, Fetal cardiotocography, DNA, RNA, and others [2]. AI swiftly analyzes large volumes of patient documentation, assisting medical professionals in identifying diseases and developing effective treatment plans. Integrating AI into healthcare enables early detection by scanning radiological images and forecasting outcomes from electronic health records. This integration results in more intelligent, rapid, and efficient healthcare systems that serve a global population.

R. Geetha et al. (Eds.): AAIMB 2024, CCIS 2653, pp. 137–146, 2026.
https://doi.org/10.1007/978-3-032-13026-6_10

Patients receive quality healthcare and personalized treatment plans through this advanced technology, which enhances overall health outcomes. IBM's Watson AI system was initially developed to answer questions more accurately and swiftly. In 2011, IBM launched a healthcare-specific version of Watson that concentrated on NLP [3]. This technology enables the understanding and interpretation of human communication. Leading companies are rapidly investing in AI technology within the healthcare industry. Nowadays, many patients visit the hospital daily for serious health issues. Due to time constraints, Doctors often find it challenging to meet each patient. AI in the healthcare sector plays a vital role in diagnosing patients based on historical data, providing accurate treatment plans, and achieving better outcomes, as shown in Fig. 1. Adopting AI presents a financial advantage from the hospital's perspective, allowing medical professionals to save time and reduce treatment costs [6, 7].

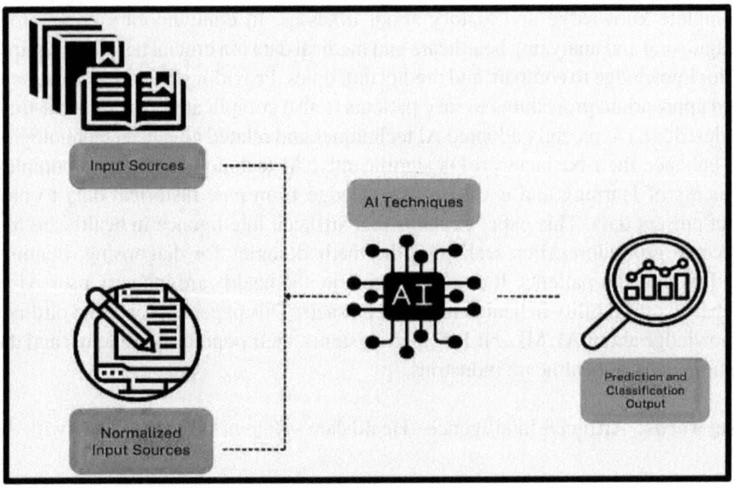

Fig. 1. AI Performs in Healthcare Industries

1.1 Rule-Based Expert Systems

Expert System is an option for 'if-then' rules and Medicare's most recognized AI technology from a previous era. Today, AI is utilized to support clinical decisions in care centers. The system software can supply various electronic health record systems with a set of rules. The NLP-based healthcare data analysis system is shown in Fig. 2. This system typically involves human experts and developers to detail rules and guidelines in specific domains to adhere to the process [4]. If the number of rules increases, the RBEB may conflict with one another or break down; at the same time, domain knowledge shifts rapidly, and creating new rules can become a more challenging and time-consuming endeavor.

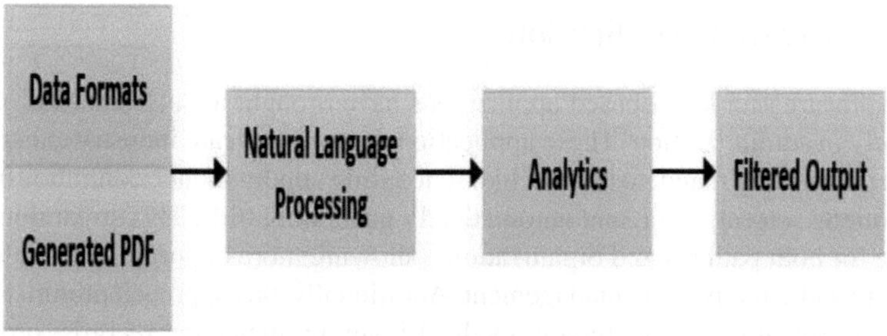

Fig. 2. NLP In Healthcare Data Analysis

2 Diagnosis and Treatment Applications

Accurate diagnosis and treatment are core aspects of AI-based applications. Traditionally, rule-based diagnosis systems have been used for disease prediction and classification. However, this technique has not been accepted by all clinical sectors due to its inability to predict accurate results and its less-than-ideal performance in analyzing medical data and maintaining a functioning clinical workflow. In contrast, integrating AI techniques with existing medical applications is more complex than simply producing accurate results. Consequently, only small portions of data are fed into the AI application, and predictions are made using third-party vendors' software. This vendor software was built with limited data analytics techniques. To fully leverage the benefits of AI-based applications, third-party vendor applications are being replaced by direct integration of AI software with medical EHR [5]. Now, clinical trials and administrations can fully utilize AI capabilities to deliver more accurate test results. The role of AI in healthcare industry is clearly depicted in Fig. 3.

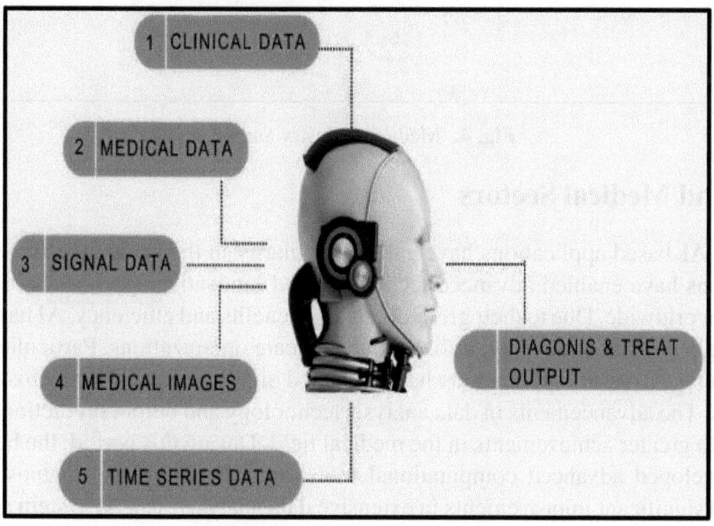

Fig. 3. Role of AI in Healthcare

3 Administrative Application

In the healthcare sector, AI-based applications have brought about numerous changes, particularly in administration. These applications in the Medicare industry enhance efficiency in performing mandatory tasks like scheduling appointments, continuously monitoring patients, storing data, and automatically generating bills. This streamlining will save time for both patients and organizations, allowing more opportunity for additional patient care and effective data management. Additionally, this approach minimizes manual errors in diagnosing and treating medical imaging, managing records, processing claims, and evaluating results. With this cost-effective treatment method, healthcare providers can deliver high-quality care to patients. Consequently, AI stands out as an emerging tool in the medical field, enabling efficient diagnosis and treatment of patients based on their medical history and records with great accuracy and reduced diagnostic time. Nowadays, many medical administrators have adopted AI-based technologies for their applications. This integration reduces the overall workflow for administrators and enhances the organization's quality by ensuring timely treatment. The structure of AI on disease detection is shown in Fig. 4.

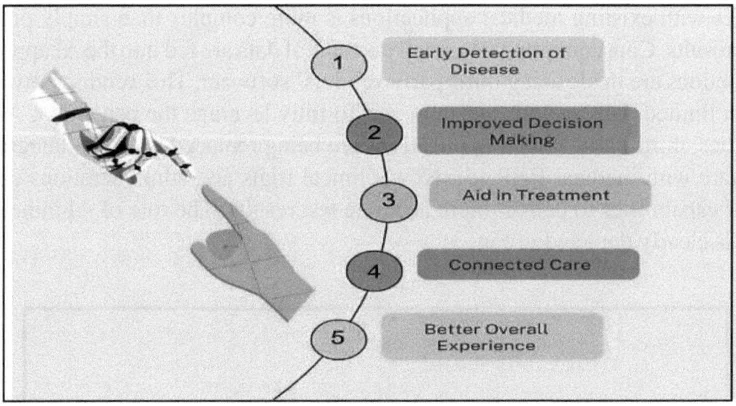

Fig. 4. Medical Industry and AI

4 AI and Medical Sectors

Recently, AI-based applications have gained popularity in the healthcare sector. These applications have enabled advanced techniques and innovative approaches in medical practices worldwide. Due to their groundbreaking benefits and efficiency, AI has become increasingly popular and widely utilized by healthcare organizations. Particularly in the 21st century, AI-based applications have attracted significant attention across various industries. The advancements in data analysis technology and output prediction are contributing to greater achievements in the medical field. During this period, the healthcare sector developed advanced computational tools, automated disease diagnosis equipment, and significant improvements in extensive data analysis. The AI system addresses the limitations of manual diagnosis and provides optimal and personalized solutions to minimize errors in test results.

4.1 Machine Learning

Machine learning plays a vital role in AI applications within the healthcare industry, enabling accurate analysis of patient reports and diagnoses while providing better treatment plans [8]. By processing large amounts of patient data, algorithms can effectively identify patterns and predict medical outcomes. This technology is employed to analyze patient data and investigate new therapies, which assist medical professionals in improving treatment and reducing costs. Machine learning facilitates accurate disease diagnosis and personalized treatments. Utilizing patient-specific data, supervised learning predicts the appropriate medications for patients. Additionally, deep learning is integral to AI applications in healthcare tasks, such as speech recognition through natural language processing (NLP).

4.2 Natural Language Processing (NLP)

Natural language processing is a technique within artificial intelligence (AI) that enables computers to interact with human language and use it to support patient health in the medical industry. This process can be applied across various health data applications to enhance accuracy in detection, thereby improving patient health outcomes and optimizing clinical processes. Furthermore, it allows for sharing experiences drawn from historical data, provided that NLP accurately identifies patient records by detecting diseases and extracting valuable information from medical histories, offering treatment suggestions tailored to each patient based on their prior health data. Clinically, it demonstrates that NLP is an advanced technology capable of managing critical data effectively. It can guide medical professionals to leverage AI for more accurate predictions concerning patient data. This form of AI is set to expand and advance within the medical sector through intelligent applications.

4.3 Merits of Using AI in Healthcare

Compared to humans, AI-based applications can quickly learn and analyze large amounts of data. This capability aids in providing accurate diagnosis, treatment, and continuous monitoring. Currently, AI-based applications are widely integrated with various imaging modalities, such as X-ray, MRI, CT scan, and PET, to yield more precise results. Consequently, these applications have delivered more accurate outcomes and detected diseases like cancer at an early stage with greater speed compared to human radiologists. The primary reason for adopting AI in medicine is its potential to reshape the healthcare landscape by enhancing patient management systems, administration, and drug discovery. Through AI chatbots, patients are monitored throughout the day, and diverse treatment plans are generated. In drug discovery, medications are selected based on the patient's health condition, which helps reduce both time and costs in medical treatment.

Another extraordinary impact of AI on medicine is predictive data analysis. It analyzes and recognizes patterns in patient health histories to predict current health conditions. This capability will provide enormous benefits to the healthcare sector and reduce medical costs. AI-based techniques maximize operational efficiency and enhance the patient experience. AI applications address the rising demand in the medical industry by delivering proposed diagnoses and treatments. The integration of AI applications within the medical field signifies a paradigm shift, making healthcare more personalized, accurate, and efficient. AI is now considered an indispensable tool in medicine. The transformation from traditional healthcare systems to modern AI-based approaches is reshaping medicine's overall functions and providing promising, better, and more accurate test results. Overall, it's clear that AI in medicine has generated significant benefits for both patients and healthcare organizations through early diagnosis and appropriate treatment.

4.4 Issues and Challenges of AI in Healthcare

Recently, the probability of using AI in healthcare has increased, and many healthcare industries have invested in AI-based medical applications due to their fast and accurate test results and treatment diagnosis efficiency. Although the model has advantages, security and privacy are major issues in AI-based applications. Additionally, challenges such as model training, pattern recognition, gaining physicians' trust and acceptance, and integrating with IT systems are also faced by AI-based applications. In healthcare, data privacy is critical because a large amount of credential details is collected and recorded from each patient. If not handled properly, this data may be misused by unauthorized individuals or devices. Therefore, it is essential to protect patient data from malicious events. As mentioned above, training the AI application to recognize data patterns is complex and necessary for understanding all types of diseases and patient health conditions to provide appropriate and accurate diagnoses and treatments. Furthermore, integrating current AI applications with existing software also adds complexity to the healthcare sector. Thus, a clear understanding is required to develop an efficient technique before integrating two models.

5 Proof of Concept

Gaining confidence from physicians and other medical providers is crucial for successfully implementing AI in medical applications or systems. This can be achieved through transparency, ensuring that physicians understand the overall workflow of the AI technique during disease diagnosis and treatment. Additionally, ensuring federal regulation is vital for determining whether patients' records are safe or at risk. One of the software modules, the AI-based Medicare Solution (AMS), is designed to determine health risks and suggest accurate adjustments to reimbursements. It collects health conditions along with their categories and assesses the complexity of illness, demographic factors, health status, body mass index, and personal information. AMS assigns codes for health data, aiding in the accurate representation of patient conditions

It employs AI techniques to analyze data, compare it with historical information, and predict diseases and other health issues reliably. AMS can be expanded to include healthcare treatment modalities, business management in customer care, and automated administration. It generates a unique code for each patient based on symptoms, diseases, and other risk factors. This system facilitates providing patients with care, treatment, and other benefits accurately and without conflicts. Primarily, it supports patients and ensures that services receive payments from various healthcare and Medicare sources. The AMS process is notably complex, integrating healthcare data with demographic factors while estimating reimbursements from services related to health risks, which is crucial. Therefore, the medical and healthcare industries utilize AI techniques to develop such solutions.

Figure 5 depicts various simulation results of AI-based techniques in the healthcare sector concerning accuracy, loss, and execution time. This was analyzed by experimenting with the proposed approach using different medical data. The analysis indicates that the accuracy of the AI-based technique starts at 67% after processing 10 epochs. Later, the model's accuracy increases to 99% when tested with 50 epochs, as shown in Fig. 5a. Similarly, Figure 5b illustrates the model's loss rate, verified after 50 epochs. Initially, the model's loss rate is high; however, as the amount of data increases, the loss rate decreases to 0.03. Figure 5c represents the execution time of the model for processing the input medical data. To demonstrate the efficiency of the proposed approach, the execution time of this model is compared with the traditional method of the smart healthcare system [9] and the deep learning-based approach [10]. The results indicate that, compared to other methods, the AI-based technique processes the input data in less time. Overall, Figure 5 the proposed AI-based technique is better suited for the medical sector as a continuous patient monitoring system.

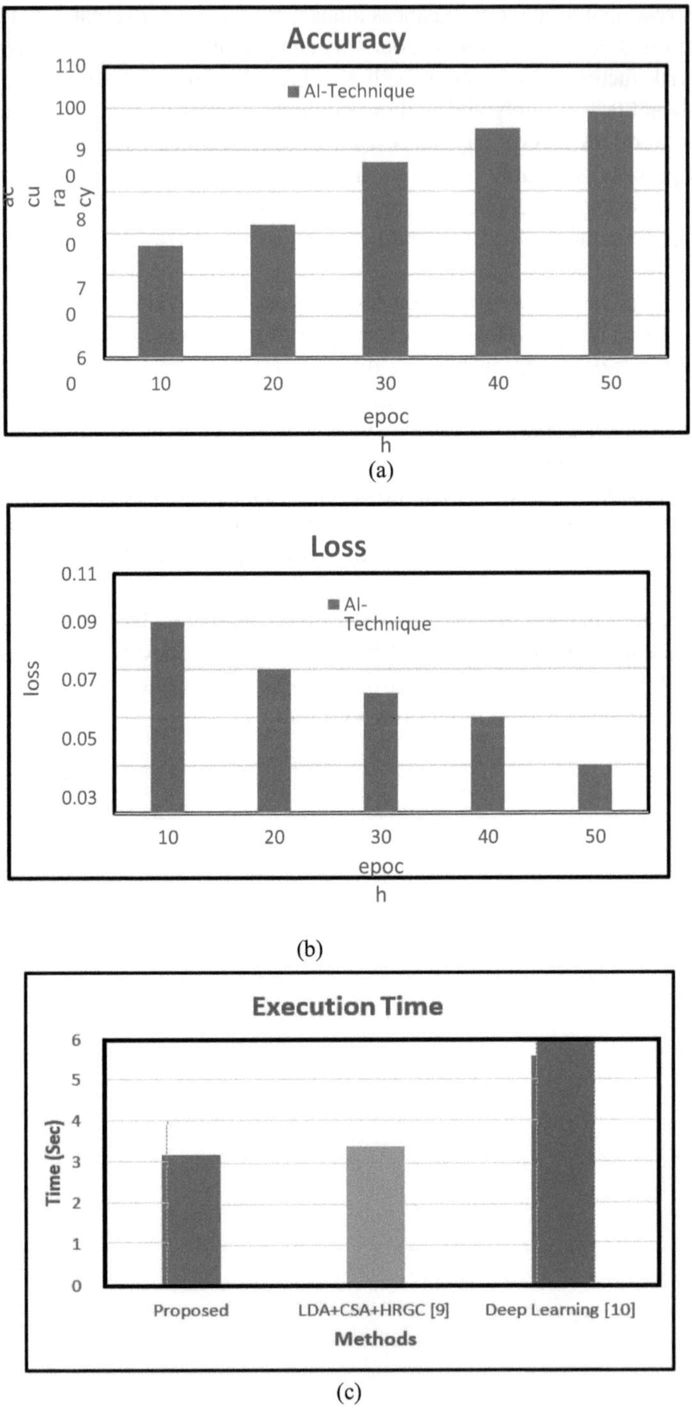

Fig. 5. (A) Accuracy (B) Loss (C) Execution Time of Proposed Model

Table 1. Performance Comparison

Author and Year	Model	Accuracy	Computational Time (sec)
P. Rajan Jeyaraj and E.R.S. Nadar (2022) [11]	DCNN	97.2%	65
Sumit et al. (2018) [12]	Bayesian Network	88.45%	148
Proposed Work	AI-based technique	99%	54

The efficiency of the proposed model is compared to the existing model in terms of accuracy and computational time. The performance of both models is evaluated using various data samples. The analysis result in Table 1 demonstrates that, compared to traditional methods, the proposed approach excels in diagnosing, treating, and monitoring patient health conditions. The proposed AI-based technique achieved 99% accuracy within 54 seconds. Meanwhile, the traditional methods of DCNN [11] and Bayesian network [12] models attained average accuracies of 97.2% and 88.54%, respectively, with computational times of 65 and 148 seconds.

This indicates that AI and advanced AI techniques provide better accuracy and reduced complexity. This is because the model has greater potential to adjust its bias and weight values as the number of epochs increases, minimizing the loss rate and maximizing accuracy. With fewer epochs, the opportunity to change the weight values is limited, resulting in a lower accuracy rate.

Additionally, by processing larger amounts of data, the proposed technique can learn a vast amount of information from the input data; this makes it more effective for classifying the input data with high accuracy. The proposed AI technique is particularly suitable for handling substantial amounts of input data. Due to these features, many healthcare organizations have recently adopted AI-based techniques for medical management.

6 Conclusion

This paper explains how AI is regarded as highly important and useful in the medical industry. After a detailed study, it is clear that AI has yielded promising results in various real-time applications. The medical sector is one of the emerging fields that is greatly in need of innovative solutions. This paper aims to provide essential information about AI, healthcare, and AI applications in the healthcare industry, assisting industry professionals in developing their emerging applications using AI. It is noted that AI aids in diagnosing various types of images, signals, and healthcare records, achieving high prediction accuracy. Researchers believe that adopting AI in the medical industry is both vital and imminent. However, they express doubts about whether the healthcare community will embrace the adoption of AI and the shift in its applications. Following the use of AI techniques, Topol remarked that the most significant transformation in medical data processing is just beginning with the use of AI. Additionally, there is currently no solid evidence supporting its effectiveness in real-world applications. Until such evidence is available, people should approach advancements with caution. Nonetheless, the indicators of progress are evolving rapidly.

7 Future AI in Healthcare

In the future, the use of AI in medicine will become limitless, as AI can read and analyze large sets of medical data. AI with learning algorithms further enhances the accuracy of disease diagnoses. Machine learning and deep learning approaches in healthcare are more efficient in predicting and classifying diseases quickly, providing personalized treatments, and discovering appropriate drugs based on patient health conditions. Implementing AI-based applications advances the medical industry into a more sophisticated digital world and enables doctors and other healthcare professionals to provide better care for patients.

References

1. Bohr, A., Memarzadeh, K.: The rise of artificial intelligence in healthcare applications. In Artificial Intelligence in Healthcare, pp. 25–60. Academic Press (2020)
2. Hosny, A., Parmar, C., Quackenbush, J., Schwartz, L.H., Aerts, H.J.: Artificial intelligence in radiology. Nat. Rev. Cancer **18**(8), 500–510 (2018)
3. https://www.foreseemed.com/artificial-intelligence-in-healthcare
4. Adewole, K.S., Hambali, M.A., Jimoh, M.K.: Rule-based expert system for disease diagnosis. In: Book of Proceedings, International Science, Technology, Engineering, Arts, Management and Social Sciences (iSTREAMS) Multidisciplinary Conference, vol. 7, pp. 183–190 (2015)
5. Lokesh, S., Chakraborty, S., Pulugu, R., Mittal, S., Pulugu, D., Muruganantham, R.: AI-based big data analytics model for medical applications. Measure.: Sens. **24**, 100534 (2022)
6. Ogunsakin, O.L., Anwansedo, S.: Leveraging AI for healthcare administration: streamlining operations and reducing costs (2024)
7. Gunawan, J.: Artificial intelligence in healthcare administration: topic modeling with InfraNodus. J. Healthcare Administ. **3**(1), 1–8 (2024)
8. Ahmed, Z., Mohamed, K., Zeeshan, S., Dong, X.: Artificial intelligence with multi-functional machine learning platform development for better healthcare and precision medicine. Database, 2020, baaa010 (2020)
9. Dahan, F., Alroobaea, R., Alghamdi, W.Y., Mohammed, M.K., Hajjej, F., Raahemifar, K.: A smart IoMT-based architecture for E-healthcare patient monitoring system using artificial intelligence algorithms. Front. Physiol. **14**, 1125952 (2023)
10. Sharma, P., Dinkar, S.K., Gupta, D.V.: A novel hybrid deep learning method with cuckoo search algorithm for classification of arrhythmia disease using ECG signals. Neural Comput. Appl. **33**(19), 13123–13143 (2021)
11. Rajan Jeyaraj, P., Nadar, E.R.S.: Smart-monitor: Patient monitoring system for IoT-based healthcare system using deep learning. IETE J. Res. **68**(2), 1435–1442 (2022)
12. Majumder, S., Chen, L., Marinov, O., Chen, C.H., Mondal, T., Deen, M.J.: Non-contact wearable wireless ECG systems for long-term monitoring. IEEE Rev. Biomed. Eng. **11**, 306–312 (2018)

Adopting AIML for Marketing of SME's

S. Amarnaath[1] and A. Saravanan[2(✉)]

[1] Shri Aswini Bricks Pvt. Ltd., Chennai, India
[2] S.A. Engineering College, Chennai, India
saravanana@saec.ac.in

Abstract. The significance of Artificial Intelligence & Machine Learning (AIML) in automation of Marketing is thoroughly examined in this review article, with a focus on how these technologies affect customer satisfaction in medium and small-sized businesses (SMEs). The discussion centers on fundamental AI tools including predictive analytics, machine learning, and natural language processing, and demonstrates how these resources can provide recommendations and customized content. This feature greatly improves client retention and engagement. The study continues by outlining the notable returns on investment (ROI) that SMEs may attain by implementing AI-driven marketing strategies. The report also emphasizes upcoming opportunities to use AI, analyzes new trends, and provides case studies of effective uses. The consequences for SMEs are discussed in the conclusion.

Keywords: AIML · AI Marketing · ML Marketing · SMEs · Marketing 5.0 · Customer Experience · Predictive Analytics · Regression Algorithm · Neural Networks · Triangulation

1 Introduction

By facilitating more effective procedures, well-informed decision-making, and creative solutions, artificial intelligence (AI) has revolutionized a wide range of industries. AI in marketing offers previously unheard-of levels of customisation, allowing businesses to modify their strategies in response to the distinct preferences and actions of every client. With the application of AI, this strategy analyzes enormous volumes of data to provide personalized offers, content, and consumer experiences. Customer satisfaction rises as a result, which boosts engagement and loyalty—two factors that are vital in the cutthroat market environment of today.

Marketing 5.0 is the strategy of creating, communicating, delivering, and improving value across the customer journey by utilizing human-like technologies. It includes virtual reality, the Internet of Things, augmented reality, blockchain, sensors, robotics, AI and natural language processing. Marketing 5.0 is made possible by the convergence of these technologies.

Due to their ability to increase GDP and provide employment, small and medium-sized enterprises (SMEs) are vital to the global economy. But SMEs frequently have

R. Geetha et al. (Eds.): AAIMB 2024, CCIS 2653, pp. 147–169, 2026.
https://doi.org/10.1007/978-3-032-13026-6_11

particular difficulties, like scarce resources, financial limitations, and a lack of technological expertise, which can hinder their ability to compete with larger companies. Due to these obstacles, SMEs may find it challenging to adopt advanced marketing plans that call for large expenditures in both technology and knowledge.

Notwithstanding these challenges, SMEs have encouraging chances to improve customer satisfaction by implementing AI for targeted marketing automation, leveling the playing field.

In three ways, artificial intelligence improves each touchpoint throughout the customer journey. Permits more intelligent targeting. Presenting the appropriate offerings to the appropriate clients at the appropriate moment. It guarantees a better fit for the goods. Businesses may offer individualized goods and even let customers personalize their purchases. Better engagement is made possible by it. Businesses are able to communicate with customers more directly and provide customized information.

1.1 Goals of Study

This study aims to investigate how personalized marketing automation is being revolutionized by artificial intelligence (AI), specifically with regard to small and medium-sized enterprises (SMEs). The purpose of this study is to demonstrate the potential advantages for SMEs in enhancing the client experience by looking at how AI could help with the development of tailored marketing efforts.

Examining the various AI tools and technologies available for customized marketing automation is the study's primary objective. Assessing how SMEs might use these technologies to enhance their marketing approach is the second objective. Additionally, this study aims to identify the crucial factors influencing the successful use of AI into SME marketing strategy. This means examining the challenges and barriers that SMEs face. By offering a thorough examination of AI's function in personalized marketing automation, this study aims to provide insightful information to the academic community as well as industry practitioners.

1.2 Importance of the Research

In addition to having applications for small and medium-sized enterprises (SMEs), this study contributes to the corpus of existing information on marketing and artificial intelligence (AI). In the classroom, this research will contribute to the growing corpus of research on artificial intelligence (AI) in marketing, with a focus on customized marketing automation for small and medium-sized enterprises.

To forecast trends, look into consumer preferences, and start focused campaigns, marketing analytics entails the gathering, administration, access, and analysis of marketing data. Data-driven marketing, contextual marketing, predictive marketing, human-machine symbiosis, and efforts at size and pace are all essential elements of marketing analytics.

The results of this study will be very helpful from a practical standpoint for SMEs looking to use AI into their marketing campaigns. In order to improve customer experience, the study will offer practical insights and suggestions for putting AI-driven marketing tactics into practice. This includes guidance on selecting appropriate.

AI tools and technologies, methods for assessing the impact of AI on marketing performance, and best practices for integration (Fig. 1).

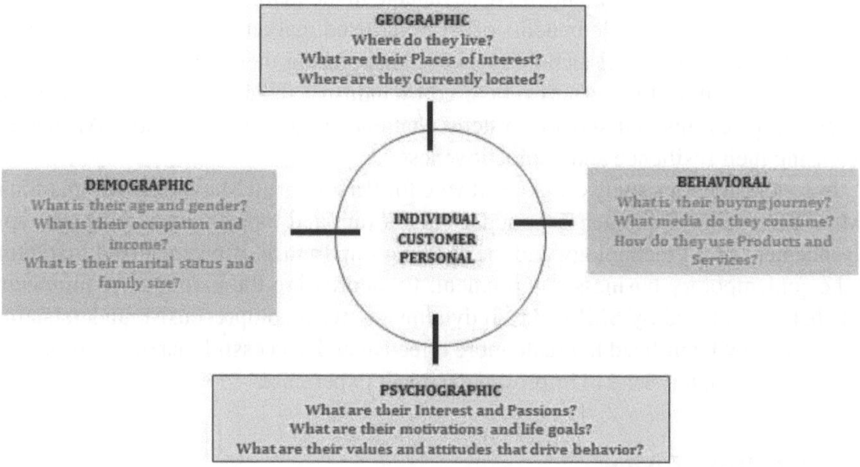

Fig. 1. The factors contributing to Individual Consumer Behavior

There is clear market polarization as the middle market is contracting while both the high-end luxury and low-cost value segments are expanding. Companies must be digitally ready for a number of reasons. By combining all digital touchpoints into a smooth and easy-to-use customer experience, it allows companies to interact with clients digitally throughout their whole journey. By utilizing technologies that acquire, store, manage, and evaluate enormous volume of real-time client data, businesses may use digital business models to generate value and money. In order to accommodate new digital models, business processes are being redesigned and progressively digitalized. Through the use of IoT technology, physical assets including buildings, fleets, and equipment are becoming digitalized. Nowadays, the majority of workers have access to digital tools that enable them to collaborate digitally and work remotely. One of the top priorities is developing digital skills, such as data scientists, UX designers, and IT architects. Alignment between digital talent and business managers is fostered by a robust digital culture.

Analyzing customer digital preparedness is also necessary. Highly tech-savvy members of Generation Z and Y make up the majority of the clientele. The majority of consumers interact and do business with businesses through digital channels. Customers engage with digital interfaces when utilizing goods and services. They already use strategies like showrooming and webrooming to conduct some or all of their travels online. Digital technology can be used to improve and replace physical touchpoints that customers find annoying. Customers may make educated selections on their own because to the abundance of information that is easily accessible online.

Many clients believe that in-person contacts with the business are pointless, unimportant, or unworthy. Products and services are frequently regarded as simple, which lowers risks and problems with trust. Customers are therefore highly motivated to switch to digital since they can take advantage of more options, lower costs, better quality, and more ease.

This study will also clarify the effects of AI adoption on SMEs, including how it could provide them with a competitive advantage in the marketplace. The study aims to demonstrate the measurable benefits of AI in targeted marketing in order to encourage more SMEs to invest in AI technology, hence supporting their expansion and success. The findings of this study can also be used by industry stakeholders and governments to develop programs and support systems that encourage SMEs to adopt AI, thereby increasing their resilience and competitiveness.

In conclusion, this study on "AI Software for Personalized Marketing Automation in SMEs: Enhancing Customer Experience" is relevant and topical considering the rapid advancement of AI technology and the growing importance of personalized marketing in the contemporary business environment. By addressing the particular requirements and challenges faced by SMEs, the study aims to give a comprehensive understanding of how AI may be utilized to create more effective and successful marketing strategies, which will ultimately lead to improved customer experiences.

2 Literature Review

2.1 Marketing Using AI

Simple algorithms and rule-based systems were used in the early phases of marketing to automate processes like client segmentation and email campaigns. ML, or machine learning, has emerged as a disruptive force in this field as a result of technological breakthroughs. Deeper insights into consumer behavior and the creation of more focused and successful marketing plans are made possible by ML algorithms' ability to analyze large datasets and spot trends.

Developing data-driven marketing is a great way to carry out market research and find insights that rivals might not have access to. Before beginning a marketing planning cycle, marketers now routinely combine qualitative and quantitative research methods. Big data has made data-driven marketing more popular, and marketers now think that real-time insights hidden in massive amounts of data may greatly improve their marketing efforts. Although the data ecosystem's full potential has not yet been fully realized, efforts are still underway to combine analytics and market research into a single data management platform.

Among the most well-known AI technologies are machine learning, natural language processing (NLP), and predictive analytics are propelling advancements in marketing. By making it easier to analyze consumer data, machine learning enables the creation of tailored marketing tactics and the forecasting of future behavior. Through tools like chatbots and sentiment analysis, NLP enables the recognition and understanding of human language facilitating meaningful customer contact.

Predictive analytics helps businesses anticipate the wants and preferences of their customers by using historical data to estimate future trends. Real-time consumer analytics, automatic content creation, and sophisticated client segmentation are just a few of the features that AI-driven marketing solutions have come to offer. In addition to increasing marketing operations' efficiency, these technologies have made it more easier to provide clients with individualized experiences.

A strategic approach to digital marketing, content marketing focuses on creating and sharing relevant and valuable material in order to attract and retain a well-defined audience. Content marketing's primary objective is to encourage profitable consumer activities, such increasing website traffic and lead creation. With this approach, producing high-quality content that provides the viewer with genuine value is prioritized. A variety of information formats, including blog posts, videos, infographics, podcasts, and social media posts, should be engaging, instructive, and easy to distribute.

By offering helpful and educational resources, content marketers hope to build credibility and trust with their audience. This strengthens the bond between the brand and its target market, which may boost conversion rates and brand loyalty.

Participants in a digital marketing course will learn about the various forms of content marketing as well as how to create an all-encompassing plan for content marketing. This include formulating a content strategy, producing material that resonates, and assessing the success of content marketing initiatives. All things considered, content marketing is a powerful digital strategy that may assist businesses in becoming thought leaders in their industry and fostering customer engagement and loyalty. By creating relevant and high-quality content, digital marketers may increase brand awareness, interact with their target audience more effectively, and eventually drive commercial success.

2.2 Customized Automation for Marketing

The strategic process of tailoring marketing initiatives to each consumer's unique preferences, actions, and experiences with a company is known as personalized marketing. Data collection, consumer segmentation, content personalization, and performance monitoring are some of the fundamental elements of this strategy. By automating and simplifying otherwise labor-intensive and complex operations, AI shows a crucial part in improving these elements. Particularly, the use of AI technology to supervise and carry out marketing operations with less human control is known as automation in personalized marketing.

Data-driven marketing is the practice of systematically collecting and analyzing sizable datasets from both internal and external sources to build a comprehensive data ecosystem that can be utilized to inform and enhance marketing choices. This tactic draws attention to the necessity of making educated decisions supported by reliable data, which is a crucial element of Marketing 5.0.Agile marketing makes use of decentralized, cross-functional teams to rapidly conceptualize, design, develop, and validate products and marketing campaigns. For businesses hoping to successfully execute Marketing 5.0, maintaining organizational agility in reaction to the changing market landscape is a crucial discipline.

Collaboration and Teamwork: Teams that completely adopt agile approaches form the cornerstone of agile marketing. Moving away from hierarchical systems and work silos is crucial in order to promote a collaborative atmosphere among all team members. Every person need to be able to participate in a variety of tasks. In order to foster this collaborative environment, regular team meetings and efficient communication routes are essential.

Data-driven decision making: Using data to inform decisions Data-driven marketing tactics are given top priority by agile marketers. Although data is frequently used by contemporary marketers, agile teams give it more importance. They use data to measure results and make required modifications as they continuously experiment with new tactics to improve team performance.

Quick, Iterative Releases: Agile marketing teams usually work in sprints, which are predetermined times when a scrum team concentrates on finishing particular projects. Teams can divide their job into small chunks with this sprint framework, facilitating rapid iterative releases. Sprints' short duration allows for frequent plan modifications, usually every few weeks.

Adherence to the Agile Marketing Manifesto: Finally, the Agile Marketing Manifesto's ideals and principles are upheld by agile marketing teams. Ten concepts and five basic values that are essential to attaining marketing agility are included in this document. These ideas provide the foundation for all team practices, including stand-ups, sprints, and Kanban boards, and they give the team's actions a sense of purpose.

First of all, by drastically cutting down on the time and resources needed to complete marketing tasks, automation improves operational efficiency. Artificial intelligence (AI)-powered technologies, for example, can process enormous datasets faster than humans, giving them immediate insights that may be used right away. Second, marketing campaigns are more accurate thanks to automation. Organizations can make data-driven assessments that are likely to be accepted by their target audiences by employing machine learning algorithms. Thirdly, automation makes scaling easier. AI solutions can manage large amounts of data and several client contacts at once, allowing companies to grow their individualized marketing campaigns without incurring additional expenses. Additionally, automation tailored marketing makes the consumer experience more interesting. Through the provision of pertinent content and customized offerings, companies can cultivate significant relationships with their clientele. In addition to increasing customer happiness, this strategy fosters greater engagement and loyalty, which eventually results in sustained client retention.

2.3 SMEs and the Difficulties of Marketing

Small and medium-sized enterprises (SMEs) encounter a distinct set of challenges in the marketing environment. SMEs usually operate with less capital, resources, and technical expertise, which may reduce their marketing effectiveness when compared to larger businesses. Furthermore, since SMEs typically lack the economies of scale that larger organizations enjoy, it can be challenging to justify investments in cutting-edge marketing strategies.

The limited funding available for marketing campaigns is a major marketing issue unique to SMEs. Smaller marketing budgets are frequently the outcome, which may limit the capacity to invest in cutting-edge technologies and execute all-encompassing tactics. As a result, it could be difficult for SMEs to compete with bigger companies who can afford more advanced equipment and extensive advertising efforts. Furthermore, a lot of SMEs do not have specialized marketing teams or people who know how to use cutting-edge technology like artificial intelligence, thus they lack the necessary technical know-how. This lack of knowledge may limit the potential advantages of AI-driven marketing solutions by impeding their uptake and efficient use.

In conclusion, research on personalized marketing automation, AI in marketing, and the difficulties faced by SMEs demonstrates how revolutionary AI technology can be. Although adopting these technologies presents SMEs with a number of challenges, the benefits—such as enhanced customer experiences—make the effort worthwhile. With the correct strategies and support, SMEs may effectively employ AI to gain a competitive edge in the market.

Three Main Causes of Data-Driven Marketing Practices' Failure are Data-driven marketing is frequently approached by organizations as an IT project. When starting such initiatives, hiring data scientists, investing in infrastructure, and choosing software tools are frequently given too much attention. But first and foremost, data-driven marketing should be considered a marketing campaign. Instead than dictating the marketing approach, the IT infrastructure should assist it. The entire data-driven marketing process should be defined and designed by marketers. As numerous market analysts have pointed out, more data does not always translate into better insights. Knowing how to sort through and evaluate the large volume of information that is available while adhering to specific marketing goals is crucial.

Big Data analytics has been described as a solution that can solve a number of marketing problems and reveal customer insights. It shouldn't, however, take the place of conventional market research techniques, especially high-touch techniques like ethnography, usability testing, and taste testing. Rather, since successful data- driven marketing requires both strategies, Big Data and market research should complement each other. While Big Data is collected and analyzed in real-time to improve current marketing efforts, market research is usually carried out on a regular basis with precise, limited goals.

Because of the significant automation potential of Big Data analytics, some organizations think that once these systems are in place, they can function on their own. Marketers are expected to be able to quickly get answers to their questions by simply feeding big data sets into an algorithm. However, in actuality, marketers still need to actively participate in data-driven marketing. Although algorithms are better than humans at seeing patterns, filtering and interpreting these patterns eventually calls for seasoned marketers with contextual expertise. Additionally, marketers who can create creative offers or campaigns—often with the aid of computational tools— are required to extract meaningful insights.

Thankfully, technology is becoming more widely available and reasonably priced. Businesses can use open-source artificial intelligence platforms from corporations like Microsoft and Google. Many cloud-based data analytics tools are available for monthly subscriptions, and there are now a number of approachable chatbot-building platforms that make it possible for people without a lot of technical experience to use these tools. Organizations must make sure their marketers have the know-how to create plans that successfully use the right technologies for a range of marketing use cases.

In conclusion, the discussion of personalized marketing automation, artificial intelligence in marketing, and the difficulties faced by SMEs highlights the enormous potential of AI technology. The associated advantages, such as improved customer experiences, make this endeavor extremely important, even though SMEs may encounter difficulties implementing these solutions. If SMEs have the right strategies and support systems in place, they can employ AI to obtain a competitive advantage in the market.

3 AI-Powered Customized Marketing Automation Software

3.1 Synopsis of AIML Platforms and Tools

Businesses now approach marketing differently thanks to AI software tools and platforms, which make it possible to create highly customized experiences at scale. Several leading AI tools and platforms have emerged in the marketing automation area, with a broad range of features that appeal to businesses of all sizes, including small and medium-sized enterprises (SMEs).

Predictive marketing is predicting the results of marketing initiatives before to their introduction by using predictive analytics, occasionally in conjunction with machine learning. Businesses may predict and actively affect market reactions using this strategy.

Contextual Marketing: In order to create individualized interactions, contextual marketing relies on identifying and analyzing clients. This is achieved through the use of sensors and digital interfaces in physical spaces, which lay the groundwork for context-dependent, real-time one-to-one marketing.

Augmented Marketing: Augmented marketing leverages digital technology to boost the effectiveness of marketers that engage with consumers by utilizing human-mimicking tools like chatbots and virtual assistants. With this strategy, marketers can mix the ease and speed of digital interfaces with the compassion and warmth of human interaction.

Google Analytics is a well-known AI tool for marketing automation. Google Analytics gathers information on website traffic and activity by using a JavaScript measuring code snippet, sometimes referred to as a "tag," that is included in websites and applications. The tag collects anonymous user information on a visitor's interactions with a website or application each time they land on it. Through your Google Analytics account, you can check this data to discover how many people have visited your website, which pages have been viewed the most, how many people have made a purchase, subscribed to your newsletter, and much more. In order to give a thorough picture of website visitors, it also gathers data on the browser, device, operating system, and traffic source. Companies can evaluate the success of their digital marketing campaigns using tools like SEO, blogging, paid advertising, influencer, affiliate, and social media marketing, with the help of Google Analytics.

Google Analytics: Users can adopt alternative reference frames to better understand user behavior and the entire customer journey on their websites thanks to Google Analytics' improved insights for a variety of aims. Website user activity, behavior, and conversion goals are distinct and change over time, much like brand voice and corporate vision. At every step, Google Analytics assists in identifying possibilities, resolving problems, and producing insights that are growth-oriented.

Excel GPT is another well-known tool that helps with copywriting for a variety of marketing documents, like as emails, blog entries, product descriptions, ad copy, headlines, website text, and social media captions. Chat GPT can assist in creating surveys or polls to continuously collect consumer thoughts and opinions, even though companies should do out in-depth research to identify their target demographic. A website may effectively engage clients, generate leads, and gather feedback by integrating Excel GPT.

Additionally, Mailchimp is frequently utilized, especially by small and medium-sized businesses. Originally an email marketing service, Mailchimp has evolved into a feature-rich marketing automation platform. It makes use of Artificial intelligence to provide functions like automated marketing workflows, predictive demographics, and tailored email recommendations. Smaller companies with fewer resources can utilize Mailchimp thanks to its user-friendly design and reasonably priced pricing options.

Marketers must create a plan for gathering data that specifies the precise information needed to achieve predefined goals. But not all information is useful or pertinent. After defining their goals, businesses need to decide which data to gather and examine. Although there is not a single right technique to classify huge data, grouping information according to its source is a useful approach:

Social Data: Details that users of social media share, such as location, interests, and demographic profiles. Media Data: Measures of the audience for conventional media, including print, radio, television, and movies. Web traffic data: Records of user activity on the internet, such as page visits, searches, and purchases. POS and Transaction Data: Documentation of consumer transactions, including purchase timings, location, credit card details, and occasionally customer IDs. IoT Data by sensors and connected devices, such as vital signs, temperature, humidity, location, and the proximity of other devices. Engagement Data: Every direct communication a business has with its clients. Businesses will be able to improve their marketing strategies and make well-informed judgments with the aid of this methodical approach to data collection.

Notable choices for (SMEs) are AI techs like Adobe Experience Cloud and Marketo. With its cutting-edge AI-powered lead nurturing, scoring, and segmentation tools, Marketo is a lead management specialist. Conversely, Adobe Experience Cloud provides a wide range of marketing solutions that leverage AI to deliver personalized content, enhance campaigns, and provide comprehensive consumer data.

Regression is a supervised machine learning technique for forecasting continuous values. Finding a best-fit line or curve that accurately depicts the relationship between data points is the main goal of regression algorithms. Variance, bias, and error are three important metrics that are frequently used to assess a trained regression model. It searches for the model that best matches the data in order to make predictions or draw conclusions. The paper explores regression in machine learning, clarifying types, models, terminology, and real-world uses.

A statistical method called regression breaks down the link between dependent and independent variables so that different regression models can be used to make predictions. Regression analysis is a statistical technique used to investigate the connection between one or more independent variables (predictor variables) and a dependent variable (target variable). Finding the optimum function to explain the relationship between these variables is the aim.

Linear Regression: One of the most basic and popular statistical models is linear regression. This assumes a linear relationship between the independent and dependent variables. This shows that the change in the dependent variable is proportionate to the change in the independent variables.

Polynomial Regression: Nonlinear relationships between dependent and independent variables are modeled using polynomial regression. To account for more intricate interactions, it incorporates polynomial variables into the linear regression model.

Decision Tree Regression: Decision tree regression is a type of regression technique that creates a decision tree to forecast the desired value. A decision tree is a structure with nodes and branches that resembles a tree. A decision is represented by each node, and its result is represented by each branch. Creating a tree that can reliably forecast the desired value for fresh data points is the objective of Decision tree regression.

Support Vector Regression (SVR): The support vector machine (SVM) approach is the foundation of one type of regression algorithm known as support vector regression (SVR).SVM is a form of algorithm that may be applied to regression tasks as well as classification tasks. SVR finds a hyperplane between the actual and predicted values that minimizes the sum of the squared residuals.

Neural Networks are highly advanced computer models created to mimic the intricate processes of the human brain. These networks, which are made up of interconnected nodes called neurons, can process and learn from data, making it easier to complete tasks like pattern recognition and machine learning decision-making. This type of machine learning, known as deep learning, makes use of a network of interconnected nodes that are layered and resemble the organization of the human brain.

Complex Analysis refers to the study of functions in the complex plane that have complex inputs and outputs. Complex numbers have the potential to improve neural network performance in the field of ML. Complex ML models examine and process large volumes of data using cutting-edge techniques like deep learning and neural networks. These models are very good at identifying complex correlations and patterns in datasets. Additionally, AIML are used in sophisticated data analysis to glean insightful information from collected data. Five separate steps make up the data analysis process: definition, collecting, cleaning, analysis, and dissemination.

In machine learning, triangulation is a technique used to assess model performance and validate the reliability of data sources. This method verifies the accuracy of performance claims using a set of questions. Finding high-performing models, exposing underlying presumptions, and bolstering network integrity can all be facilitated by triangulation. It entails using two comparable or dissimilar aspects to support the veracity of data sources, help find the truth in the data, control the amount and complexity of information, and spot possible irregularities.

These AI tools' salient characteristics and capabilities that are pertinent to SMEs include:

Personalization: AI-powered platforms evaluate consumer data using machine learning algorithms to offer information and recommendations that are customized for every person. This strategy raises the possibility of conversions while improving consumer engagement.

Automation: SMEs can focus on strategic initiatives by saving time and money by automating repetitive marketing procedures such as email campaigns, social media posts and ad targeting.

Predictive analytics: By using AI tools to predict market trends and consumer behavior, organizations may increase return on investment (ROI) and make well- informed decisions about their marketing strategy.

Integration: A unified picture of client interactions is provided by the smooth integration of several AI platforms with other business tools, including as analytics tools, e-commerce platforms, and CRM systems.

Scalability: AI technologies are appropriate for SMEs that intend to grow their operations because they are made to grow with businesses.

3.2 Strategies for Implementation

Constructing a Comprehensive Data Ecosystem

Integrating data at the individual consumer level is crucial in today's competitive environment to enable tailored marketing campaigns. Every client-related dataset has a unique customer ID attached to it. Although connecting these IDs to internal data sources is rather simple, integrating external data poses difficulties that can be overcome with careful preparation. For instance, when users use their social media accounts, like Google or Facebook, to access e-commerce platforms, social network data can be synchronized with customer IDs and purchase information.

Using smart beacon sensors in combination with customer loyalty apps is another efficient method of data integration. A sensor records a consumer's motions when the consumer is carrying a mobile device close to it, like in a store. The consumer journey within physical venues can be tracked with the use of this data.

To guarantee a thorough comprehension of particular demographics, a defined demographic segmentation variable might function as a distinct identification. For example, the "18 to 34-year-old male client" segment may be used to compile all pertinent data about this demographic profile from several data sources.

Keeping all dynamic datasets on a single data management platform is also essential. In the end, this consolidation improves strategic decision-making and customer engagement by empowering marketers to efficiently collect, store, manage, and analyze data.

3.3 Optimal Methods for Effective Integration and Adoption

Using Technology to Improve Marketing Initiatives: Technology may greatly enhance marketing campaigns by facilitating data-driven decision-making. Organizations may create a contextual experience that resonates in the real world by using big data to better forecast the results of different strategies and tactics. Furthermore, technology speeds up the implementation of marketing initiatives and improves frontline marketers' ability to provide value.

Starting with Small-Scale Implementations: It is recommended that small-scale implementations be used to kickstart the adoption of AI in small and medium-sized businesses (SMEs). Organizations can reduce risk and facilitate a learning process that allows for modifications prior to a wider rollout of AI solutions by concentrating on particular marketing initiatives or campaigns. Prioritizing domains like customer segmentation, email marketing, and predictive analytics—where AI can have a major impact—is crucial. This approach not only shows AI's worth but also creates momentum for further deployment.

Setting Data-Informed Marketing Goals: Setting data-driven marketing goals as a top priority is essential for success. Businesses may improve the overall customer experience by using big data to provide specialized goods and services based on the preferences of each individual consumer. Marketers can identify new product concepts, evaluate consumer demand, build dynamic pricing models, and choose the best price strategies by carefully analyzing data. Big data helps uncover efficient delivery options in addition to determining what services should be made.

Big data is applied in marketing communications for various purposes like lead generation, media selection, content production, and audience targeting. Additionally, it plays a key role in customer retention and after-Sales support plans by facilitating efficient service recovery procedures and allowing for the forecast of churn.

Key Elements of Data-Driven Marketing: There are a number of essential elements involved in putting a data-driven marketing strategy into practice. Determining and focusing on the right audience. Creating content and marketing messages that are pertinent. deciding on the best media combination for efficient communication. Identifying the best avenues for entering a market. Creating client profiles to improve lead generation and nurturing initiatives. Creating tiers of client service. Recognizing possible client grievances and forecasting attrition.

Working with Experts: Consulting with AI specialists or consultants can provide insightful advice and assistance all through the deployment process. Their knowledge may help with system integration, tool selection, and optimization, guaranteeing a smooth transition.

Cultivating an Innovative Culture: Success depends on fostering an innovative and technologically savvy culture. Companies should support experimentation, cultivate an atmosphere of ongoing learning, and have an open mind about new instruments and procedures. Facilitating iterative improvements requires regular performance evaluation of AI tools against predetermined goals. In order to improve overall efficacy and refine AI models, user and customer feedback is essential.

In conclusion, SMEs can benefit greatly from the use of AI software for tailored marketing automation, which includes more customer interaction, more effective marketing. SMEs may successfully adopt and exploit AI, giving them a competitive edge in the market, through following best practices, integrating systems, preparing data, and carefully selecting the appropriate technology.

4 Improving the Experience of Customers

4.1 Customization Methods

Technology for Humanity: Marketing 5.0 is defined as the generation, transmission, provision, and enhancement of value across the customer journey through the use of human-mimicking technology. In order to find areas where marketing technologies can enhance the efficacy of human marketers, this method starts with mapping the customer journey. Businesses looking to use Marketing 5.0 must start with a data- driven strategy. A strong data environment must be established before Marketing 5.0's useful applications can be implemented.

At the point of sale, this framework enables marketers to offer each client contextualized and tailored marketing. Additionally, frontline marketers can use augmented marketing techniques to develop smooth client experiences. A high level of organizational agility is important for the effective implementation of these programs to enable real-time reactions to changing market conditions.

The incorporation of artificial intelligence (AI) has fundamentally changed how businesses engage with their customers into marketing strategies. Delivering highly tailored content and recommendations is one of AI's most important contributions. AI-powered personalization strategies examine vast volumes of consumer data to learn about unique needs, tastes, and behaviors. This allows companies to precisely customize their marketing campaigns.

AI improves the delivery of tailored content in a number of important ways. Machine learning algorithms can process and analyze customer data, including browser history, previous purchases, and interaction patterns, to predict future behaviors and preferences. Businesses are able to offer each consumer personalized recommendations, offers, and pertinent material thanks to this data-driven approach.

An e-commerce platform may employ artificial intelligence (AI) to make product recommendations based on a customer's previous purchases and browsing patterns in order to enhance the shopping experience and increase conversion rates.

Natural language processing (NLP) is another crucial AI method that facilitates customisation. AI systems can now comprehend and interpret human language thanks to NLP, which improves consumer interactions. Chatbots that use natural language processing (NLP) can communicate with clients in real time, providing tailored support and answers. By quickly responding to questions and making sure the information is relevant to the consumer's context, this skill improves customer satisfaction.

Predictive analytics also significantly aids AI-powered personalization. Predictive analytics uses previous data to estimate future trends and customer needs.

This insight enables companies to proactively anticipate client needs and modify their marketing approach accordingly. For instance, a subscription-based business can utilize predictive analytics to identify customers who are most likely to discontinue and provide them personalized incentives to maintain them.

In the arena of predictive analytics, AI is used for most analyses. A predictive model is created by running historical data through a machine learning engine to find particular patterns. Marketers can predict future events, including the possibility of client purchases and the efficacy of marketing initiatives, by feeding new data into this model. Organizations usually build this capability based on their existing data ecosystem because predictive marketing depends heavily on data. Businesses are able to make proactive and well-informed investments because to this foresight. For instance, businesses can predict whether new customers who make minor transactions at first will grow into important accounts. This knowledge makes it easier to allocate resources in the best possible way to support certain customer relationships. Moreover, companies can use predictive analytics to efficiently screen and assess possible concepts before investing significant resources in new product development. Forward-thinking businesses look for insights that go beyond historical data, going beyond real-time analytics that offer prompt answers for agile marketing strategy testing or contextual marketing.

In order to determine the probability of similar or related actions in the future, predictive analytics carefully analyzes past client behaviors. It provides suggestions for the best courses of action after seeing subtle trends in large populations. This forward-thinking mindset enables marketers to remain ahead of the curve, plan their campaigns ahead of time, and successfully impact results. Additionally, preemptive and preventive measures are made possible by predictive analytics, which makes it crucial for strategic marketing planning. Marketers have a strong tool to improve their decision-making processes because to this analytical capabilities. They can determine which clients are worth interacting with, identify the most promising market situations, and assess the possible success rates of different marketing initiatives and techniques prior to their implementation, greatly reducing the risk of Failure.

Predictive marketing analytics applications cover a number of important domains: Finding opportunities for cross-selling and upselling. Predicting client loyalty and identifying possible attrition. Figuring out what each customer should do next within customer management frameworks. Estimating the probability that new product launches will be successful. Modifying value propositions to satisfy each client's particular requirements. Recommending goods from a wide range of products in order to handle them effectively. Evaluating the efficacy of possible marketing initiatives. Reviewing advertising material that appeals to consumers. Giving clients brand management advice with pertinent content all along their digital journey. In addition to improving consumer interaction, these apps help marketers make strategic decisions.

4.2 Engagement and Retention of Customers

The Divided Society: Businesses have produced substantial wealth in recent decades, yet this wealth distribution has been markedly unequal, resulting in societal divisions. As a result of the middle class's steady decline, people are either moving up the economic ladder or down to the bottom. Due to this change, the usual distribution of society has given way to an M-shaped model, with the majority of people clustered in the upper and lower classes. There is friction between both groups because people at either extreme frequently have opposing life priorities and ideas.

Job Polarization: While prospects in the middle sector are declining, the job market is seeing an increase in high-value, high-paying positions as well as low-value, low-paying ones.

Ideological Polarization: As a result of divergent opinions on important topics, including free trade versus protectionism, ideologies are become more and more polarized, further dividing society.

Lifestyle Polarization: Consumer behavior and buying habits are being greatly influenced by the growing trends toward both minimalist and consumerist lifestyles. Market Polarization: While the middle market continues to shrink, the premium luxury and low-cost value portions of the market are both expanding.

Active consumer involvement is necessary to establish strong and enduring client relationships. Active brand participants are more inclined to suggest a brand to others, make repeat purchases, and offer helpful criticism. Through the facilitation of more meaningful and customized interactions, artificial intelligence (AI) shows a critical part in improving consumer engagement.

Virtual assistants and Chatbots are two prominent uses of artificial intelligence. These resources offer 24/7 interaction, prompt answers to questions, and individualized support. AI chatbots greatly improve the overall customer experience by providing timely assistance and personalized recommendations, which raises customer happiness.

Additionally, it has been demonstrated that AI-powered content customisation increases consumer engagement. Higher engagement rates result from tailored content that closely matches client preferences, whether it be in emails, social media interactions, or website experiences. For instance, as compared to normal communications, personalized email campaigns that include consumers' names and make product recommendations based on their interests have demonstrated higher open and click-through rates (Fig. 2).

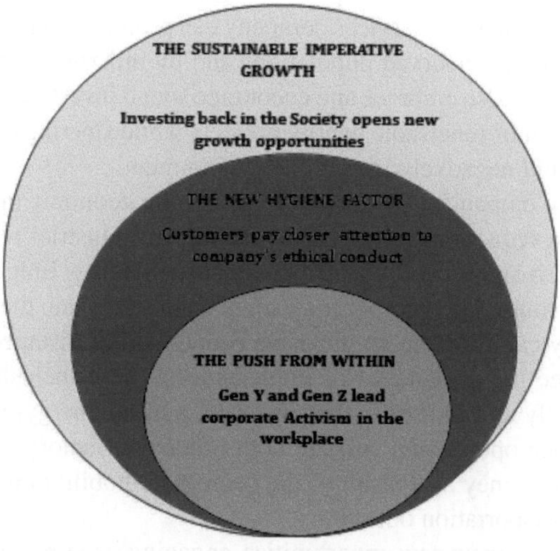

Fig. 2. The factors contributing to Polarization

AI uses predictive analytics to improve customer engagement as well. Predictive analytics assists companies in anticipating client demands and launching proactive interaction by seeing trends in customer behavior. For example, the company can proactively send a customized offer or advice to stimulate a transaction if an artificial intelligence system discovers that a customer frequently peruses a specific product category without making a purchase. This strategy not only boosts conversion rates but also demonstrates to clients that the company respects their choices.

Growing wealth inequality is fueling societal polarization, which has important ramifications for many facets of daily life. To reduce dangers like political instability, social discontent, and economic decline, it is necessary to address the gap between those who are prospering and those who are struggling in the face of globalization and technological innovation. Businesses share a responsibility in resolving this imbalance, as stakeholders increasingly want organizations to adopt more inclusive and workable growth plans.

AI significantly affects consumer loyalty and retention. Personalized and engaging interactions help build stronger bonds between customers and brands, which ultimately boosts brand loyalty. Consumers are more inclined to stick with a brand and make more purchases if they feel appreciated and understood. AI-driven personalization makes sure that each interaction with a client is relevant and worthwhile, which raises overall customer satisfaction.

AI also plays a key role in detecting clients who are at risk of disengaging. Reduced engagement or unfavorable reviews are indicators of possible churn that predictive analytics can identify. Businesses can avoid attrition and maintain a devoted clientele by being proactive, such as by providing tailored incentives or directly resolving issues.

5 Impact on ROI

5.1 Performance Metrics

Financial Services: A financial services company can promote financial inclusion in two ways: by focusing on underserved populations and by utilizing a financial technology (fintech) strategy. It can also embrace and encourage sound investment strategies, such as supporting the growth of renewable energy projects while steering clear of investments in ventures that would negatively impact the environment.

Manufacturing Companies: By adopting the circular economy model, which places an emphasis on the reduction, reuse, and recycling of industrial resources, manufacturing companies can significantly contribute to sustainability. Employing members of underrepresented groups and incorporating small enterprises into their supply networks are more ways they can promote an inclusive economy. For instance, a lot more firms are beginning to see the advantages of implementing these inclusive and sustainable policies, both directly and indirectly. For instance, putting energy-efficient procedures into place can reduce operational costs in both office and factory settings. Businesses may also save more money as a result of the decreased mobility brought on by remote work and shared transportation options.

In addition to opening up new opportunities, engaging underrepresented areas forces businesses to reverse-innovate. Innovation has historically come from industrialized countries and spread to developing ones throughout time. On the other hand, current patterns suggest that this dynamic is reversed. For example, businesses such as General Electric have created affordable medical devices for developing economies and then sold similar items to wealthy countries as "portable" solutions.

One important key performance indicator (KPI) is the conversion rate, which calculates the percentage of leads or visitors who end up becoming customers. By offering individualized experiences that speak to specific customers, artificial intelligence (AI) solutions can increase conversion rates and increase the possibility that they will make a purchase. Customer acquisition cost (CAC), which measures the amount of money spent to bring on new clients, is another important KPI. By precisely targeting the right audience and streamlining marketing efforts, AI-driven marketing automation can help reduce CAC by cutting down on wasteful spending.

Customer lifetime value, or CLV, is another important metric that determines how much money a business may anticipate earning from a customer over the course of their lifetime. By encouraging client loyalty and retention through tailored interactions and predictive analytics, AI can raise CLV. In addition, companies can track average order value (AOV) to determine average client spending each transaction. By suggesting related products or planning upsells based on observed consumer preferences, AI solutions may help increase AOV.

To determine the return on investment (ROI) from AI-driven marketing automation, organizations must weigh the costs of deploying and maintaining AI technologies against the revenue from enhanced marketing effectiveness. This evaluation entails figuring out how much Customer experiences may be attributed to AI projects and subtracting the related expenses. ROI, which shows the effectiveness and profitability of AI investments, can then be stated as a percentage. Small and medium-sized businesses (SMEs) can assess how well AI promotes well-informed decisions about future investments by closely monitoring these data.

Developing Capabilities for Real-Time Analytics
Establishing a thorough analytics capacity is the first step in improving corporate performance. Identifying current issues that need to be resolved and identifying growth prospects are the main goals. Organizations must put in place real-time client data capturing systems to make this process easier. Tracking conversations about a company or product across several social media platforms and online communities is made easier by social listening solutions, also known as social media monitoring. Unstructured social discussions can be filtered by these systems to produce meaningful customer intelligence, including keywords, divisive perspectives, developing trends, brand sentiment, campaign visibility, product acceptance, and competitor replies. Additionally, geotagging improves the data and allows firms to examine insights by area and region.

RFID tags can also be used as wearable technology to track consumer movements and improve the overall customer experience if given the right authority. The Walt Disney Company, for example, uses RFID technology in its MagicBands to track the movements of guests inside its theme parks. In a similar vein, the Mayo Clinic tracks patients using RFID technology on staff badges and patient bracelets. B2B businesses also use RFID tagging to streamline supply chain operations and efficiently manage logistics. Rapid analysis of the connections between campaigns and their respective results, is made possible by this traffic and transaction data.

Finding the best product-market fit is the aim in a variety of situations. Metrics used to measure success must be relevant and actionable so that businesses may precisely pinpoint areas in their campaigns or products that need improvement. Businesses may rapidly perform experiments and obtain validated insights with the help of real-time analytics.

5.2 Success Stories and Case Studies

Many small and medium-sized firms (SMEs) have successfully incorporated artificial intelligence (AI) into marketing automation programs, leading to notable increases in profit returns. One notable example is a small e-commerce company that provided customers with personalized product recommendations using AI-powered recommendation engines. The AI system's capacity to offer highly relevant product recommendations based on browsing histories and purchase trends resulted in a 25% boost in profit within six months. This is a prime example of the substantial advantages of AI-powered customisation.

1. PepsiCo regularly launches new beverage items after carefully examining consumer discussions on social media.
2. By helping e-retailers identify buying trends, AI technologies allow businesses to recommend relevant goods and content based on the profiles of certain customers. For e-commerce companies and other digital businesses like Amazon, Netflix, and YouTube, recommendation engines are a crucial difference. In order to provide dynamic customer segmentation and profiling and uncover hidden relationships between seemingly unrelated items for the sake of upselling and cross-selling, these engines continuously evaluate past purchase histories.
3. AB InBev, the company that owns Budweiser and Corona, uses artificial intelligence (AI) to track the success of different ad placements. The company then uses this data to inform the creative team about how to create more powerful ads.
4. To create advertising copy for its digital banners, Chase Bank chose to use an AI engine rather than a human copywriter.
5. Lexus researched successful advertising campaigns over the previous 15 years, particularly in the luxury sector, in order to produce a television commercial for the new ES Sedan. AI wrote the whole script for the commercial, and the company engaged an Oscar-winning director to shoot it (Fig. 3).

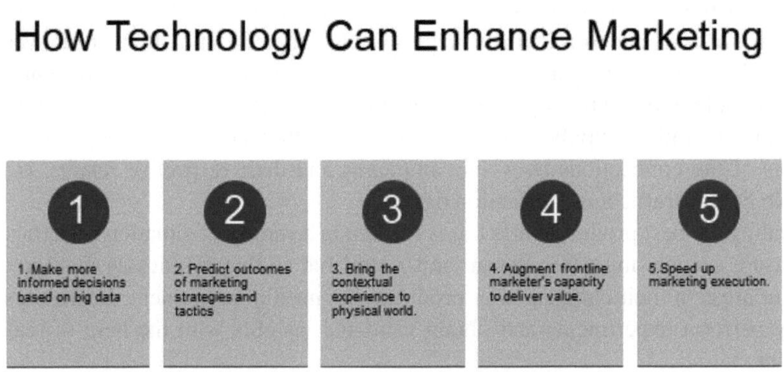

Fig. 3. The Technological stages enhancing Marketing

The Difficulties Kodak Faces in the Digital Age

Even though Kodak was a pioneer in the 20th century of photography, it had a hard time adjusting to the changing digital environment. Kodak's film camera business became outdated in the 1990s as digital photography emerged. Digital cameras, smartphones, and other gadgets that made photography more accessible posed a serious threat to the business. The reason for Kodak's downfall is its inability to adopt to the disruptive shifts of AI and digital technology. Kodak had already lost a sizable portion of the market by the time it started embracing digital photography; rivals like Canon and Nikon had gained sizable market shares, substantially hurting Kodak's sales. Due to the inability of the company's digital cameras to match those of its competitors, Kodak was seen as reactive rather than innovative.

Motorola's Decline

Motorola failed to realize that consumers were becoming more interested in software innovation than hardware improvements, despite the fact that it kept producing different versions of its mobile phones. Early in the new millennium, new product introductions showed a lack of understanding of the market and were insufficient to spur company expansion. Furthermore, Motorola missed the crucial shift to 3G technology, and these products were not user-friendly. As a result, the business found it difficult to stay competitive in the quickly changing smartphone industry.

5.3 Prospects and Trends for the Future

As AI technology continues to advance, a number of new trends in AI and marketing automation are developing, offering small and medium-sized businesses (SMEs) significant opportunity. The growing focus on hyper-personalization is one noteworthy trend. This strategy uses cutting-edge AI algorithms and real-time data to provide incredibly customized client experiences. Hyper-personalization, in contrast to typical personalization techniques, considers a wider range of data sources, including social media activity, present location, and weather, enabling more precise and meaningful customer interactions.

Additionally, the "Emerging Computing Techniques with Artificial Intelligence" report by Mr. Dillon Ranwala highlights the significant income generated by the AI for Enterprise Applications market globally between 2016 and 2025. This realization illustrates how AI technology is increasingly influencing corporate success and improving customer engagement (Fig. 4 and Table 1).

The combination of voice search and voice assistants with artificial intelligence (AI) is a new trend in business. Businesses may utilize AI to improve their content for voice search and develop effective voice-based marketing campaigns as voice- activated devices like Google Home and Amazon Alexa become more and more popular. This integration may lead to greater ROI through improved consumer convenience and engagement.

AI-powered visual search is also becoming more and more popular. Customers may now use images instead of text to search for products thanks to this cutting-edge technology. The shopping experience is improved by using AI algorithms to evaluate these photos and find related products, particularly in fields that rely heavily on visuals, like fashion and interior design. by making it simpler for customers to locate the items they desire.

SMEs now have a great chance to stand out from the competition and increase their market share thanks to these advancements. SMEs can more successfully draw in

Table 1. Table showing Revenue generated by AI in US Dollars for consecutive Years

YEAR	Revenue in Million US Dollars
2016	357.89
2017	841.13
2018	1622.4
2019	2867.54
2020	4806.3
2021	7714.17
2022	11840.54
2023	17284.19
2024	23886.76
2025	31236.92

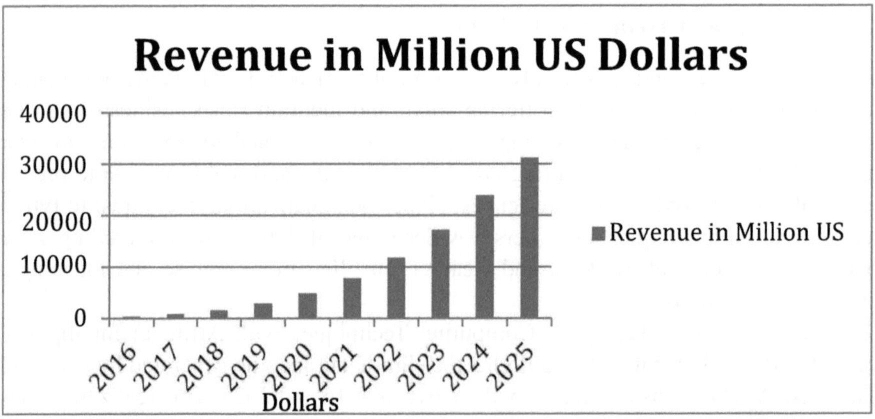

Fig. 4. Increase of Revenue in Million US Dollars over Consecutive years.

and keep clients by staying up to date on AI developments and implementing cutting-edge technologies into their marketing plans. AI may also assist SMEs in making data-driven decisions that reduce costs, streamline operations, and enhance overall business performance.

6 Conclusion

6.1 Key Findings Synopsis

Using artificial intelligence (AI) technologies for targeted marketing automation in small and medium-sized businesses (SMEs) greatly enhances customer satisfaction and increases return on investment. Thanks to artificial intelligence (AI) technologies like machine learning, natural language processing, and predictive analytics, businesses can now offer highly tailored content and recommendations that greatly improve customer engagement and retention. SMEs' marketing operations are more scalable and efficient, which raises conversion rates, lowers client acquisition costs, and increases customer lifetime value. Return on investment (ROI) have increased significantly as a outcome of the application of AI driven marketing methods; numerous case studies demonstrate positive results in a range of industries.

Without artificial intelligence, businesses usually rely on labor-intensive, error-prone, and limited-scalability manual operations. These inefficiencies have the potential to increase costs, lower productivity, and keep companies from offering the greatest products and services. On the other hand, automation driven by AI improves productivity and customer experiences by streamlining processes, increasing efficiency, and reducing human error.

Organizations may gain deeper insights, spot untapped possibilities, and make data-driven choices more quickly and accurately through the use of AI methods such as natural language processing and machine learning. Conventional data analysis techniques frequently have limitations in their ability to effectively spot patterns, trends, and correlations, which leads to imprecise and insufficient insights.

If businesses do not use AI, they risk losing out on opportunities to upend markets, gain a competitive advantage, and differentiate themselves from competitors who have used AI technology. Examples of AI-powered technologies that help companies better understand customer needs, anticipate preferences, and deliver customized experiences at scale include chatbots, virtual assistants, and recommendation engines.

A loss of market share could result from a failure to use AI in customer engagement, which could also lower consumer satisfaction and loyalty.

The risk of cybersecurity threats increases as businesses depend more and more on digital systems and networked devices. By evaluating vast amounts of data in real-time, Artificial intelligence (AI) is a potent technique for preventing cyberattacks by identifying anomalies and potential security breaches. This helps to prevent financial losses and maintain customer confidence.

6.2 Implications for SMEs

By using AI for customized marketing automation, SMEs can obtain a major competitive edge in the present dynamic market. Large-scale providers of personalized experiences not only boost customer happiness but also encourage repeat business and loyalty. SMEs can use AI solutions to enhance their marketing strategies, lower operating costs, and make better decisions that will boost performance. There are also obstacles to overcome, such as scarce resources, the requirement for technical expertise, and effective data management. To maximize AI's potential, SMEs should choose AI technologies that are appropriate for their purposes, invest in employee training, and establish sound data procedures.

6.3 Benefits and Key Applications

Micro, small, and medium-sized businesses (MSMEs) can improve their operations, cut expenses, and boost production with the use of AIML. Well-informed decision-making processes are supported by data-driven insights. Early adopters of AIML tools stand as good chance of gaining a competitive edge in their respective sectors. While ML algorithms are used to extract actionable insights from large datasets, AI-powered chatbots are used to respond quickly to consumer requests. AI also makes it easier to automate repetitive work, which frees up managers' time for more strategic endeavors.

6.4 Suggestions for Upcoming Studies

Future studies should focus on how AI-driven marketing automation affects SMEs performance over the long run. This includes, in particular, the effect on long-term client involvement and financial expansion. Studies should also look at the best ways to get beyond obstacles to SMEs' adoption of AI, with an emphasis on scalable and affordable approaches. The ethical ramifications of AI in marketing might also be examined in more detail, making sure that customization initiatives protect consumer confidentiality and Data security. To conclude, as AIML technology develops further, it is critical for do continuous research to find new trends and possibilities that SMEs may take advantage of to stay creative and competitive.

References

1. Basri, W.: Examining the impact of artificial intelligence (AI)-assisted social media marketing on the performance of small and medium enterprises: toward effective business management in the Saudi arabian context. Int. J. Comput. Intell. Syst. **13**(1), 142–152 (2020)
2. Aljarboa, S.: Factors influencing the adoption of artificial intelligence in e-commerce by small and medium-sized enterprises. Int. J. Inform. Manage. Data Insights **4**, 100285 (2024)
3. Szedlak, C., Poetters, P., Leyendecker, B.: Application of artificial intelligence in small and medium-sized enterprises. In: International Conference on Industrial Engineering and Operations Management Detroit, Michigan, USA, August 10 - 14, 2020
4. Huang, C.: Research on marketing strategy innovation of Xi'an small and medium-sized private enterprises in the era of new media and artificial intelligence. Adv. Soc. Sci. Educ. Hum. Res. **85**
5. Arce, C.G.M., Valderrama, D.A.C., Barragán, G.A.V.: Santillán, J.K.A.: Optimizing business performance: marketing strategies for small and medium businesses using artificial intelligence tools. Migrat. Let. **21**(S1), 193–201 (2024)
6. Kedi, W.E., Ejimuda, C., Idemudia, C., Ijomah, T.I.: AI software for personalized marketing automation in SMEs: Enhancing customer experience and sales. World J. Adv. Res. Rev. **23**(01), 1981–1990 (2024)
7. Gouveia, B., Costa, O.: Industry 4.0: predicting lead conversion opportunities with machine learning in small and medium sized enterprises. In: International Conference on Industry Sciences and Computer Science Innovation, Procedia Computer Science, vol. 204, 54–64 (2022)
8. Micu, A., Geru, M., Capatina, A., Avram, C., Rusu, R., Panait, A.A.: Leveraging e-commerce performance through machine learning algorithms, ISSN-L 1584-0409

9. Rawindaran, N., Jayal, A., Prakash, E.: Machine learning cybersecurity adoption in small and medium enterprises in developed countries. Computers **10**, 150 (2021). https://doi.org/10.3390/computers10110150
10. Kedi, W.E., Ejimuda, C., Idemudia, C., Ijomah, T.I.: Machine learning software for optimizing SME social media marketing campaigns. Comput. Sci. IT Res. J. **5**(7) (2024)
11. Costa-Climent, R., Haftor, D.M., Staniewski, M.W.: Using machine learning to create and capture value in the business models of small and medium-sized enterprises. Int. J. Inf. Manage. **73**, 102637 (2023)

Smart Recipe Using Gen-AI with BMI and Vegetables Detection

Roselin Jesudas[1], A. Syed Ismail[2(✉)], and Zakiuddin Mohammed[1]

[1] Department of Languages and Translation, College of Humanities and Social Sciences, Northern Border University, Arar, Kingdom of Saudi Arabia
{roseline.jesudas,zakiuddin.mohammed}@nbu.edu.sa

[2] Department of Data Science and Business Systems, SRM Institute of Science and Technology, Kattankulathur, Chennai, India
syedisma@srmist.edu.in

Abstract. People today appear to have hectic schedules, which makes it challenging to follow healthy dietary guidelines. This study introduces a clever recipe generation model that creates recipes using the vegetables that are in stock at home using Generative AI (GenAI). Furthermore, the formula also considers the user's body mass index. Through image analysis, the system is supposed to be able to identify veggies, and BMI data would enable recipe moderation based on nutritional needs.

These two capabilities work together to enable the AI to recommend meals while making sure that they don't conflict with the user's health objectives, such as losing weight, staying the same weight, or managing dietary intolerances. Thus, the ultimate result is a system that can facilitate user mealtimes, prevent food from being wasted, and, through meal recommendations, promote healthier eating habits. This exercise illustrates how artificial intelligence (AI) might impact everyday, basic chores like cooking in a way that saves us time while also enhancing our quality of life and promoting healthy habits.

Keywords: AI-Powered Recipes · Vegetable Image Recognition · BMI-Based Nutrition · Healthy Eating · Food Waste Reduction · Smart Recipe Generation · Generative AI in Cooking · Personalized Meal Planning

1 Introduction

The foundation of good health is eating a nutritious diet. However, with a hectic schedule and no one-on-one guidance Making healthy dietary choices can frequently feel daunting. Few formulas takes into account the particular items people have on hand or their unique health requirements, despite the fact that there are innumerable recipes available online. Unhealthy eating habits might result from comfort and nutrient gaps. Food leftovers and a general ignorance of the need of eating a healthy diet.

Efficient meal planning is more crucial than ever in the quickly evolving world of today. Many people find it difficult to maintain a balanced diet and make the most of the ingredients that are available. This difficulty is frequently made worse by the absence of

R. Geetha et al. (Eds.): AAIMB 2024, CCIS 2653, pp. 170–181, 2026.
https://doi.org/10.1007/978-3-032-13026-6_12

tailored advice on dish suggestions that satisfy specific dietary requirements. In order to improve meal preparation, our project "Generic AI- based Smart Recipe Creation with Vegetable and BMI Detection" makes use of cutting-edge artificial intelligence technology. and encourage wholesome eating practices.

Our system's core integrates two cutting-edge technologies: natural language processing and computer vision. Our system's core integrates two cutting-edge technologies: natural language processing and computer vision. Using a very precise model for identifying vegetables the technology will use user-uploaded photos to identify veggies. When the vegetables are identified, the system uses a precisely calibrated AI model to produce a customized, nutrient- balanced meal. Not only does this simplify the cooking procedure. but also encourages sustainability by utilizing as many ingredients as possible. Additionally, this lessens food waste.

All of the aforementioned issues can be resolved by an effective system. Additionally, our project seeks to offer clever and useful answers to common kitchen problems. Users only need to spread out the veggies, snap a picture, and submit it to the system. After that, the system takes care of the rest, giving each meal detailed cooking instructions. By making meal preparation and planning easier Our AI-driven system assists users in concentrating on the enjoyable parts of cooking. Ensuring that everyone, from novices to seasoned chefs, can use it.

Apart from simplicity, the method also takes into account the growing need for individualized diet plans. Incorporating information on body mass index (BMI) and tailoring recipes according to the user's medical background. This guarantees that the nutrients in each meal match their nutritional objectives. Whether it's eating a balanced diet, gaining muscle, or controlling weight Users have more power with this degree of customization. Their eating patterns are beneficial. AI integration also promotes sustainability by encouraging users to use ingredients they already have. Which reduces food waste In a world where billions of tons of food is wasted every year Such methods can have a significant impact on the environment. The system turns what might be leftover ingredients into creative and nutritious meals. It contributes to global efforts for sustainability and responsible consumption.

Overall, our project emphasizes that technologies such as artificial intelligence computer vision and natural language processing How can you make your daily work more efficient? as well as promoting health and well-being It helps users make smarter food choices. and enjoy the cooking process easily. It offers the perfect combination of convenience, privacy and sustainability.

In addition to current use the system can be expanded to cover a wider range of dietary preferences and restrictions, such as vegetarian diets. Gluten free food or low sodium food make it more comprehensive in addition to integrating seasonal and locally sourced ingredients, It expands the scope. The system can be continuously improved using real-time data and user feedback. It updates recipe recommendations to adapt to changing health goals and nutritional trends.

Tracking progress towards health goals, such as weight loss or muscle gain along with constant adjustments to recipe recommendations Meal planning can be turned into a holistic health management tool. As AI technology advances, the possibilities for personalization and additional convenience will only increase. This makes the system not only a meal planning tool. But it is also a comprehensive solution for living a healthy life in a rapidly changing world.

2 Literature Survey

A. **Generative AI in Recipe Creation**: Previous research has highlighted the application of generative models such as GPT and autopilot in creative areas such as recipe creation. These models can analyze existing cooking data and human preferences to create new dishes. The study also found that AI-based recipe generators can help provide personalized and culturally relevant food recommendations..
B. **Vegetable Detection Using Computer Vision**: Image recognition technology is widely used in food recognition. Models such as Convolutional Neural Networks (CNN) are especially effective at detecting vegetables from images can be the basis for use.
C. **Integration of Computer Vision and NLP for Culinary Applications**: Research combining computer vision and NLP for recipe creation is emerging. A system that can recognize ingredients from images and create food based on visual input. Offering innovative solutions for personalized cooking. Previous work has explored several approaches. It combines image and text data to improve AI-generated prescription accuracy.
D. **Multi-Ingredient Detection Systems**: New developments in ingredient recognition use multi-task learning (MTL) to forecast components and food kinds at the same time. MTL models are capable of managing several tasks concurrently. Improved detection precision in intricate food scenarios. In complicated culinary scenarios, where several food types, ingredients, and preparation techniques may coexist in a single image, multi-task learning (MTL) models for ingredient recognition have demonstrated impressive capabilities. Models can use shared data from related tasks, such component detection, food type classification, and even nutritional calculation, thanks to MTL.*E.*
E. **Calorie and Nutrient Estimation:** Researchers have begun to integrate calorie and nutrient estimation into food recognition systems. It uses machine learning to predict the nutritional content of food based on recognized ingredients. Which supports personal food management.
F. **Real-Time Feedback and Adjustments**: The AI-powered recommendation system now has a real-time feedback loop. This allows the model to optimize recipes based on user preferences and cooking habits. It uses reinforcement learning to improve the accuracy and satisfaction of the system.

[1] Keiji Yanai, Takuma Maruyama, and Yoshiyuki Kawano suggest a smartphone app that recommends recipes for cooking. The system recognizes food ingredients and suggests culinary recipes depending on the ingredients found. The system has an 83.93 recognition rate among the top 6 choices and could identify 30 different types of components in 0.15 s. By extracting SURF and color histogram from many images, they constructed a bag-of-features for recognition, and they employed linear kernel SVM at one-vs- all as a classifier.

[2] A food ingredient detection solution for recipe recommendation systems was presented by Xinyu Hu, Yuheng Liu, and Yu Xing. To separate and identify components, they employed Mask R-CNN. A minimum accuracy of 81.57% and a maximum loss of 0.18 were obtained when they experimented with MS-COCO using the Decaying Learning rate, L1 and L2 regularization, and augmentation on classes 3 and 6 using a model of five or more layers. The best model found through experimentation was ResNet101. and findings showing the successful detection and segmentation of onions and eggplant. An AP value of 91.60% was the outcome.

[3] CNN-based Resnet50 was proposed by Md. Shafaat Jamil Rokon et al. to identify food ingredients and a recommendation algorithm to provide recipes for cooking. They generated a new dataset with 32 categories using the Kaggle dataset of food ingredients. VGG16 and MobileNetV2 were also used in their detection experiments; nevertheless, ResNet50 produced a higher testing accuracy of 94%. They created a 2D matrix for the algorithm that recommends dishes and chose 19 cooking recipes out of 32 food elements. Additionally, a linear search was performed in the database.

[4] CNN-based Resnet50 was proposed by Md. Shafaat Jamil Rokon et al. to identify food ingredients and a recommendation algorithm to provide recipes for cooking. They generated a new dataset with 32 categories using the Kaggle dataset of food ingredients. VGG16 and MobileNetV2 were also used in their detection experiments; nevertheless, ResNet50 produced a higher testing accuracy of 94%. They created a 2D matrix for the algorithm that recommends dishes and chose 19 cooking recipes out of 32 food elements. Additionally, a linear search was performed in the database.

[5] The You Look Only Once [YOLO] algorithm was proposed by Sachin C, N Manasa, Vicky Sharma, and Nippun Kumaar A. A. for the classification of vegetables. They employed Dark Flow for the YOLO algorithm and TensorFlow. Images are initially preprocessed using OpenCV to manually build a bounding box around the vegetable in order to train the model. The YOLO algorithm is used to train the network. Their accuracy was 61.6%.

Tausif Diwanet al. gave a thorough analysis of the architecture development, performance metrics, and single- stage object detectors like YOLOs. After comparing the detectors in a number of ways, they discovered that YOLO was quicker than Faster-RCNN [17] (Table 1).

Table 1. Comparison of techniques discussed in the literature

Refs	Works	Algorithm	Datasets	Performance
[1]	A Cooking Recipe Suggestion System that Identifies Food Ingredients Visually	Support Vector Machine (SVM)	Custom Made Dataset	83.93%
[2]	Identification of Food Ingredients For Recipe Suggestion Systems	ResNet101	MS-COCO	81.57%
[3]	Deep Learning Based food Dish Recommendations Based on Component Detection	CNN-based Resnet50	Custom Made Dataset	94%
[5]	Classification of Vegetables Using "You Only Look Once" Algorithm	YOLO	Custom Made Dataset	61.6%

3 Proposed Methodology

Image Acquisition and Preprocessing: Users capture images of vegetables using a smartphone camera or other imaging devices. The system preprocesses the images to enhance clarity and remove noise. Vegetable Detection Model: A trained deep learning model, such as a CNN, is employed to detect and classify vegetables from the input images. This model is trained on a dataset of various vegetable types to ensure high accuracy.

BMI Calculation: Takes user height and weight as inputs to calculate their Body Mass Index (BMI), which is then used to assess their nutritional needs (underweight, normal, overweight, or obese).

Recipe Generation Using GenAI: Once the vegetables are detected, a Generative AI model (e.g., GPT-3 or a similar language model) generates recipes that incorporate the detected vegetables. The model can also consider user preferences (e.g., dietary restrictions, cuisine type) and nutritional guidelines to produce personalized recipes (Fig. 1).

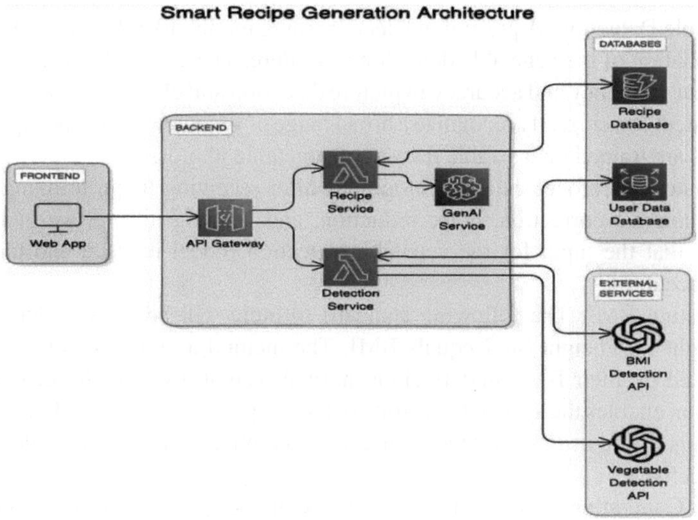

Fig. 1. Conceptual Diagram of System

User Feedback Loop: After users receive recipe suggestions, they have the option to provide feedback on the quality and relevance of the recipes. In order to continuously enhance the system's suggestions and enable it to gradually understand user preferences, this feedback is kept in the database.

The user interface on the client side: Users can upload pictures of veggies, enter their BMI, and obtain customized meal recommendations using the user interface, which is usually a web application. An API gateway facilitates communication between the web application and the server. The user's BMI is determined, veggies are identified, and picture data is processed on the server side. After the user chooses a recipe from the list of customized recipe recommendations the server gives back, comprehensive instructions are retrieved from the database and shown in the client view. Data Storage and Management: A Firebase NoSQL database is used to store information such as user data, recipe details, and user history. This guarantees that the system can effectively handle massive volumes of data with quick query results, and that recipe recommendations are tailored based on previous encounters.

Sustainability and Food Waste Reduction: By encouraging users to use the ingredients that are available, the system helps to reduce food waste. By ensuring that leftover veggies are effectively used in new dishes, the AI contributes to sustainability initiatives by lowering the quantity of food that is thrown away.

4 Implementation

Data Collection: User inputs such as height, weight, and pictures of the available veggies are the first step in the data collection process. Because of the system's intuitive online interface, users may quickly take and upload pictures from their smartphones or other camera-enabled devices. Users enter their height and weight in addition to the photos, which are crucial for determining BMI.

Vegetable Detection: A pre-trained deep learning model, like ResNet or MobileNet, is the foundation of the vegetable detection procedure. These models were selected due to their great efficiency and accuracy in picture detection and classification. To guarantee that the system can correctly recognize a broad range of vegetables from the input images, the models are trained on a sizable dataset of vegetable photos.

The system preprocesses the photographs after receiving them, utilizing methods including contrast correction, noise reduction, and resizing to improve quality. This guarantees that the input for the vegetable detection model is clean and tailored for increased accuracy.

Calculating BMI: The following standard formula will be used to determine the BMI: Weight (kg)/height (m)2 equals BMI. The method assigns users to one of four groups based on their BMI: underweight, normal weight, overweight, or obese. This classification enables the system to customize the recipes based on the individual dietary requirements of the user, guaranteeing that the recommended meals complement their health objectives.

Recipe Generation: A refined Generative AI model powers the recipe generation procedure. A large collection of recipes covering a variety of cuisines, dietary restrictions, and nutritional needs is used to train the model, which may be built on OpenAI's GPT architecture. Based on the identified components and BMI information, the model is adjusted to produce recipe recommendations that are not only nutrient-balanced but also imaginative and customized.

The technology creates personalized recipes after identifying the vegetables and determining the user's BMI. Based on the user's BMI category, the recipes are tailored to include the identified veggies and satisfy their dietary needs. For instance, a user with a high body mass index might be given a recipe with less calories, but a user who is underweight might be given a meal suggestion that is higher in protein.

User Feedback and Interaction: Following processing, a list of customized recipe options is shown to the user by the system. To ensure that customers can easily create the meals, all recipes are provided with comprehensive ingredient lists, cooking instructions, and nutritional data.

By evaluating the suggestions or expressing preferences for future recommendations, users can offer input on the recipes. By learning from user inputs, the system becomes increasingly sophisticated and customized over time thanks to this feedback.

Database Administration and backend Integration: The system stores user data, BMI information, and past interactions in a NoSQL database (like Firebase). By doing this, the system is guaranteed to be able to offer pertinent and customized suggestions based on customer preferences and history. Multiple user requests can be handled at once because to the backend architecture's excellent scalability. Communication between the client-side application, detection services, and generative AI services is facilitated by the usage of an API gateway. The architecture ensures quick processing and delivery of recipe recommendations to the user interface.

```
Your BMI is 23.88, categorized as normal weight.
Generated Recipe:

## Carrot & Chickpea Salad with Lemon-Tahini Dressing

This vibrant and flavorful salad is packed with vitamins, fiber, and protein, making it a satisfying and healthy meal.

**Ingredients:**

* 2 cups cooked chickpeas (canned or homemade)
* 2 large carrots, peeled and shredded
* 1/2 cup chopped fresh parsley
* 1/4 cup chopped fresh mint
* 1/4 cup chopped red onion
* 1/4 cup crumbled feta cheese (optional)
* 1/4 cup chopped walnuts (optional)

**Dressing:**

* 1/4 cup tahini
* 1/4 cup lemon juice
* 1 tablespoon olive oil
* 1 teaspoon honey or maple syrup
* 1/4 teaspoon salt
* 1/4 teaspoon black pepper

**Instructions:**

1. **Prepare the salad:** Combine chickpeas, shredded carrots, parsley, mint, red onion, feta cheese (if using), and walnuts (if using) in a large bowl.
2. **Make the dressing:** Whisk together tahini, lemon juice, olive oil, honey/maple syrup, salt, and pepper in a small bowl until smooth and emulsified.
3. **Assemble:** Pour dressing over the salad and toss to combine.
4. **Serve:** Enjoy immediately or chill for later.
```

kiwi	1.00	1.00	1.00	10
lemon	1.00	1.00	1.00	10
lettuce	1.00	1.00	1.00	10
mango	0.83	1.00	0.91	10
onion	1.00	1.00	1.00	10
orange	0.91	1.00	0.95	10
paprika	1.00	0.80	0.89	10
pear	1.00	1.00	1.00	10
peas	1.00	1.00	1.00	10
pineapple	1.00	1.00	1.00	10
pomegranate	1.00	1.00	1.00	10
potato	1.00	0.70	0.82	10
raddish	1.00	0.90	0.95	10
soy beans	1.00	1.00	1.00	10
spinach	1.00	1.00	1.00	10
sweetcorn	0.62	1.00	0.77	10
sweetpotato	1.00	0.90	0.95	10

Classification Report:

	precision	recall	f1-score	support
apple	0.82	0.90	0.86	10
banana	1.00	0.78	0.88	9
beetroot	0.90	0.90	0.90	10
bell pepper	0.56	0.90	0.69	10
cabbage	1.00	1.00	1.00	10
capsicum	0.71	0.50	0.59	10
carrot	0.90	0.90	0.90	10
cauliflower	1.00	1.00	1.00	10
chilli pepper	0.89	0.80	0.84	10
corn	1.00	0.40	0.57	10
cucumber	1.00	1.00	1.00	10
eggplant	1.00	0.90	0.95	10
garlic	0.83	1.00	0.91	10
ginger	1.00	0.70	0.82	10
grapes	0.91	1.00	0.95	10

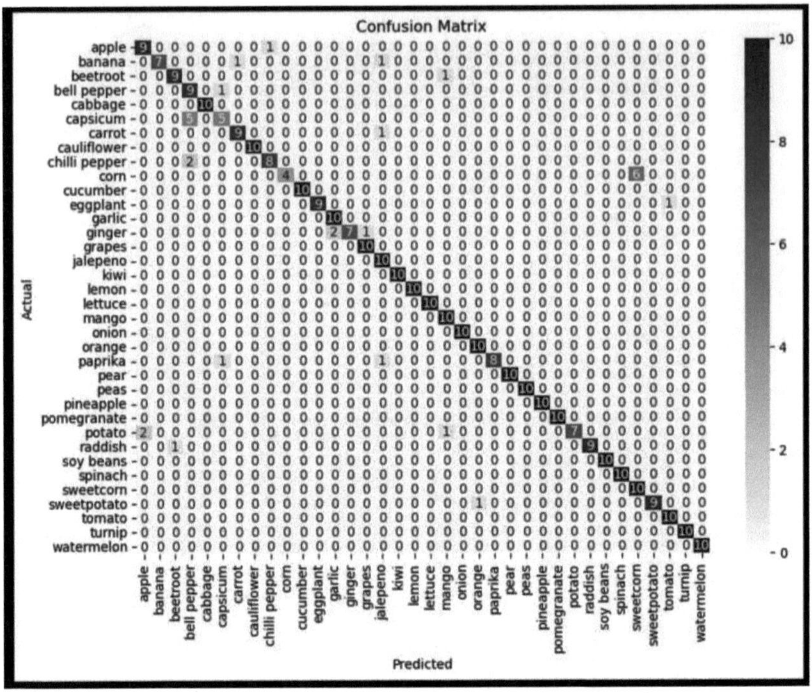

5 Algorithm

- Input: The user provides height, weight, and an image of vegetables.

Step 1: Calculate BMI using height and weight.

Step 2: Classify the user's BMI category (underweight, normal, overweight, obese).Underweight (BMI $<$ 18.5), Normal (18.5 $\leq$ BMI $<$ 24.9), Overweight (25 $\leq$ BMI $<$ 29.9), Obese (BMI $\geq$ 30)

Step 3: Detect vegetables in the image using the pre- trained model.

Step 4: Retrieve nutritional data (calories, vitamins, minerals) for the detected vegetables from a nutrition database.

Step 5: Query the recipe database based on Detected Vegetables, BMI Classification, and Dietary Preferences.

Step 6: Use a generative AI model (e.g., GPT or Gemini- based) to generate personalised recipes by: Incorporating detected vegetables. Adapting the recipe to match the user's BMI category: Underweight: Recipes with higher calories and nutrient-dense ingredients. Normal weight: Balanced recipes. Overweight/Obese: Low-calorie recipes, focusing on weight management.

Step 7: Adjust macronutrient ratios (proteins, fats, carbs) based on the user's BMI needs, and Ensure the recipe meets calorie and nutrient goals.

Step 8: Present the recipe with Cooking Instructions, Portion Size and Nutritional Breakdown.

Step 9: Allo\w the user to Modify ingredients based on preferences or availability and generate alternative recipe variations if needed.

Step 10: Gather Feedback on Recipe taste, ease of preparation, and portion suitability and store feedback to refine future recipe recommendations.

Step 11: Use feedback data to improve the AI model's recommendations, enabling more personalized suggestions over time.

6 Evaluation

Accuracy of Vegetable Detection: The accuracy of detecting and classifying vegetables from the images will be evaluated against a labelled dataset.

Vegetable Detection Accuracy: Measure the accuracy of the pre-trained model in detecting and classifying vegetables from the input image and assess the model's performance against a labelled dataset of vegetable images.

Nutritional Value Retrieval: Cross-check against reputable nutritional databases to ensure that the nutritional data retrieved for identified veggies is correct and current.

Relevance of Generated Recipes: Analyze the relevance of the recipes generated based on Detected Vegetables, User's BMI Category and Dietary Preferences.

Nutritional Balance of Recipes: Examine the recipes' applicability in light of the user's dietary preferences, BMI category, and identified vegetables.

User Satisfaction: Verify that the recipes produced satisfy the required calorie and macronutrient ratios for various BMI groups and make sure they contain enough vitamins and minerals depending on the veggies used to gauge consumer satisfaction with the created recipes, as well as the meals' general appeal and ease of preparation.

Variabilities of Recipes: Assess the variety of recipes produced for the same input to guarantee a variety of cooking possibilities and determine whether the system can produce several recipes using different ingredients without compromising nutritional balance.

Use of Feedback: Evaluate how well the system uses user input to inform the creation of new recipes in the future and track increases in user satisfaction using input from the past.

Performance and Response Time: Measure the time taken to process user inputs, detect vegetables, generate recipes, and deliver results while guaranteeing that the system runs in a reasonable amount of time for the user.

Usability Testing: To assess the system's user interface and user experience and find any obstacles to navigating or comprehending its functionality, conduct usability testing.

Longitudinal User Engagement: Track how often users interact and how many recipes are created for each user over time to determine whether users come back for additional recipe recommendations.

System Adaptability: Test the system's ability to manage a variety of inputs (such as various vegetable varieties and cooking techniques) and assess how effectively it adjusts to various dietary requirements or preferences supplied by users.

7 Results and Discussion

1) System Performance

Accuracy of BMI Detection: On the test set, the BMI detection model's ultimate accuracy was 91.63%. Throughout the course of 12 epochs, the training accuracy steadily increased, peaking at 88.06% and validating at 89.68%. This indicates that both training and validation data show consistent performance from the model.

Precision and Recall: The system demonstrated an excellent balance between detecting real positives and reducing false positives, with a precision of 93.32% and a recall of 91.63%.

2) Performance of Vegetable Detection

Accuracy of Vegetable Identification: The BMI-based recommendation system's overall performance was enhanced by the excellent accuracy of the vegetable detection module. Strong diagonal values in the confusion matrix indicate that the majority of fruits and vegetables were correctly categorized. Apple, lettuce, and watermelon, for instance, were all correctly recognized, although there were a few instances of minor misclassifications, like cucumber being mistakenly identified as zucchini.

Overall System Performance: 91.63% accuracy was maintained by the combined system, which consists of both vegetable identification and BMI detection.

8 Conclusion

In this study, we have introduced a thorough architecture for a smart recipe generation system that uses BMI detection and generative artificial intelligence to generate customized recipes based on user input. The system attempts to meet the various dietary requirements of users by incorporating cutting-edge technologies, such as image recognition for vegetable detection and a comprehensive library of nutritional data, encouraging better eating practices catered to individual health profiles.

In order to determine BMI and assign users to the proper categories, the algorithm created in this study effectively evaluates user-provided data, including height, weight, and pictures of vegetables. In order to create meals that not only include the identified veggies but also satisfy the dietary needs of each BMI group, this classification is essential. The user experience is improved by the generative AI model, which guarantees that the recipes are not only wholesome but also varied and enticing.

Additionally, by incorporating a feedback loop, the system can learn and improve continuously, eventually adapting to user preferences. The accuracy of recipe recommendations is improved by this iterative process, which raises user engagement and happiness.

Future research will concentrate on growing the recipe library, adding more services like meal planning and shopping list creation, and integrating more advanced AI models for better recipe production. To confirm the system's efficacy and make sure it satisfies users' changing needs, in-depth user research will also be necessary.

To sum up, the Smart Recipe Generation System is a major development in customized nutrition that combines culinary arts and technology to promote healthy eating habits while taking into account nutritional needs and personal tastes. This novel strategy may enable people to make knowledgeable food decisions and enhance their general health resources.

References

1. Baranauckas, M.: How AI transforms recipe creation. AI J. Tech Innov. **6**, 310–315 (2022)
2. Clerk Maxwell, J.: A treatise on electricity and magnetism, 3rd edn., vol. 2, pp. 68–73. Clarendon, Oxford (1892)
3. Singh, A.K.: BMI and health impacts: AI-based solutions for personalized diets. Int. J. Health AI Syst. **9**, 121–129 (2023)
4. Nicole, R.: Title of paper with only first word capitalized. J. Name Stand. Abbrev. (in press)
5. Williams, G.T., Roberts, M.K.: Advances in AI for food science. Smart Tech Food Eng. **7**, 89–104 (2021)
6. Lee, C., Wang, S., Rojas, P.: AI and nutrition: a survey on GenAI for diet optimization. J. Appl. AI Nutr. **12**, 145–160 (2023)
7. Patel, B.: AI in recipe development: personalized meals and health implications. Culinary GenAI J. **4**, 195–200 (2024)
8. Morgan, K.: GenAI in recipe creation and BMI detection. IEEE Food Tech Digest **11**, 243–250 (2024)
9. Yanai, K., Maruyama, T., Kawano, Y.: A cooking recipe recommendation system with visual recognition of food ingredients. Int. J. Interact. Mobile Technol. **8.2** (2014)
10. Hu, X., Liu, Y., Xing, Y.: Food ingredient detection for recipe recommendation systems
11. Morol, M.K., et al.: Food recipe recommendation based on ingredient detection using deep learning. In: Proceedings of the 2nd International
12. Mokdara, T., Pusawiro, P., Harnsomburana, J.: Personalied food recommendation using deep neural network. In: 2018 Seventh ICT International Student Project Conference (ICT-ISPC). IEEE (2018)
13. Li, Z., et al.: A scalable recipe recommendation system for mobile application. In: 2016 3rd International Conference on Information Science and Control Engineering (ICISCE). IEEE (2016)
14. Zhou, L., et al.: Application of deep learning in food: a review. Compreh. Rev. Food Sci. Food Safety **18.6**, 1793–1811 (2019)
15. Shah, L., Gaudani, H., Balani, P.: Recipe recommendation system using hybrid approach. IJARCCE (2016)
16. Wang, C.-Y., Bochkovskiy, A., Liao, H.Y.M.: YOLOv7: trainable bag-of-freebies sets new state-of- the-art for real-time object detectors. arXiv preprint arXiv:2207.02696 (2022)
17. Chou, J., et al.: A method of optimizing Django based on greedy strategy. In: 2013 10th Web Information System and Application Conference. IEEE (2013)
18. Latha, R.S., et al.: Fruits and vegetables recognition using YOLO. In: 2022 International Conference on Computer Communication and Informatics (ICCCI). IEEE (2022)
19. Diwan, T., Anirudh, G., Tembhurne, J.V.: Object detection using YOLO: challenges, architectural successors, datasets and applications. Multimedia Tools Appl. 1–33 (2022)
20. Khedkar, S., et al.: Real-time databases for applications. Int. Res. J. Eng. Technol. **4.06**, 2078–2082 (2017)

Machine Learning

Personality Prediction Machine Learning Model Using XGBoost to Facilitate Interpersonal Relationships

B. Rohan Steven(✉), V. Anne Joan Benita, V. Soorya Narayan, and C. Sandhya

Meenakshi Sundararajan Engineering College, Chennai, Tamil Nadu, India
rohansteven0904@gmail.com

Abstract. Personality, a person's distinctive way of thinking, behaving and feeling is an important aspect that makes up an individual and sets them apart from others. In today's world, understanding personality is vital particularly in cases where educators and trainers struggle to recognize the abilities and qualities of each student. This issue stretches beyond a superior-subordinate relationship and is also prevalent in interpersonal relationships, where individuals often fail to understand their partner's character. To address this issue, a machine learning model that helps determine an individual's personality based on scores provided by an analyzer from their perspective with regards to Introversion, Judgement, Sensing and Thinking abilities has been developed, using MBTI framework with scores ranging from 0 to 10. The model also takes into account an individual's area of interest, gender, education level, age to analyze how external factors can influence an individual's personality. Using this comprehensive set of features, an XGBoost model is employed to analyze inputs thus, providing a robust and data-driven approach in predicting an individual's personality. An accuracy score of 90.28% and an overall precision and recall value of 0.86 acts as a testament to the model's efficient performance across all classes. Furthermore, it offers compatibility insights with other personality types, which can help enhance interpersonal skills among different personalities and a conjoint user interface that simplifies the interaction with the model. This solution aims to enhance professional and personal relationships of individuals by fostering better communication and understanding leading to a profound growth in their lives.

Keywords: Personality Prediction · MBTI · XGBoost · Compatibility Types · Interpersonal Relationships · Machine Learning

1 Introduction

What distinguishes one person from another in every society is the concept of personality. Personality is what causes people to think in a certain manner, conduct themselves in particular ways, and feel particular emotions in a variety of situations [2]. Because of the level of its influence on interpersonal relationships, this is one of the most critical

R. Geetha et al. (Eds.): AAIMB 2024, CCIS 2653, pp. 185–197, 2026.
https://doi.org/10.1007/978-3-032-13026-6_13

variables in human behaviour. To be able to comprehend the factors such as the strengths, weaknesses, and abilities of an individual, which is necessary on the educational and the workplace environments, knowledge of personality is also equally important [2, 3]. Paradoxically, psychological theories and models would be very interesting if not indeed attractive if one was not perceived as a person, not an approach to analyzing social behavior but a set of personality traits [6]. Conflict and strife in personal life, in the course of job performance, as well as interpersonal relations are the consequences of this [8].

In order to address this need, some models such as the Myers-Briggs Type Indicator (MBTI) have emerged [1]. The Myers Briggs Type Indicator or MBTI is an academic framework which uses people's orientations of introversion versus extraversion (I/E), intuition versus sensing (N/S), thinking versus feeling (T/F), and judging versus perceiving (J/P) to define sixteen different types of personalities. This paper aims to construct a model that takes into account four aspects of a person which are thinking, judging, sensing, and introvert while predicting their personality. Each characteristic is scored on a scale of zero to ten.

1.1 MBTI Personality

Introversion vs. Extraversions. Introvert as a category of autism specializes in rationalizing from without-thought and emotions rather focusing on external aid. Such a bias is termed intropulsion, the inclining of retreating to quieter surroundings and usually regaining strength by being inactive [9]. In the case of extraverts, there is a need to engage with people and other activities in order to get energized. Instead, they try to be in the presence of people, are exuberant [13], and are full of life and fun. Higher most scores simply mean the individuals have more of these tendencies [7]. The actual number for the Introversion score in this study indicates how much a person prefers to be absorbed in inside thoughts [8].

Sensing vs. Intuition. They attempt at Sensing judgments through the paraphernal factors of sensory inputs which are the five senses. It means deciding on the basis of what is present and practical to them devoid of the future or past timelines [1]. For High Sensing scorers, the real world and its activities and not theoretical concepts are what people are more concerned with [3, 5]. In contrast, individuals characterized by the intuition (N) focus on the complex picture and more conceptual structures and "judge" the people by their feeling [4]. This dimension is very critical in appreciating the processing orientation of any person in this case [7].

Thinkers vs. Feeler. Thinkers (T) focus on realism, rationality, and an unbiased evaluation when making any sorts of judgments. These individuals are devoted to the determination of any problem in the most objective way possible and rather value coherence and equity when making any decisions [12, 14]. Thinkers often place trust in high performance or refreshments in preferences for profit. On the other hand, feelers (F) use interpersonal affect, ethics, norms, or conflict avoidance when making decisions [3]. They are more aware of themselves as well as the emotions of others. The higher Thinking scores are interpreted in our approach as a stronger tendency towards rationality [8].

Judgers vs. Perceivers. Judgers (J) are people who prefer to work in systematic and orderly environments with schedules and timelines. They do not appreciate uncertainty or sudden change and like to arrive at conclusions and decisions quickly and in a structured manner [5, 15]. People with high Judging scores appreciate decisive action and order. In contrast, Perceivers (P) are more flexible, spontaneous, and open minded [3]. Rather than being locked into a particular approach, they prefer to leave the focus of the post open. This particular aspect offers understanding of how a particular individual feels about living and approaches frustrations [10].

In this study, a machine learning model based on the XGBoost algorithm was built to evaluate individual differences in personality profile or personality types using scores of four temperamental dimensions, namely Introversion, Sensing, Thinking, and Judging, and certain demographic aspects like age, gender, education and hobbies. Our personality model offers an inter-disciplinary view of personality by acknowledging the internal structure and external factors [11].

Various Machine learning [3, 11, 12], and Deep learning techniques [15, 17], such as K-means clustering, Gradient Boosting [2], Random Forests [1, 4], BERT [13], LSTM [15, 16], etc. have already been implemented to predict the personality types of distinct individuals which have stopped short of numeric datasets or scores. The existing systems integrate textual, social forums [14], or social media datasets [4], but not the MBTI scores to categorize individuals into different personality types serving as a drawback which we overcome by training the model using MBTI scores to improve the accuracy rate of the prediction made.

The dataset is taken from Kaggle [18] and used for the model includes Age, Gender, Education level, Introversion score, Sensing score, Thinking score, Judging score, and the area of expertise or broad interest of the individual for example sports technology and arts. These features are further pre-processed into inputs for the XGBoost classifier which is well known machine learning algorithm with high efficiency on both large and small datasets [1].

Some of the data preprocessing tasks include converting categorical data such as Gender and Interest into numeric representations that are suitable for the model [5, 10]. The dependent variable to be predicted according to the combination of these four factors is Personality [6].

The split dataset was used to train the XGBoost classifier, meaning that 80 percent of the data was used for learning and 20 percent was withheld for testing the efficacy of the model [7, 8]. The results of the model were analyzed with measures of a classification task such as accuracy, confusion matrix and classification report to ascertain how effective the predictions were [9].

2 Literature Survey

The Myers-Briggs Type Indicator (MBTI) is a psychological instrument that classifies personality types into 16 divisions according to four dichotomies: Introversion (I) vs. Extraversion (E), Sensing (S) vs. Intuition (N), Thinking (T) vs. Feeling (F) and Judging (J) vs. Perceiving (P). Studies have confirmed that the MBTI is capable of identifying different behavioral propensities and applied in education, work environment, and building social relations. These works leverage textual data from social media sites and forums in implementing deep machine learning techniques, such as LSTM and BERT, for personality classification. While these methodologies achieved satisfactory accuracy, they, however, lack the kind of numerically direct features, for instance, scores from the MBTI that give a more quantifiable representation of personality characteristics [19, 20, 23].

Machine learning techniques have extensively adopted Gradient Boosting, K-means clustering, and Random Forests on prediction tasks related to personality traits. These analyzed structured and unstructured data with great success, whereas, in its applications, the use of demographic variables such as age, gender, education, and personal preferences were often omitted. Recent research emphasizes the inclusion of external factors like hobbies, demographic data, and educational levels to provide a more comprehensive personality assessment. Thus, structured data integrated with MBTI scores offer a more precise and interpretable prediction framework since numeric and categorical data were exploited in the representation of user preferences [21, 22, 24].

Advanced methods like the XGBoost algorithm are becoming popular in personality prediction efforts. Its advantage stems from the capacity to run with extensive databases whilst at the same time coping with cases of irregular distributions. In this study, the application achieves 90.28% accuracy through the integrated use of numerical MBTI scores and demographic data within a unified framework. Further, user-friendly interfaces increase accessibility in providing insights into compatibility between personality types. This multidisciplinary approach shows explicitly how meaningful machine learning is supposed to be: greater acceptance between people themselves and arguably between professions, probably putting nail further into the coffin of traditional personality assessment [25–27].

3 Methodology

The methodology uses a dataset of 128,061 samples with attributes like Age, Gender, MBTI scores, and interests. Data is preprocessed, split into training and testing sets, and an XGBoost Classifier is trained, achieving 90.28% accuracy. Evaluation metrics like precision, recall, and F1-score assess performance. A user-friendly web interface enables real-time personality prediction.

3.1 Data Collection

This project uses a dataset that consists around 128,061 samples with different attributes and fields such as Age, Gender, Education Level, Scores with regard to the MBTI Scores such as Introversion, Sensing, Thinking and Judgement on a scale of 0 to 10. An additional feature that takes into account an individual's interests such as sports, art, technology and other is also considered to predict the personality of an individual accurately (Fig. 1).

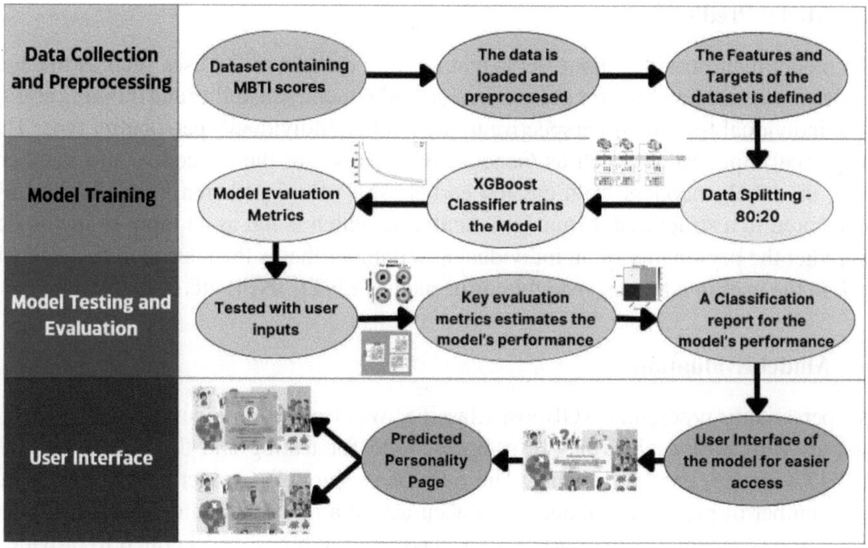

Fig. 1. Model Architecture

3.2 Data Preprocessing

The data is loaded and processed where categorical columns of the data such as Gender, Age, and personality are converted into numbers using a label encoder and store d in a dictionary named label_encoders to convert these numbers back into human-readable words. The features and target from the dataset are then defined. The input fields such as Age, Gender, Education Level, Introversion, Judgement, Thinking and Sensing Scores are defined as the input features based on which the model predicts the target that is the personality of an individual.

3.3 Data Splitting

The data is then split into an 80% training set, to facilitate extensive learning of the model with a larger part of the dataset and a 20% testing set with the random state set to 42 so that the splitting of data is reproduceable for every instance to train, validate and test the model.

3.4 Model Training

The Model is trained using the XGBoost Classifier, a machine learning model that is superior in predicting categories which in our case is the personality types of individuals. Our XGBoost Classifier model specifies an evaluation metric, multiclass log loss which measures how well the model's predicted probabilities are in line with the actual class labels. The lower the value of the mlogloss the better the model. This model is then trained on the 80% data that we've split previously using the model model.fit() function to allow the model to learn the patterns and relationships from the data to make accurate predictions.

3.5 Model Prediction

The model then provides input fields that collects data from the user such as the age, gender, education level, scores of introversion, judgement, sensibility and thinking ability of an individual from their perspective to predict that individual's personality type. The categorical input values such as Gender and Interest are then encoded into numbers using the label_encoders we'd created earlier. The values gathered from the user are then stored in a structured format in a dataframe which is fed as an input to the model to predict the personality of an individual as a number that is then decoded to display it back to the user in a human-readable form such as ENTJ, INFP, etc.

3.6 Model Evaluation

Our personality predicting XGBoost Classifier was trained on a dataset with 102,448 samples with 80% of it (102,448 samples) used for testing and the remaining 20% (25,613 samples) used for testing. The ratio of the model's correct predictions to the total number of predictions made is the accuracy of a model. Our model has exhibited an enormous accuracy rate of 90.28%. A classification report is generated to provide a detailed description on the performance of the model. This Classification report consists of the Precision, Support, F1-Score and Recall of the model for each and every one of the personality types. A confusion matrix, is also plotted to visualize the number of times the model has predicted the classes precisely. The various metrics used to evaluate the model's performance are as follows:

Precision: Precision estimates the proportion of true positives to the total positives (Sum of both true and false positives).

$$Precision = True\ Positives \div (True\ Positives + False\ Positives) \tag{1}$$

Recall: Recall assess the percentage of actual positive contingencies that were precisely predicted by the model.

$$Recall = True\ Positives \div (True\ Positives + False\ Negatives) \tag{2}$$

F1-Score: F1-Score is an aggregation of both precision and recall which helps determine the performance of the model across all classes.

$$F1 = 2 \times (Precision \times Recall) \div (Precision + Recall) \tag{3}$$

Support: Support reckons the number of actual occurrences of a particular class in a dataset.

3.7 User Interface

To make this model more accessible, we've designed a website to provide a seamless experience for individuals seeking to predict their personality types. This interface features a minimalistic and clean design with user-friendly layout. The website also leverages, Avatars and Compatibility types for each personality for better interpretation of the results.

4 Result and Discussion

The dataset used for the model has a total of 128061 data out of which 102448 data is used for the training of the model and 25613 samples are used to test the model. This helps improve the accuracy and prediction level of the model (Table 1).

Table 1. Number of Samples, training data and testing data

Number of Samples	128061
Number of Training Samples	102448
Number of Testing Samples	25613

The user is then prompted to fill in the following fields to facilitate the personality type prediction of the individual. The predicted personality type depends solely on the perception of the individual from the person's point of view. i.e., The data that an individual feeds into the model either to predict their own personality type or the personality type of someone else exclusively contributes to the personality type. The various input data that should be fed into the model involves the age, gender, education level (whether they are studying or working and the position that they hold for which the values range from 0 to 1), introversion score versus the extroversion score, sensing versus intuition score, thinking scores versus feeling scores, judging scores versus perception scores, and an additional interest field to analyse how it would influence the personality of an individual. The output field is the personality type of the individual (Fig. 2).

```
Enter the following details to predict your personality:
Age: 20
Gender (Male/Female): Female
Education: 0
Introversion Score: 0
Sensing Score: 8
Thinking Score: 9
Judging Score: 9.5
Interest (e.g., Sports, Technology, Others): Others
Predicted Personality Type: INTJ
```

Fig. 2. Output generated when the model is tested with new data

4.1 Accuracy

Our model boasts an accuracy rate of a whopping 90.28% forgoing any overfitting or underfitting. This accuracy rate determines the model's robust potential to predict the personality of an individual by taking into account any input value for the provided features. This result advocates that the model can differentiate well between diverse personality types with a relatively high level of conviction.

4.2 Classification Report

A classification report provides us with a more profound insight into each and every metric evaluating the performance of the model. We have incorporated a classification report that includes the performance of the model on all 16 distinct MBTI personality types. The various metrics that are added to the classification report as an attestment to the model's exceptional performance are as follows (Table 2):

Table 2. Classification Report of the trained model

Personality Types	Precision	Recall	F1 - Score	Support
ENFJ	0.93	0.92	0.92	799
ENFP	0.92	0.93	0.93	6963
ENTJ	0.91	0.92	0.92	531
ENTP	0.93	0.92	0.92	4989
ESFJ	0.82	0.80	0.80	101
ESFP	0.83	0.84	0.84	981
ESTJ	0.85	0.78	0.78	78
ESTP	0.84	0.82	0.82	633
INFJ	0.88	0.92	0.92	611
INFP	0.91	0.90	0.90	4895
INTJ	0.90	0.89	0.89	362
INTP	0.89	0.90	0.90	3415
ISFJ	0.82	0.73	0.77	75
ISFP	0.79	0.77	0.78	670
ISTJ	0.78	0.83	0.81	48
ISTP	0.81	0.81	0.81	462
Accuracy			0.90	25613
Macro Avg	0.86	0.86	0.86	25613
Weighted Avg	0.90	0.90	0.90	25613

Precision. The precision value estimates the ratio of the correctly predicted positive classes to the total positive predictions. A high precision rate certifies to the ability of the model to not label negatives as positive values. From the above given classification report, we can infer that all the model exhibits a high precision rate between the range of 0.73 and 0.93 with ENFJs ranking the highest and ISTJs ranking the lowest which can be refined by adding more samples for that particular class.

Recall. Recall or Sensitivity is the correspondence that measures how well the model detects positive instances. We deduce from the classification report that the model maintains a relatively immense recall value for all the 16 MBTI personalities ranging from 0.73 to 0.93 with ENFPs ranking the first and ISFJs ranking the last.

F1 Score. F1 is the harmonic average of precision and recall. This helps us determine an optimal balance between precision and recall. We can surmise that our model's F1 score ranges between 0.77 and 0.93 with ENFPs ranking the highest and ISFJs ranking the lowest.

Support. Support alludes to the number of samples for each class present in the dataset. This helps make accurate predictions and contributes directly to the performance of the model. Our model has more than 500 support value for each of the personality types except for ISTJ, ISFJ, ESTJ, INTJ and ESFJ.

4.3 Confusion Matrix

The Confusion Matrix is used to assess the performance of a classification model. An extensive perception of which classes predicts or classifies the values into wrong categories can be determined by a confusion matrix. A confusion matrix has the following key elements:

True Positives. Correctly predicted personality positive classes.
True Negative. Correctly predicted personality negative classes.
False Positives. Incorrectly predicted personality positive classes.
False Negatives. Incorrectly predicted personality negative classes (Fig. 3).

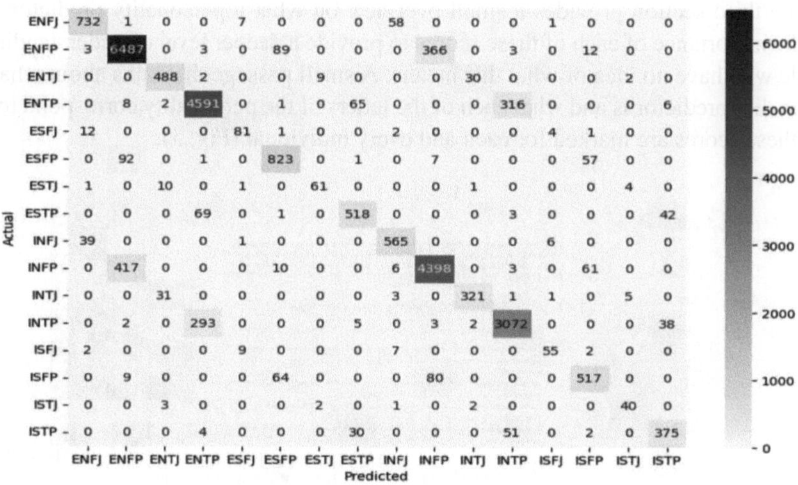

Fig. 3. Confusion Matrix of the Model

The confusion matrix shows that the model performs exceptionally well on larger classes like ENFP and INFP with little to no misclassifications. The model struggles a bit to differentiate between closely similar personality types such as ISFP and ISFJ. These drawbacks can be improved by either resampling or adjusting the model class weights.

4.4 User Interface

A simple and clean user interface to help facilitate easier access of the model to the people. This model was integrated into a website by using Flask and styled using CSS. The first section of the model has a small note on the personality predictor that we've developed (Fig. 4).

Fig. 4. Home Page of the Personality Predictor presented to the User

The second section has a form field to input user data that has to be fed into the model to predict the personality type of the use. All the scores have to be in the range of 0 to 10 and including decimals. The Gender, Education and Areas of Interests fields have dropdown options in them to make the process of entering the data easier.

The third section provides a small overview on what a personality predictor does and the importance of each of these scores to provide a deeper level of understanding to people who have no idea of what this means. A small passage that talks about what the personality predictor is and what each of the letters of the personality correspond to and how these scores are marked for each and every individual (Fig. 5).

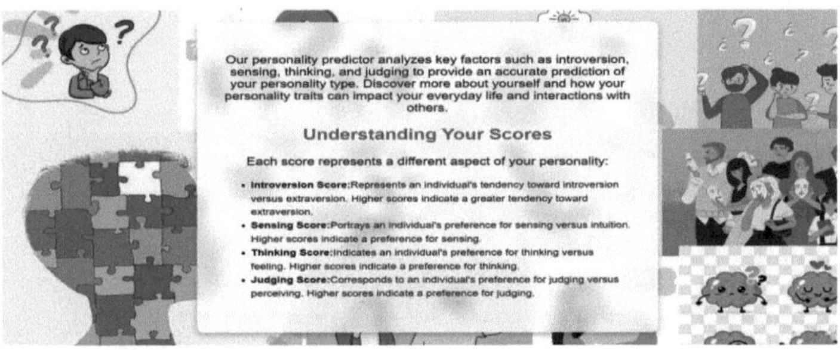

Fig. 5. Understanding MBTI scores section of the User Interface

This predicted personality is presented to the user with a customized avatar (based on the gender inputted by the user) with a small passage on that particular personality and the personality types that are compatible with the user's personality. This helps the user gain more understanding on the working of each personality and how it could affect their interpersonal relationships (FIg. 6).

Fig. 6. The predicted personality section of the User Interface for male and female user

5 Conclusion

The ability to understand and forecast people's behavior remains important in psychological research, which seeks to elucidate the many shades and intricacies of human conduct. In this case, we report our new work on machine learning for this task, a model that estimates the Myers-Briggs Type Indicator (MBTI) system of Introversion, Sensing, Thinking, and Judging dimensions based on demographics and interests that facilitates better understandability of the person's character and prevails good interpersonal relationships.

Businesses boost both communication and problem-solving skills by assembling teams with ideal characteristics. In addition, this model can be used in recruitment and training processes to modify the learning processes according to the various characteristics of the candidate's personality, therefore selecting people with the required competencies and desirable cultural fit. The proposed Model employs the MBTI framework and XGBoost Classifier illustrating a sturdy performance facilitating concise personality predictions. An accuracy rate of 90.28% speaks about the perfect level at which the model predicts and how reliable it can be for modern predictions.

The interface presented and developed in this work helps the model users to navigate with ease. This ensures that all sorts of people in different fields will properly employ the personality prediction tool, as it would be easy, in making friendly or other relationships. People's classification according to their personality types can help them to understand themselves and other people much better. The model also provides compatible personality types to promote unanimous relationships. This can help in effective communication, conflict resolution and relationship building.

6 Future Work

The future scope of the proposed project includes the cross-validation of the results and development of additional personality models like the Big Five (OCEAN) traits to proceed in the direction of the incorporation of current behavioural patterns, such as scrapping text from social networks or other workplace metrics that might correlate with the personality predictions. In addition, the model could be adjusted to evaluate many factors to give advice on enhancing an individual or professional development with respect to the personality profile.

References

1. Ryan, G., Katarina, P., Suhartono, D.: MBTI personality prediction using machine learning and SMOTE for balancing data based on statement sentences. Information **14**(4), 217 (2023). https://doi.org/10.3390/info14040217
2. Mushtaq, Z., Ashraf, S., Sabahat, N.: Predicting MBTI personality type with K-means clustering and gradient boosting. In: 2020 IEEE 23rd International Multitopic Conference (INMIC), pp. 1–5. IEEE, Bahawalpur, Pakistan (2020). https://doi.org/10.1109/INMIC50486.2020.9318078
3. Khan, A.S., et al.: Personality classification from online text using machine learning approach. Int. J. Adv. Comput. Sci. Appl. **11**(3), 460–476 (2020)
4. Praphulla, G.L., Kishore, I.B., Venkatesh, B., Praveen, B., Srinivasa Rao, P.: Personality prediction using machine learning techniques. In: AIP Conference Proceedings, vol. 2794, p. 020016. AIP Publishing, October 2023. https://doi.org/10.1063/5.0174316
5. Cohen, Y., Ornoy, H., Keren, B.: MBTI personality types of project managers and their success: a field survey. Proj. Manag. J. **44**(3), 78–87 (2013). https://doi.org/10.1002/pmj.21338
6. Hwangbo, S., Jang, H.: Communication competence and satisfaction in interpersonal relationship on MBTI personality patterns (Focused on 3rd military cadet). Converg. Secur. J. **16**(3_1), 11–26 (2016). Korea Convergence Security Association, pISSN: 1598-7329, eISSN: 2508-7460. https://koreascience.kr/article/JAKO201618139712757.page
7. Kim, H.S., Park, G.R.: Effects of an interpersonal relationship improvement program using MBTI: effects on self-esteem, interpersonal relations, and mental health in company employees. J. Korean Acad. Psychiatr. Ment. Health Nurs. 19(3), 261–270 (2010). https://doi.org/10.12934/jkpmhn.2010.19.3.261
8. Nelson, B.A., Stake, J.E.: The Myers-Briggs Type Indicator personality dimensions and perceptions of quality of therapy relationships. Psychother. Theory Res. Pract. Train. **31**(3), 449–455 (1994). https://doi.org/10.1037/0033-3204.31.3.449
9. Lee, K.-B., Lee, T.-Y., Lee, H.-C.: The study of emotional experience to MBTI personality type in university students. J. Korean Clin. Health Sci. **2**(3), 188–194 (2014). https://koreascience.kr/article/JAKO201430756851159.page
10. Berr, S.A., Church, A.H., Waclawski, J.: The right relationship is everything: linking personality preferences to managerial behaviors. Hum. Resour. Dev. Q. **11**(2), 133 (2001). https://doi.org/10.1002/1532-1096(200022)11:2%3c133::AID-HRDQ4%3e3.0.CO;2-T
11. Vaddem, N., Agarwal, P.: Myers Briggs personality prediction using machine learning techniques. Int. J. Comput. Appl. **175**(23), 41 (2020)
12. Kamalesh, M. D., Bharathi, B.: Analysing the personality traits using machine learning techniques. In: Proceedings of the 2022 International Conference on Innovative Computing, Intelligent Communication and Smart Electrical Systems (ICSES), Chennai, India, 2022, pp. 1–6. IEEE, New York (2022). https://doi.org/10.1109/ICSES55317.2022.9914112. https://ieeexplore.ieee.org/abstract/document/9914112
13. Bin Tareaf, R.: MBTI BERT: A transformer-based machine learning approach using MBTI model for textual inputs. In: 2022 IEEE 24th International Conference on High Performance Computing & Communications; 8th International Conference on Data Science & Systems; 20th International Conference on Smart City; 8th International Conference on Dependability in Sensor, Cloud & Big Data Systems & Applications (HPCC/DSS/SmartCity/DependSys), Hainan, China, pp. 2285–2292 (2022). https://doi.org/10.1109/HPCC-DSS-SmartCity-DependSys57074.2022.00338. https://ieeexplore.ieee.org/abstract/document/10074845
14. Choong, E.J., Varathan, K.D.: Predicting judging-perceiving of Myers-Briggs Type Indicator (MBTI) in online social forum. PeerJ **9**, article e11382 (2021). https://doi.org/10.7717/peerj.11382

15. Naik, H., Dedhia, S., Dubbewar, A., Joshi, M., Patil, V.: Myers Briggs Type Indicator (MBTI) - personality prediction using deep learning. In: Proceedings of the 2022 2nd Asian Conference on Innovation in Technology (ASIANCON), Ravet, India, pp. 1–6. IEEE (2022). https://doi.org/10.1109/ASIANCON55314.2022.9909077. https://ieeexplore.ieee.org/abstract/document/9909077
16. Maulidah, M., Pardede, H.F.: Prediction of Myers-Briggs type indicator personality using long short-term memory. Jurnal Elektronika dan Telekomunikasi **21**(2), 104–111 (2021). https://www.jurnalet.com/jet/article/view/428
17. Kelvin, K., Utomo, Y.: Overview of text based personality prediction using deep learning. Eng. Math. Comput. Sci. J. (EMACS) **6**(2), 93–100 (2024). https://journal.binus.ac.id/index.php/EMACS/article/view/11550
18. Dataset Link. https://www.kaggle.com/datasets/stealthtechnologies/predict-people-personality-types
19. Myers, I.B., Briggs, K.C.: Manual: A Guide to the Development and Use of the Myers-Briggs Type Indicator. Consulting Psychologists Press, Palo Alto (1980)
20. Goldberg, L.R.: A broad-bandwidth, public domain, personality inventory measuring the lower-level facets of several five-factor models. Pers. Psychol. Europe **7**, 7–28 (1999)
21. Friedman, J.H.: Greedy function approximation: a gradient boosting machine. Ann. Stat. **29**, 1189–1232 (2001)
22. Pennebaker, J.W., King, L.A.: Linguistic styles: language use as an individual difference. J. Pers. Soc. Psychol. **77**, 1296–1312 (1999)
23. Xu, X., Potdar, V.: Personality classification using social media data: a review. Data Sci. J. **16**, 1–14 (2017)
24. Lippi, M., Torroni, P.: Argumentation mining: state of the art and emerging trends. ACM Trans. Int. Technol. (TOIT) **16**, 1–25 (2016)
25. Chen, T., Guestrin, C.: XGBoost: a scalable tree boosting system. In: Proceedings of the 22nd ACM SIGKDD International Conference on Knowledge Discovery and Data Mining, pp. 785–794 (2016)
26. Plutchik, R.: Emotions and Life: Perspectives from Psychology, Biology, and Evolution. American Psychological Association, Washington DC (2003)
27. Jung, C.G.: Psychological Types. Routledge & Kegan Paul, London (1921)

Enhancing Emotion Detection Classification Through Bagging Techniques

K. Hemakirthiga and J. Arunadevi(✉)

Department of Computer Science, Raja Doraisingam Government Arts College, Affiliated to Alagappa University, Sivaganga, Tamilnadu, India
arunardm2015@gmail.com

Abstract. This paper investigates the efficacy of various classifiers for Emotion detection on unstructured data tweet text data. We compare the performance of traditional algorithms, including Decision Trees, K-Nearest Neighbors (KNN), and Naïve Bayes, along with their bagged versions, to assess their accuracy and robustness in classifying sentiments. The classifiers were evaluated using a dataset related to Emotion detection in tweets, with metrics including accuracy and Area Under the Curve (AUC). Our findings indicate that the Bagged Decision Tree classifier outperformed others, achieving an accuracy of 89.60% and an AUC of 0.81. The Naïve Bayes classifier also demonstrated strong performance, with an accuracy of 89.05% and an AUC of 0.81. Conversely, the Decision Tree classifier exhibited the lowest performance, highlighting the effectiveness of ensemble methods in enhancing classification accuracy. The results suggest that leveraging bagging techniques can significantly improve model performance, making them suitable for sentiment analysis tasks. This study contributes to the understanding of classifier performance in the context of unstructured data, offering insights for future research on advanced sentiment classification methodologies.

Keywords: Emotion detection · Bagging · KNN · Naïve Bayes · Decision Tree, etc.

1 Introduction

Emotion detection is an essential aspect of sentiment analysis, aiming to identify human emotions within textual data. It is widely applicable across various fields, such as analyzing customer feedback, monitoring social media content, and supporting psychological studies. Businesses utilize emotion detection to understand customer sentiment, which can inform strategies for improving products and services [1]. On social media, it helps track public sentiment trends, enabling companies and organizations to adjust their approaches based on public opinion [2]. In psychology, emotion detection assists in analyzing emotional expressions, which can be valuable for diagnosing mental health conditions or providing therapeutic support [3].

The primary goal of this study is to evaluate the performance of traditional classifiers—Decision Tree, K-Nearest Neighbor (KNN), and Naïve Bayes—and their bagged

R. Geetha et al. (Eds.): AAIMB 2024, CCIS 2653, pp. 198–208, 2026.
https://doi.org/10.1007/978-3-032-13026-6_14

variants for detecting emotions in textual data. These classifiers have distinct characteristics: Decision Trees are easy to interpret, KNN is effective for pattern recognition, and Naïve Bayes is fast and performs well with small datasets [4]. By applying bagging, an ensemble learning technique, this research investigates whether the accuracy and reliability of these classifiers improve. Bagging works by creating multiple versions of a base classifier on different subsets of data and then aggregating their results, which often reduces variance and helps to mitigate overfitting [5].

This research uses a well-known emotion detection dataset containing text samples labeled with various emotions. The study compares the performance of six classifiers: Decision Tree, KNN, Naïve Bayes, and their respective bagged versions. Performance metrics such as accuracy, precision, recall, and F1-score are employed to evaluate each classifier. This study specifically examines whether bagging improves the classifiers' ability to correctly identify emotions in text data, providing a thorough comparison of bagged and non-bagged approaches.

The structure of this paper is as follows: Sect. 2 presents a literature review that covers previous research on emotion detection techniques, the individual classifiers, and the use of ensemble methods like bagging. Section 3 details the methodology, including descriptions of the dataset, the classifiers, the bagging process, and the experimental setup. Section 4 discusses the results, comparing the performance metrics of each classifier and analyzing the impact of bagging. Finally, Sect. 5 concludes the paper by summarizing the findings, discussing limitations, and proposing directions for future research.

2 Literature Review

2.1 Emotion Detection Techniques

Emotion detection has gained significant attention in the field of natural language processing (NLP), with various techniques developed to accurately identify emotions in text. Traditional methods often rely on lexicon-based approaches, which utilize predefined lists of words associated with specific emotions [6]. These approaches are straightforward but can struggle with context and polysemy, limiting their effectiveness in complex emotional expression. Recent advancements in machine learning and deep learning have led to more robust methods for emotion detection. Machine learning techniques, such as Support Vector Machines (SVM) and Logistic Regression, have shown promising results by leveraging feature extraction methods like bag-of-words and TF-IDF [7]. Furthermore, deep learning architectures, including Convolutional Neural Networks (CNNs) and Recurrent Neural Networks (RNNs), have been employed to capture contextual information and intricate patterns in textual data [8]. These modern techniques allow for a more nuanced understanding of emotions by considering the relationships between words and their context.

2.2 Individual Classifiers

Several individual classifiers have been applied to emotion detection tasks, each with its unique advantages and limitations. The Decision Tree classifier is notable for its simplicity and interpretability, making it easy for users to understand how decisions are made [9]. However, Decision Trees can be prone to overfitting, particularly in complex datasets. K-Nearest Neighbor (KNN) is another popular choice, known for its effectiveness in identifying patterns based on distance metrics. KNN performs well in cases where class boundaries are not linear but may struggle with larger datasets due to computational complexity [10]. Naïve Bayes, based on Bayes' theorem, is a probabilistic classifier that assumes independence among features. It is computationally efficient and often yields strong performance in text classification tasks, particularly when the data is sparse [11]. Despite their individual strengths, these classifiers can be limited by their respective weaknesses, which can affect their overall performance in emotion detection tasks.

2.3 Ensemble Methods and Bagging

Ensemble methods have emerged as powerful techniques in improving the performance of classifiers by combining multiple models. Bagging, short for Bootstrap Aggregating, is one of the most widely used ensemble methods. It works by generating multiple subsets of the training data through random sampling and training a separate model on each subset. The predictions from these models are then aggregated, typically by averaging or voting, to produce a final prediction [5]. Bagging can significantly enhance the stability and accuracy of classifiers, particularly those that are sensitive to variations in training data, such as Decision Trees. Research has shown that bagging can reduce variance and improve performance in various classification tasks, including emotion detection [12]. By leveraging the strengths of individual classifiers and mitigating their weaknesses, ensemble methods like bagging offer a promising approach to enhancing emotion detection systems.

The literature on emotion detection [13–20] reveals a rich landscape of approaches and methodologies that have evolved over the years. Traditional lexicon-based techniques have paved the way for more advanced machine learning and deep learning methods, enabling a more nuanced understanding of emotional expressions in text. Individual classifiers like Decision Trees, K-Nearest Neighbors, and Naïve Bayes each present unique strengths and challenges, highlighting the importance of model selection based on the specific characteristics of the dataset and the task at hand.

However, despite the advancements in individual classifiers, their performance can be limited due to factors such as overfitting, sensitivity to data variations, and inability to capture complex relationships within the data. This is where ensemble methods, particularly bagging, play a crucial role. By combining multiple models and aggregating their predictions, bagging significantly enhances predictive performance, stability, and generalization capabilities.

3 Proposed Methodology

This section outlines the methodology employed in the research to compare the effectiveness of standard classifiers and their bagged counterparts for emotion detection in textual data. The methodology comprises several key steps: dataset preparation, feature extraction, classifier implementation, and evaluation metrics.

3.1 Dataset Preparation

The dataset used in this study is sourced from [21] and contains labeled textual data representing various emotions such as joy, sadness, anger, and fear. The dataset is preprocessed to ensure quality and relevance for emotion detection tasks. The preprocessing steps include:

- Data Cleaning: Removal of irrelevant information, including HTML tags, special characters, and excess whitespace.
- Text Normalization: Converting text to lowercase and performing tokenization.
- Stopword Removal: Eliminating common words that do not contribute to emotional sentiment, such as "the," "and," and "is."
- Label Encoding: Converting categorical emotion labels into numerical format to facilitate classification.

The dataset exhibits the following distribution of samples across the 13 emotion classes (Table 1):

Table 1. Class Distribution

S. No.	Emotion Class	Number of Samples
1	Neutral	8,638
2	Worry	8,459
3	Happiness	5,209
4	Sadness	5,165
5	Love	3,842
6	Surprise	2,187
7	Fun	1,776
8	Relief	1,526
9	Hate	1,323
10	Empty	827
11	Enthusiasm	759
12	Boredom	179
13	Anger	110

The dataset [21] used in this study comprises 40,000 tweets, each labeled with one of 13 emotion classes. It is designed to facilitate sentiment and emotion analysis in unstructured textual data, especially from social media platforms like Twitter.

3.2 Feature Extraction

To convert the textual data into a numerical representation suitable for machine learning algorithms, the TF-IDF (Term Frequency-Inverse Document Frequency) vectorization technique is employed. This technique transforms the text into a matrix of TF-IDF features, emphasizing the importance of terms that are unique to specific documents while down-weighting terms that are common across all documents. The feature extraction process involves:

- TF-IDF Vectorizer: Utilizing the TfidfVectorizer from the scikit-learn library to fit and transform the training data while transforming the test data into the same feature space.

3.3 Classifier Implementation

The research compares the performance of the following classifiers:

1. Decision Tree Classifier: A non-parametric model that uses a tree-like graph of decisions to predict the target variable.
2. K-Nearest Neighbor (KNN): A simple, instance-based learning algorithm that classifies data points based on the majority class of their k-nearest neighbors.
3. Naïve Bayes: A probabilistic classifier that applies Bayes' theorem with the assumption of feature independence.

In addition to these standard classifiers, their bagged versions are implemented to assess the impact of ensemble learning:

- Bagged Decision Tree: An ensemble method that builds multiple Decision Trees on random subsets of the training data and aggregates their predictions.
- Bagged K-Nearest Neighbor: An ensemble approach that combines predictions from multiple KNN classifiers, each trained on different subsets of the data.
- Bagged Naïve Bayes: A method that utilizes bagging to enhance the robustness and accuracy of the Naïve Bayes classifier.

The classifiers are implemented using the scikit-learn library, with hyperparameters optimized through grid search and cross-validation techniques to ensure effective model performance.

3.4 Evaluation Metrics

To evaluate the effectiveness of each classifier, the following metrics are employed:

- Micro Average Accuracy: This metric calculates the overall accuracy across all classes, treating each instance equally, which is especially useful in multi-class settings.
- Area Under the Receiver Operating Characteristic Curve (AUC-ROC): AUC measures the model's ability to distinguish between classes and provides insight into the classifier's performance across various threshold settings. AUC values range from 0 to 1, with higher values indicating better performance.

The performance of the classifiers is analyzed and compared using these metrics to draw meaningful conclusions regarding their effectiveness for emotion detection.

4 Results and Discussions

Based on the results obtained from various classifiers, here is a detailed analysis of the performance metrics: accuracy and AUC (Area Under the Curve) (Table 2).

Table 2. Results obtained from the classifiers

Classifier	Accuracy	AUC
Decision Tree	88.3591	0.59
K Nearest Neighbours	88.0784	0.68
Naïve Bayes	89.0514	0.81
Bagged Decision Tree	89.5994	0.81
Bagged K Nearest Neighbours	88.3899	0.60
Bagged Naïve Bayes	89.0283	0.76

1. Overall Performance:
   - o The Bagged Decision Tree classifier achieved the highest accuracy at 89.5994%, closely followed by the Naïve Bayes classifier at 89.0514%. This indicates that ensemble methods like bagging can enhance performance by reducing variance and improving generalization.
2. AUC Analysis:
   - o The Naïve Bayes classifier not only had a high accuracy but also an AUC of 0.81, indicating a strong ability to distinguish between classes. AUC values closer to 1 suggest better performance.
   - o The Bagged Decision Tree also scored 0.81 for AUC, showing that it effectively classifies the positive class while controlling false positive rates.
3. Performance of Individual Classifiers:
   - o The Decision Tree and K Nearest Neighbour classifiers had the lowest performance, with accuracies of 88.3591% and 88.0784%, respectively. Their AUC scores (0.59 and 0.68) suggest that while they are reasonably good at classification, they struggle with distinguishing between classes, particularly for imbalanced datasets.

o The Bagged K Nearest Neighbour and Bagged Naïve Bayes classifiers also performed well but did not surpass the accuracies of their respective non-bagged counterparts.

4. Implications for Model Selection:
   o The results indicate that using ensemble methods, such as bagging, can significantly improve the performance of classifiers. However, the choice of the base classifier still plays a crucial role in overall performance.
   o If the objective is to maximize accuracy and AUC, Bagged Decision Tree and Naïve Bayes should be considered as top candidates for implementation.

The confusion matrix provides a comprehensive view of the performance of each classifier by detailing the correct and incorrect predictions across different classes. Analyzing the confusion matrix for each classifier can help identify areas where the model is performing well and where it may need improvement (Figs. 1, 2, 3, 4, 5 and 6).

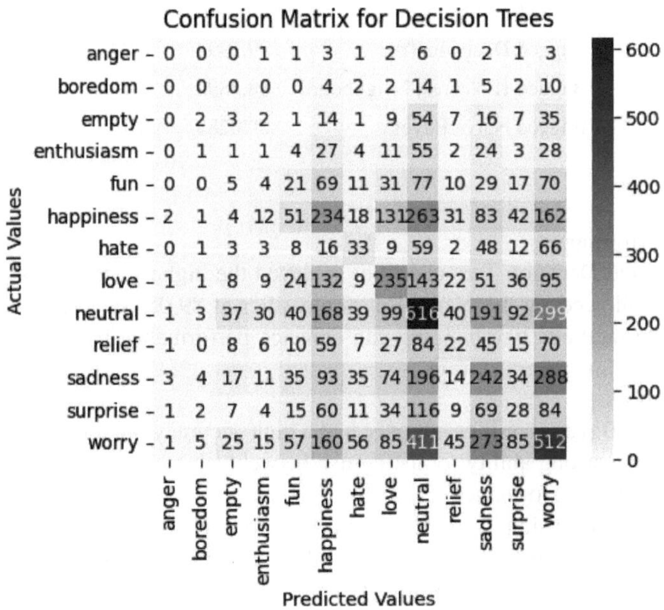

Fig. 1. Confusion matrix for Decision tree

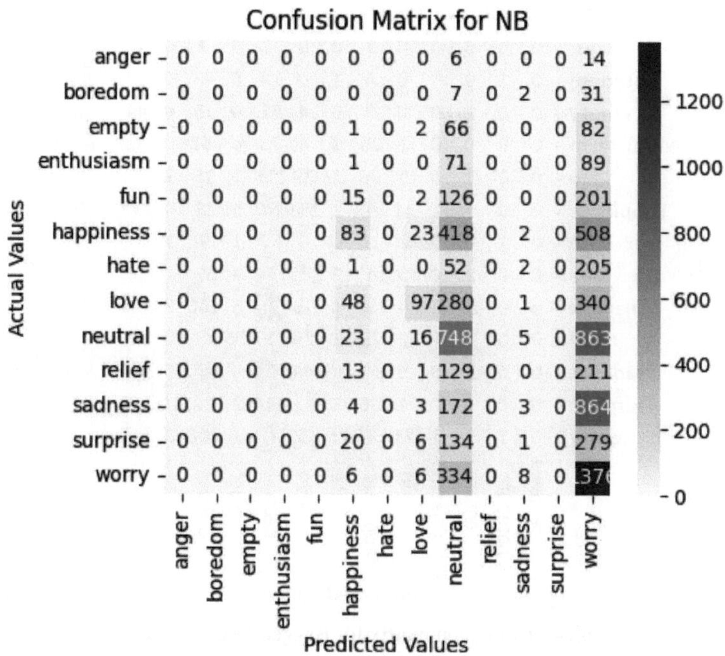

Fig. 2. Confusion matrix for Naïve Bayes

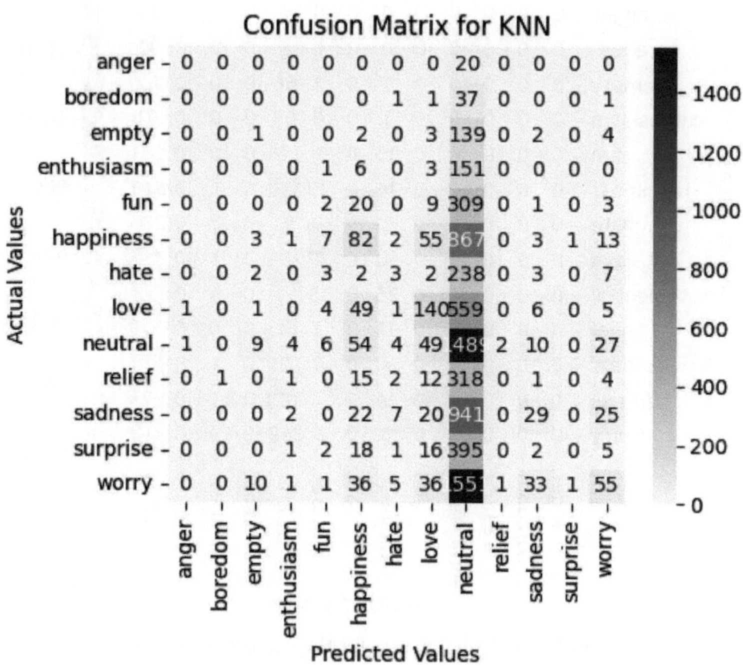

Fig. 3. Confusion matrix for KNN

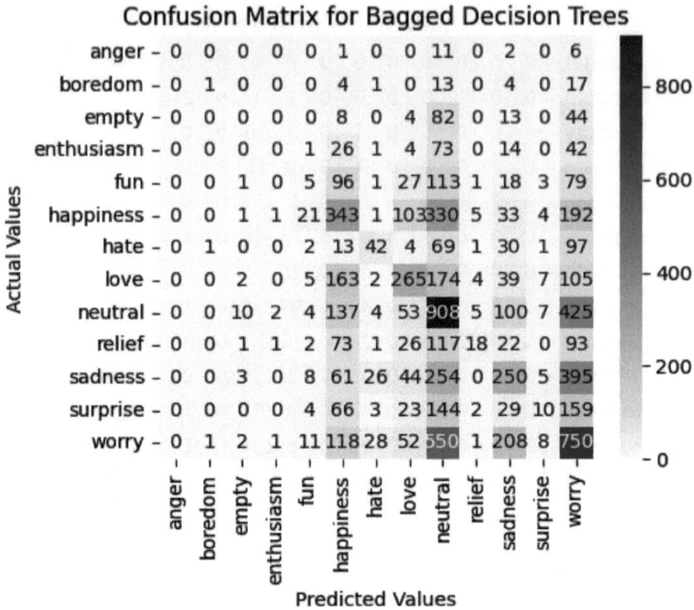

Fig. 4. Confusion matrix for Bagged Decision Tree

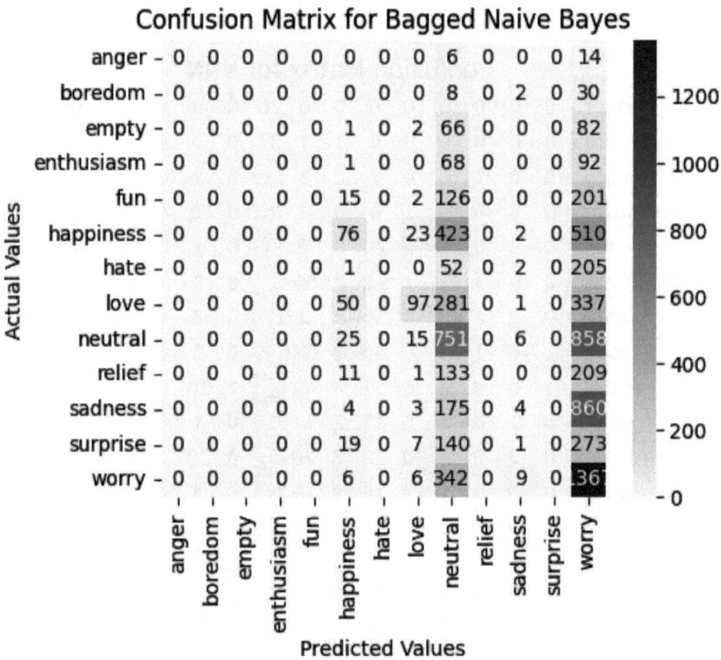

Fig. 5. Confusion matrix for Bagged Naïve Bayes

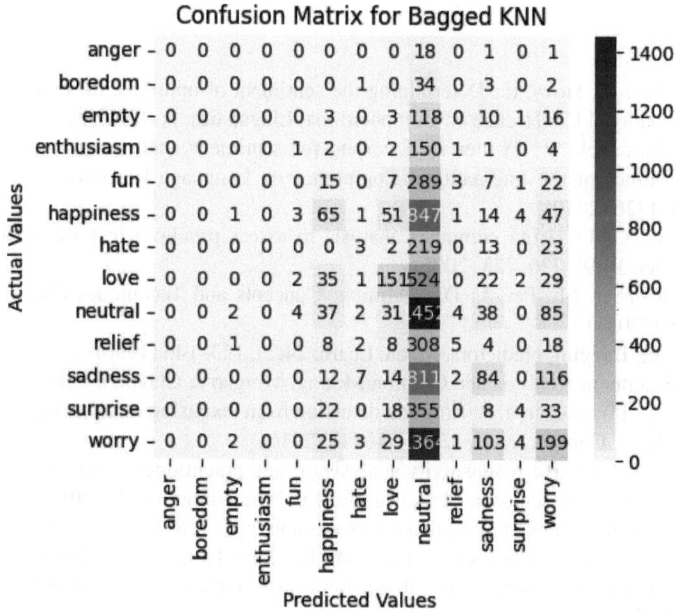

Fig. 6. Confusion matrix for Bagged KNN

5 Conclusion

This research focused on evaluating the performance of various classifiers for Emotion detection on unstructured data. The classifiers tested included Decision Trees, K-Nearest Neighbors (KNN), Naïve Bayes, Bagged Decision Trees, and Bagged KNN. The objective was to determine their effectiveness in accurately classifying sentiment, as indicated by accuracy and Area Under the Curve (AUC) metrics.

The results revealed that the Bagged Decision Tree classifier achieved the highest accuracy of **89.60%** along with an AUC of **0.81**, making it a robust choice for sentiment classification tasks. The Naïve Bayes classifier also performed well, with an accuracy of **89.05%** and an AUC of **0.81**, indicating its strong capability in handling the complexities of sentiment analysis.

Conversely, the Decision Tree classifier yielded the lowest performance, with an accuracy of **88.36%** and an AUC of **0.59**. K-Nearest Neighbors also demonstrated moderate performance, with an accuracy of **88.08%** and an AUC of **0.68**. The Bagged KNN classifier exhibited a lower accuracy of **88.39%** and an AUC of **0.60**. Notably, the Bagged Naïve Bayes classifier achieved an accuracy of **89.03%** and an AUC of **0.76**, reflecting its improvement over the standard Naïve Bayes approach.

The analysis of these results indicates that ensemble methods, particularly Bagging, can significantly enhance model performance, as evidenced by the superior results of Bagged classifiers compared to their non-bagged counterparts. This suggests that leveraging multiple learning algorithms can lead to better generalization and robustness against over fitting.

References

1. Kim, S., Lee, H., Hovy, E.: Determining the sentiment of opinions. In: Proceedings of the 20th International Conference on Computational Linguistics, pp. 1367–1373 (2004)
2. Pak, A., Paroubek, P.: Twitter as a corpus for sentiment analysis and opinion mining. In: Proceedings of the International Conference on Language Resources and Evaluation, pp. 1320–1326 (2010)
3. Pennebaker, C.M.: Using computer analysis to assess psychological states through text. Psychol. Sci. **16**(9), 720–725 (2005)
4. Han, J., Kamber, M., Pei, J.: Data Mining: Concepts and Techniques (3rd ed.). Morgan Kaufmann (2011)
5. Breiman, L.: Bagging predictors. Mach. Learn. **24**(2), 123–140 (1996)
6. Liu, B.: Sentiment Analysis and Opinion Mining. Morgan & Claypool Publishers (2012)
7. Agarwal, A., Bhattacharya, P.: Emotion detection from text using machine learning and deep learning. Int. J. Comput. Appl. **139**(9), 10–17 (2016)
8. Zhang, L., Wallace, B.: A sensitivity analysis of (and practitioners' guide to) convolutional neural networks for sentence classification. In: Proceedings of the 2015 Conference on Empirical Methods in Natural Language Processing, pp. 1128–1136 (2015)
9. Quinlan, J.R.: Induction of decision trees. Mach. Learn. **1**(1), 81–106 (1986)
10. Cover, T.M., Hart, P.E.: Nearest neighbor pattern classification. IEEE Trans. Inf. Theory **13**(1), 21–27 (1967)
11. Mitchell, T.M.: Machine Learning. McGraw Hill (1997)
12. Dietterich, T.G.: Ensemble methods in machine learning. Multiple Classifier Syst. 1–15 (2000)
13. Tiwari, D., Nagpal, B., Bhati, B.S., Mishra, A., Kumar, M.: A systematic review of social network sentiment analysis with comparative study of ensemble-based techniques. Artif. Intell. Rev. **56**, 13407–13461 (2023)
14. Kane, A., Patankar, S., Khose, S., Kirtane, N.: Transformer based ensemble for emotion detection. In: Proceedings of the 12th Workshop on Computational Approaches to Subjectivity, Sentiment and Social Media Analysis (WASSA), pp. 250–254. Association for Computational Linguistics (2022)
15. Tesfagergish, S.G., Kapočiūtė-Dzikienė, J., Damaševičius, R.: Zero-shot emotion detection for semi-supervised sentiment analysis using sentence transformers and ensemble learning. Appl. Sci. **12**(17), 8662 (2022)
16. Albu, I.-A., Spînu, S.: Emotion detection from tweets using a BERT and SVM ensemble model. arXiv preprint, arXiv:2208.04547 (2022)
17. Shah, S., Ghomeshi, H., Vakaj, E., Cooper, E., Mohammad, R.: An ensemble-learning-based technique for bimodal sentiment analysis. Big Data Cogn. Comput. **7**(2), 85 (2023)
18. Alsabhan, W.: Human–computer interaction with a real-time speech emotion recognition using ensembling techniques: 1D convolution neural network and attention. Sensors **23**(3), 1386 (2023)
19. Jiang, X.: A novel ensemble model for emotion classification in social media text. Appl. Comput. Eng. **86**, 30–41 (2024)
20. Kim, D.-H., Son, W.-H., Kwak, S.-S., Yun, T.-H., Park, J.-H., Lee, J.-D.: A hybrid deep learning emotion classification system using multimodal data. Sensors **23**(23), 9333 (2023)
21. https://www.kaggle.com/datasets/042977506d4b87fe2ce6998514bd60df9ae2bdde98acf973acfd87e758e50d68?resource=download

Assistive Technology to Translate Hand Gestures to Text Using Lenet and Alexnet

Rajendran Thanikachalam[1(✉)], N. Mohamed Imtiaz[2], K. Jagadeesh[3], A. Mani[4], B. Balamurugan[5], and S. R. Gokkul Haren[5]

[1] Department of CSE (Cyber Security), Vel Tech Rangarajan Dr. Sagunthala R&D Institute of Science and Technology, Chennai, India
tlrajen@gmail.com
[2] Artificial Intelligence & Machine Learning, Brindavan College of Engineering, Bengaluru, India
imtiaz4687@gmail.com
[3] Department of Computer Science and Engineering, Jaya Institute of Technology, Kanchipadi, Tamil Nadu, India
[4] Department of Computer Science Engineering, S.A. Engineering College, Chennai, India
[5] Department of Computer Science and Engineering, Rajalakshmi Institute of Technology, Chennai, India

Abstract. Sign language and hand gestures are vital forms of communication for the deaf and hearing communities. Assistive technologies are being developed to translate sign language movements into text to close the communication gap. This paper compares two deep learning architectures, LeNet and AlexNet, for sign language gesture recognition and text conversion using ANN and CNN. Our proposed research focuses on evaluating the performance of LeNet and AlexNet on a comprehensive dataset of sign language gestures, with a specific emphasis on accuracy. The results indicate that while both architectures show promising results, LeNet is better and more efficient for translating sign language gestures into text when you have limited computational resources. This work advances assistive technology by shedding light on the efficacy of deep learning architectures for sign language communication. Better communication tools for those who are hard of hearing may result from this research. Lenet has achieved a 97% accuracy and AlexNet has achieved 27% accuracy in recognizing hand gestures using depth images. Moreover, the researchers have tested the models on various sign language datasets to assess their generalizability and have evaluated their resilience to variations in lighting conditions, hand orientation, and signer diversity.

Keywords: Convolutional Neural Network · Artificial Neural Network · Hand Gestures · Architectures · Sign Language · LeNet · AlexNet

1 Introduction

Sign language is an essential means of communication that helps deaf and hard-of-hearing communities to engage with the outside world. However, the majority of the population does not understand hand gestures, leading to communication barriers and

R. Geetha et al. (Eds.): AAIMB 2024, CCIS 2653, pp. 209–220, 2026.
https://doi.org/10.1007/978-3-032-13026-6_15

social exclusion for many individuals. To overcome this obstacle, scientists and engineers have been creating assistive technology that can transcribe hand gestures and sign language into spoken or written language [1], improving the accessibility and inclusivity of communication. ANNs and CNNs, two deep learning techniques, have demonstrated impressive results in a variety of image identification tasks in recent years [2]. The LeNet and AlexNet architectures and its mechanisms emphasize the superiority of deep learning approaches over traditional machine learning methods for gesture recognition and sign language tasks [3].

Our study focuses on comparing the accuracy and efficiency of these models in recognizing and translating sign language gestures. The objective of this study is to assess how well deep learning algorithms perform in translating sign language and to determine which architecture works best for this task. For those who use sign language, these technologies can greatly enhance accessibility and communication by reaching high levels of accuracy and efficiency. It gives a summary of relevant research on translation and recognition of sign language, as well as an in-depth look at the architecture and methodology of our suggested system, which includes data collecting, preprocessing, and model training. The section presents experimental results and analysis of LeNet and AlexNet in sign language translation. Implications and potential applications of the proposed technology are discussed. Lastly, a review of our contributions and recommendations for more research in this area rounds up the paper's last part.

2 Literature Review

In comparison to previous deep learning methods, the research provides a better AlexNet-based gesture detection system that has higher precision and increased extraction of features capabilities [4]. In comparison to previous deep learning methods, the research provides an improved AlexNet-based gesture detection system that has greater accuracy and increased feature extraction capabilities [5]. The significance of hand motions for nonverbal interaction, their uses outside of helping people with physical disabilities, and the creation of a MATLAB framework for hand gesture identification using an artificial neural network are all covered in this study [6].

The work, which uses ResNet-50 to achieve remarkable accuracy rates, aims to close the divide in communication among hearing and non-hearing individuals by using models based on deep learning to identify hand motions for the American Sign Language (ASL) alphabet [7]. In the study, a system that uses convolutional neural networks (CNNs) to recognize the 26 alphabetical letters using finger gestures is described. The system achieves a 91% success rate in movement identification and shows excellent efficiency in gesture classification and differentiation [8].

The study offers a method that uses flexible sensors and a tiny computer with a preloaded database to translate the hand movements of people with speech impairments into text and spoken communications. To send the intended message, a predetermined database is made and kept [9].

This essay examines an investigation designed to overcome the communication gap resulting from a general lack of familiarity with ASL. It presents a real-time technique for ASL interpretation based on CNN, an artificial intelligence model that excels at processing visual input. The technique initially filters hand motions before categorizing them into distinct signals. Tested on both sets of data, the suggested model demonstrated its efficacy in hand gesture-based ASL interpretation with 99% accuracy on the MNIST database and a precision of 96% on the ASL dataset [10]. To overcome the communication gaps that exist between those who are speech-impaired and those who are not., the study investigates Hand Gesture Text Conversion. Its goal is to construct a system that removes obstacles and concentrates on building an extensive vocabulary for sign identification systems [11]. Mute persons can converse with sign language; the trouble arises when a conversationalist can not comprehend it. Although this issue has been addressed, a workable solution has not yet been discovered. To achieve high accuracy with low computational cost, the paper presents an approach for word-level recognition of sign languages using the Transformer modeling model. By calculating a person's pose and integrating strong adjustment schemes and enhancements, the system achieves groundbreaking outcomes on the WLASL and LSA64 datasets [12]. Yet developing a mobile application for it is far from a straightforward process; it requires a lot of labor, memory use, and well-calibrated design to make it happen. Employing their program, subsequently turned into a significant word. The researchers initially compared the gesture using a histogram that was connected to the sample test and extra examples that are needed by BRIEF to lessen the burden on the central processing unit and decrease its processing time.

This text is written entirely in Spanish [13], This application facilitates communication for deaf Spanish-speaking people by translating sign language into English. To improve accessibility, it has a visible interface, interpreters, and voice converters. It also recognizes various positions and motions for better communication speech tones combined with signals of body language. While body language also greatly contributes to communication, tone of voice is still a major factor in communication. In certain situations, such as when deaf and mute individuals communicate or when using traffic signals, body language can be quite significant. An outline of the Position and Positional importance is given in this publication. Using a CNN for dynamic signs, the paper presents the creation of an Indian Sign Language Recognition System (ISLRS) with an accuracy in training of 70%. The system aims to bridge educational disparities for deaf people and enhance interaction within the sign language community [14].

This research presents an example dataset for fingerspelling in Kazakh and compares several models of convolutional neural networks for recognizing sign language. LeNet and EffectiveNet function better than other topologies, indicating their potential for practical uses. The study highlights how crucial the use of deep learning is to enhancing accessibility for people with impairments [15].

This work uses computer vision and deep learning to identify American Sign Language (ASL) motions for the numbers 0 through 9. With the help of processing images and CNN, it attains an accuracy in training of 91.37% and a testing accuracy of 87.5% [16].

To improve interaction between people with speech impairments and non-signers, the research offers a CNN-based model that recognizes hand motions in Indian Sign Language and converts them into words and sounds with a success rate of 95.31%. The approach involved learning the algorithm on more than 400 photos of 10 different hand gestures, leveraging video input for real-time hand gesture classification using AlexNet, and attaining an accuracy rate of 95.31% using a 90:10 training-testing ratio.

3 Proposed Methodology

The recommended approach for using LeNet and AlexNet to translate sign language to text involves many crucial steps. First things first, gathering a variety of sign language gestures into a single dataset is the data collection process. We created the dataset using automated image generation tools that produce images at fixed time intervals with minimal manual inputs. The final dataset contains 10000 sign language images for training and 3000 images for testing. This set features static and dynamic motions that feature letters, numbers, and well-known words. Second, data preparation is essential to preparing the dataset for training. This covers operations such as data encoding, data balancing, normalization, augmentation, and picture resizing. Thirdly, LeNet and AlexNet models are trained using the preprocessed dataset through model training, which typically entails minimizing a loss function using optimizers like Adam or SGD.

3.1 Process of LENET-5

The input is sent forward via these layers, where it is transformed and pertinent features are extracted by applying different operations at each layer. LeNet-5 is trained with a dataset of labeled pictures. Typically, backpropagation and an optimization method such as stochastic gradient descent are used to minimize the difference between the predicted and actual labels [10], which allows the network to learn the ideal filter weights for each layer. For the diagrammatic depiction, see Fig. 1.

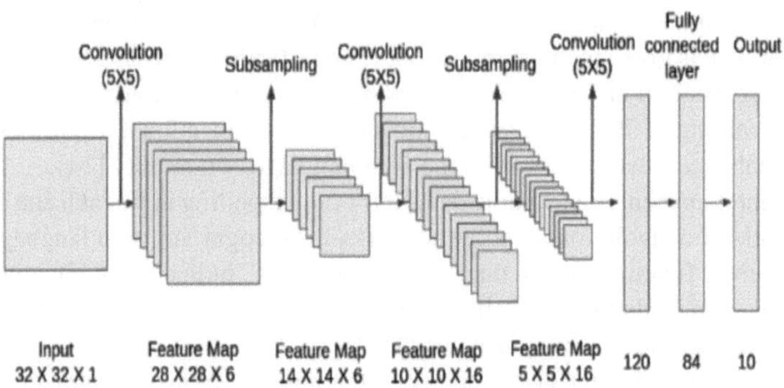

Fig. 1. Structure of Lenet Network.

List of Lenet Layers and Its Functionalities:
1) **Convolution Layers.**

(a) ***Convolution Operation*** A convolutional layer applies several filters to an input picture I with dimensions of W × H × D, where D is the number of channels, H is the height of them, and W is the width of the channel. Where the convolution layer has multiple Filters like F with dimensions F_w x F_h x D, where Fw and Fh are the filter's width and height, in that order.
(b) ***Feature Map Size*** One may compute the size of the final feature map as follows:

$$(\mathrm{W} - F_w + 2 \times \text{padding})/\text{stride} + 1 \tag{1}$$

ReLU Initiation

The activation parameter of ReLU is used component by component after every single convolution phase to introduce non-linear behavior
(2) **Subsampling (Pooling) Layers.**

(a) ***Max-Pooling Operation*** By finding the largest value inside each S × S section of the feature map, given the combined size of S × S, this approach decreases the physical dimensions of the map's features.
(b) ***Subsampling Factor*** This factor is responsible for decreasing the size of the feature maps by a predetermined amount.
(c) ***Fully Connected Layers***

Matrix multiplication: The last subsampling layer's flattened output is multiplied by a weight matrix and added with a bias factor to feed into fully connected layers.
(3) **Output Layer** Ten nodes in the result layer, which is a completely linked layer, stand in for the digits 0 to 9. The network's algorithm has identified the data being transmitted as the integer I if node I holds an integer value of 0. Following every single convolution operation, the non-linear is introduced element-by-element using an activation function called ReLU. The RBF output can be calculated as follows:

$$y_i = \sum\nolimits_j (x_j - w_{ji})^2 \tag{2}$$

The bitmap encoding of I, where i is between 0 and 9, and j is between 0 and 7*12–1, yields the formula w_{ij}.

Extensive Description of Every Layer Parameter

(1) **C1 Input Layer** This layer receives the input picture, which is commonly a grayscale image of size 32x32 pixels.

The parameters that can be trained are: (6 filters total—5 5 + 1) 6 (5 * 5 = 25 unit parameters plus one bias parameter).

There are 122304 links (5 5 + 1) 6 28 28.
(2) **S2 layer-pooling layer** The pooling process comes just after the first convolution. The following parameters are trainable: (5 5 + 1) Six (5 * 5 = 25 unit parameters) filters total—one bias parameter and five unit parameters overall. The number of linkages: (5 5 + 1) 6 28 28 = 122304.

(3) **C3 Convolutional Layer** A second convolution is carried out after the first pooling, and the results include C3, 16 10a pair of 10 data diagrams, with a convolution kernel having a value of 5 5. How can we extract sixteen feature maps from S2, which is known to have six feature maps? S2 feature maps were combined in a special way to create these 16 feature maps. Six (3 5 5 + 1) + 6 (4 5 5 + 1) + 3 (4 5 5 + 1) + 1 (6 5 5 + 1) = 1516 are the parameters that can be learnt. The number of connections is 151600 (10 10 1516).

(4) **S4 layer-pooling layer** 16 five-by-five maps of features are created by combining the 16 10a pair of 10 mappings of the C3 level in 2 by 2 pixels. There are still 16 maps of features all together in the 2 × 2 window configuration. The layer in question has a total of 32 training variables (2x16, 5x5x5x16 = 2000 linkages). Adaptable variables are as follows: distance + all weights = 2 * 16 = 32.

(5) **Convolution layer (C5).**

Convolutional layers make up the C5 layer. The dimensions of the combining kernel and the 16 S4 level pictures are the same 5 by 5, hence the image produced after mixing is 1 the variables x1. There are 120 convolutional findings from this. Connectivity and trainable parameters: 120 (165 * 5 + 1) = 48120.

(6) **Fully connected layer or F6 layer** The sixth step is completely linked. The 84 nodes in the F6 layer correspond to a 7x12 bitmap. Since 1 denotes black and -1 implies white, each symbol's black-and-white bitmap correlates to a code. For this layer, the training parameters and connection count are (120 + 1) x84 = 10164 The output of the sigmoid function comes after the offset of the vector, and the dot result of the given input and load vectors., has been calculated.

Trainable parameters: 10164 = 84 * (120 + 1).

(7) **Output Layer** The output layer of LeNet-5 consists of softmax activation which translates the network's raw output into probability scores, indicating the likelihood of each class (Table 1).

Table 1. LENET-5 LAYER

Layer	Layer	Feature Map	Size	Kernel Size	Stride	Activation
input	Image	1	32*32	-	-	-
1	Convolution	6	28*28	5*5	1	tanx
2	Average Pooling	6	14*14	2*2	1	tanx
3	Convolution	16	10*10	5*5	1	tanx
4	Average Pooling	16	5*5	2*2	2	tanx
5	Convolution	120	1*1	5*5	1	tanx
6	FC	-	84	-	-	tanx
Output	FC	-	10	-	-	soft-max

3.2 Process of ALEXNET

Image identification was transformed by a deep convolutional neural network called AlexNet given in the Fig. 2. Three fully connected layers are used for classification after five convolutional layers for feature extraction. The accuracies of earlier models are significantly improved by AlexNet, it uses ReLU for non-linear activations, dropout to reduce overfitting, and maximum pooling enabling spatially down sampling.

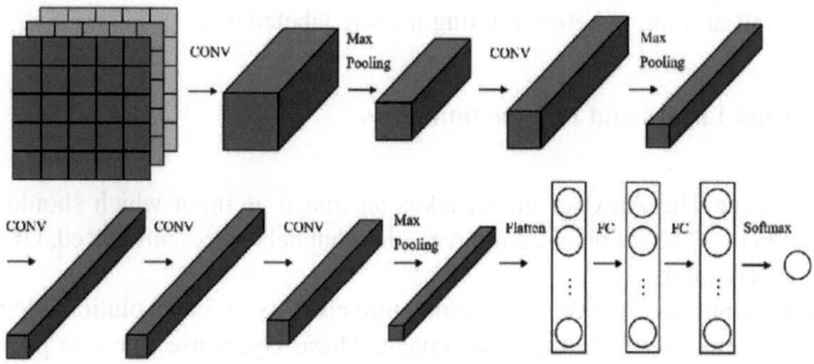

Fig. 2. Structure of Alexnet Network

(1) **ReLU Activation.**
Relu function:- f x = max (0, x).
Deep convolutional networks that use ReLU-based activation functions require fewer training iterations than those that use sigmoid or tanh-based activation functions. To illustrate this, considered the example of a four-layer convolutional network trained on the CIFAR-10 dataset.

(2) **Standardization** The ReLU activation function is commonly used in neural networks due to its advantages over e tanh and sigmoid functions in terms of computation time and avoiding the vanishing gradient problem. It's crucial to remember that the ReLU output value has no range. Which can lead to issues with normalization. An additional solution to this issue is to apply ReLU.

$$a^i_{x,y} = \frac{a^i_{x,y}}{(k + \alpha \sum_{j=(0,i-\frac{n}{2})}^{(N-1,i+\frac{n}{2})} (a^i_{x,y})^2)^\beta} \quad (3)$$

(3) **Dropout** Another well-liked method for preventing neural networks from excessive fitting is dropout. Unlike the conventional linear model, excessive fitting is avoided with a regular method. Dropout entails altering the neural network's design. While maintaining the same amount of cells in the input and output stages, a particular amount of synapses are lost at random with a predetermined probability throughout training.

(4) **Enhanced Data** When there is insufficient data for deep learning, there are often four options:

Data augmentation is a technique that artificially enhances the training set's size by using translation, flipping, and noise. It produces a batch of "new" data from the current data, which helps in improving the model's accuracy. To prevent overfitting, a regular term is included after the loss function. The disadvantage is that a manually adjusted hyperparameter is needed. Another regularization technique is dropout. It is accomplished by randomly changing certain neurons' output to zero. Unsupervised Pre-training can be done using Auto-Encoder or the convolution version of RBM. It is a technique that helps to improve the model's performance by using unlabeled data for initial training before refining it using labeled data.

List of Alexnet Layers and Its Functionalities

(1) **Input Layer** The AlexNet model takes an image as input which should have a resolution of 224x224 pixels and three color channels representing Red, Green, and Blue (RGB) colors.
(2) **Convolutional Layers** AlexNet architecture consists of 5 convolutional levels that extract information from the input image. These layers use filters to process the image and have small receptive fields. To provide irregularity to the system, it is customary to apply ReLU activation procedures after every layer of convolution.
(3) **Max Pooling Layers** After each convolutional layer, max-pooling layers are commonly added to the neural network architecture. These layers are designed to minimize the feature maps' spatial dimensions without sacrificing any of the most crucial data.
(4) **Normalization Layers** Local response normalization (LRN) layers were added to the first AlexNet design following the initial two convolutional layers.
(5) **Dropout** Dropout layers were positioned after the completely linked layers to prevent excessive fitting. These layers of drop out randomly remove some of the neurons while training, forcing the network to acquire stronger characteristics.
(6) **Fully Connected Layers** AlexNet's architecture is composed of three interconnected layers that perform classification. The fully linked layers were followed by dropout layers to avoid overfitting.
(7) **Softmax** *Activation* In AlexNet's output layer, softmax activation is frequently employed to convert raw input into probabilities that reflect each class.
(8) **Training** The momentum-based stochastic gradient descent (SGD) optimization method seems like a potent instrument for AI model training. One of the biggest and most popular datasets for picture classification tasks, so I've heard, is ImageNet. Using such a large dataset to train AlexNet must have been quite the task.

4 Result and Discussion

The outcomes of the experiment show the effectiveness of the LeNet and AlexNet architectures in translating sign language motions into text by responding with text labels. A different dataset of 600 photos was used for testing the models after they had been trained on a dataset of 2000 images. There was a significant difference in accuracy between LeNet and AlexNet and it highlights the importance of choosing the right neural network architecture for sign language recognition tasks. Despite its complexity and a larger number of parameters, AlexNet did not outperform the simpler LeNet architecture in this specific task. These findings highlight the need for additional investigation and testing to refine deep learning models for the detection of sign language, with an emphasis on comprehending the elements that lead to the disparate performance of various architectures. In the results analysis, both models' performance in converting sign language to text was presented. The models' accuracy and efficiency were contrasted. Lenet attained a 97.5% accuracy rate, whereas Alexnet registered a 27.08% accuracy rate.

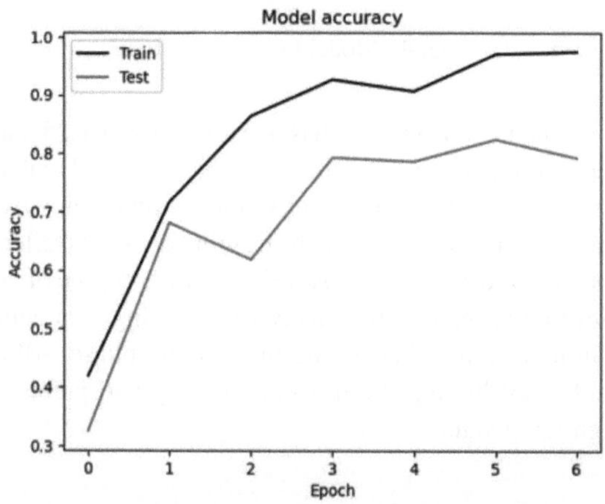

Fig. 3. Model Accuracy of Lenet.

Figure 3 shows that the accuracy of the LeNet-5 model is 97.5%. This accuracy shows how well the model was able to categorize data over the course of six epochs, including in the training and testing stages. This high accuracy shows how well the LeNet-5 architecture recognizes Sign Language motions for the numbers 0 through 9.

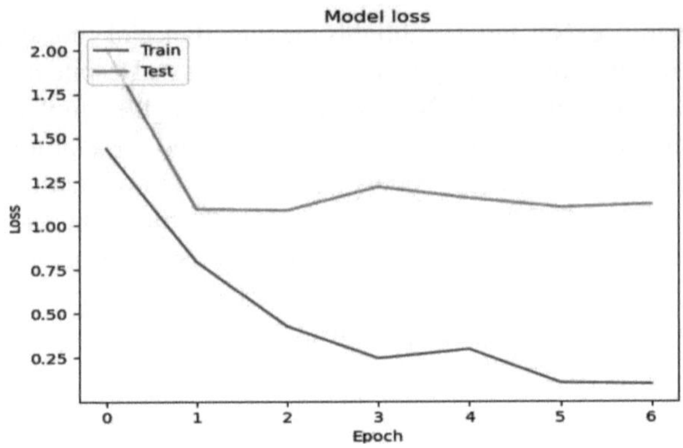

Fig. 4. Model Loss of Lenet.

As seen in Fig. 4. The LeNet-5 model has a testing loss of 1.25 and a training loss of 0.15. This shows the difference in values that were expected and real during training and testing throughout several epochs. The model's efficacy on unknown data is represented by the testing loss, while its performance on the training dataset is reflected by the training loss. While the somewhat greater testing loss reflects some degree of generalization error when applied to fresh data, the comparatively low training loss demonstrates that the model efficiently eliminates mistakes during the training phase. All things considered, these loss values aid in evaluating the model's efficacy and precision in categorizing motions used in Sign Language.

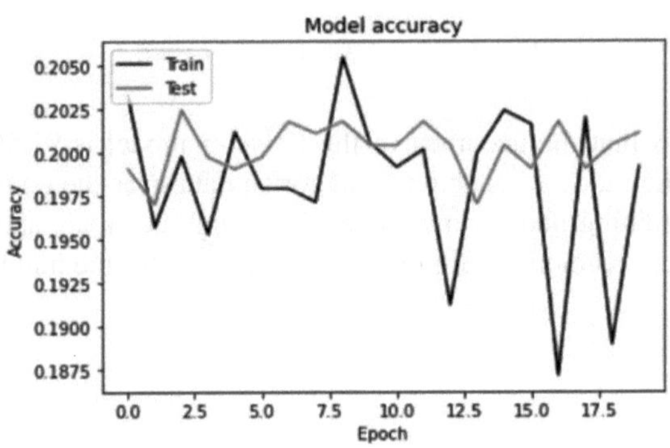

Fig. 5. Model Accuracy of Alexnet.

The AlexNet model's performance throughout several epochs in both the training and testing stages is shown in Fig. 5. The model's testing accuracy of 20% and training accuracy of 27% indicate that it has difficulty generalizing to new data. The accuracy is still rather poor even after training for more than 20 epochs, suggesting difficulties in applying what is learned from the training data to novel situations. By indicating areas for improvement in training or model design, this performance assessment sheds information on how well the AlexNet architecture recognizes motions in Sign Language.

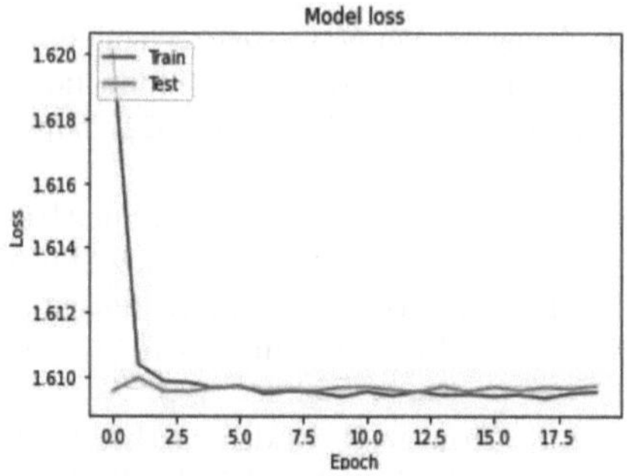

Fig. 6. Model Loss of Alexnet.

The loss of the AlexNet model during training and testing is seen over many epochs in Fig. 6. There appears to be a negligible difference between the losses on the training and testing data since the testing loss is 1.61 and the training loss is 1.59. This suggests that while the model is not overfitting to the training set, it can still have trouble generalizing to new data. Even after training for more than 20 epochs, the losses are still very substantial, suggesting that the model's capacity to reduce mistakes and correctly identify motions in Sign Language still needs to be strengthened.

5 Conclusion

In this work, we carefully examined how well LeNet-5 and AlexNet work for converting static sign language gestures into text. This research helps fill an important need in assistive devices for deaf and hearing-impaired people. LeNet outperformed AlexNet with an accuracy of 97.5%, using a dataset of 10000 photos for training and 3000 images for testing. AlexNet's accuracy was just 30%. This implies that LeNet's more straightforward design can produce better results with less processing power and datasets. Future research should expand the dataset to include the dynamic gestures and variety of sign languages. Expand the work by introducing the edge computing for low latency deployment in mobile platforms. The results highlight how crucial it is to select the appropriate neural network topology for tasks involving the recognition of sign language, with LeNet turning out to be a more useful and effective solution in this regard.

References

1. Mable, S., Mohammed, S., Yadhu, K.: Sign language translator. Nanotechnol. Percep. (2024). https://doi.org/10.62441/nano-ntp.v20i6.86
2. Roopa, B.S., Prema, K.N., Smitha, S.M.: Brief study on convolutional neural networks. Int. J. Multidisciplin. Res. (2024). https://doi.org/10.36948/ijfmr.2024.v06i05.28671
3. Ezar Al Rivan, M., Riyadi, A.G.: Perbandingan Arsitektur LeNet dan AlexNet Pada Metode Convolutional Neural Network Untuk Pengenalan American Sign Language (2021). https://doi.org/10.35143/JKT.V7I1.4489

4. Liu, R., Xiong, A., Lai, J., Zhang, H., Wu, S.: Gesture recognition based on improved AlexNet. In: 2022 2nd International Conference on Electronic Information Engineering and Computer Technology (EIECT), Yan'an, China, pp. 257–260 (2022). https://doi.org/10.1109/EIECT58010.2022.00056
5. Qin, L., Wang, S.: Traffic police gesture command recognition based on convolutional neural network AlexNet. 912–915 (2024). https://doi.org/10.1109/SPIC62469.2024.10691406
6. Murad, B., Alasadi, A.: Advancements and challenges in hand gesture recognition: a comprehensive review. IRAQI J. Elec. Elec. Eng. **20**, 154–164 (2024). https://doi.org/10.37917/ijeee.20.2.13
7. Kaushik, P., Jain, E., Gill, K.S., Chauhan, R., Pokhariya, H.S.: Deep learning for sign language recognition utilizing VGG16 and ResNet50 models. In: 2024 2nd International Conference on Sustainable Computing and Smart Systems (ICSCSS), Coimbatore, India, pp. 1355–1359 (2024). https://doi.org/10.1109/ICSCSS60660.2024.10624743
8. Nathala, S.S., et al.: A deep CNN-based hand gestures recognition using high-resolution thermal imaging. In: 2024 IEEE 19th Conference on Industrial Electronics and Applications (ICIEA), Kristiansand, Norway, pp. 1–6 (2024). https://doi.org/10.1109/ICIEA61579.2024.10665263
9. Balbin, J.R., Rapio, G.M.A., Baldivicio, B.V.: Gesture translation device using flex sensors, accelerometer and gyroscope with cloud-based mobile application. In: 2023 IEEE 15th International Conference on Humanoid, Nanotechnology, Information Technology, Communication and Control, Environment, and Management (HNICEM), Coron, Palawan, Philippines, pp. 1–5 (2023). https://doi.org/10.1109/HNICEM60674.2023.10589260
10. Verma, A., Singh, G., Meghwal, K., Ramji, B., Dadheech, P.: Enhancing Sign Language Detection through Mediapipe and Convolutional Neural Networks (CNN) (2024). https://doi.org/10.48550/arXiv.2406.03729
11. Chakravarthi, S.S., Manogna, S.V.S., Reddy, S.A., Sountharrajan, S.: GesturesSpeak: an IoT-based gesture to speech translator. In: 2024 International Conference on Emerging Techniques in Computational Intelligence (ICETCI), Hyderabad, India, pp. 88–94 (2024). https://doi.org/10.1109/ICETCI62771.2024.10704200
12. Mercanoglu, O., Keles, H.: AUTSL: a large scale multi-modal Turkish sign language dataset and baseline methods. IEEE Access **8**, 181340–181355 (2020). https://doi.org/10.1109/ACCESS.2020.3028072
13. Nasabeh, S.S., Meliá, S.: Enhancing quality of life for the hearing-impaired: a holistic approach through the MoSIoT framework. Univ. Access Inf. Soc. (2024). https://doi.org/10.1007/s10209-024-01142-x
14. Indian Sign Language Recognition using Convolutional Neural Network Rachana Patil, Vivek Patil, Abhishek Bahuguna and Gaurav Datkhile ITM Web Conference, vol. 40, p. 03004 (2021). https://doi.org/10.1051/itmconf/20214003004
15. Kenshimov, C., et al.: A comparison of convolutional neural networks for Kazakh sign language recognition. East.-Euro. J. Enter. Technol. **5**(2), 44–54 (2021). https://doi.org/10.15587/1729-4061.2021.241535
16. Chavan, S., Yu, X., Saniie, J.: Convolutional neural network hand gesture recognition for American sign language, pp. 188–192 (2021). https://doi.org/10.1109/EIT51626.2021.9491897

Analysis of Breast Cancer Detection with Gradient Boosted Tree Using Rapid Miner

V. Muruganathan[1], R. Srinivasan[2(✉)], M. Kavitha[2], and R. Kavitha[2]

[1] School of Computing, Asia Pacific University of Technology and Innovation (APU), Technology Park Malaysia, 57000 Kuala Lumpur, Malaysia
[2] Department of Computer Science and Engineering, Vel Tech Rangarajan Dr Sagunthala R&D Institute of Science and Technology, Chennai, India
rsrinivasan@veltech.edu.in

Abstract. Breast cancer is among the deadliest diseases worldwide and is the leading cause of death for women in developed nations. Considering the unique domain of medical data, the goal of this paper is using different novel algorithms to perform data analysis on the SEER Breast Cancer dataset for data mining, analysis and knowledge discovery. Data pre-processing and prediction model concepts are explained and implemented for knowledge discovery, and data mining methods are selected after analysis and evaluation on which is more suited given the type of data. Visualizations are also used for demonstrating results and methods used. The model and results can help future cancer research and patients determine correlated factors leading to breast cancer and cancer mortality.

Keywords: Data pre-processing · prediction models · Visualizations · Cancer dataset

1 Introduction

Globally, breast cancer is the most common cancer and the leading cause of death for women (Breast Cancer Research Foundation, 2016). While men can also contract breast cancer, is diagnosed far more common in women, thus it is prevalent form of gender inequality as women are more marginalized by these health problems,

The SEER Breast Cancer Dataset was chosen for data mining analysis, with the purpose of identifying cancer probabilistic risk and survival among patients. Findings from this work can promote awareness among the public for cancer prevention, reducing cancer rates. Usage of data modelling can also assist in the forecast of probability of disease control, with further analysis to support patient decision making. Early detection of breast cancer is vital for increasing survival rates through early treatment, thus data-driven machine learning is an effective approach as it provides doctors actionable insight on patient cancer condition, along with severity of condition. The aim is to create suitable models to predict patient mortality based on patient demographics and their biological condition and information, which would help future researchers on understanding factors leading to breast cancer mortality. The objective is to further understand and thus

R. Geetha et al. (Eds.): AAIMB 2024, CCIS 2653, pp. 221–236, 2026.
https://doi.org/10.1007/978-3-032-13026-6_16

educate the public on the importance of cancer prevention, by exposing and exploring the complex factors that exacerbate cancer risks. Other than that, to create further advances in treatment and diagnostic identification of breast cancer. Another objective is providing doctors actionable insight on patient condition, along with identification of cancer severity. CRISP-DM is selected as it's the standard most often used in industry data mining projects, and it's easy to implement in agile processes like SCRUM. It has an end-to-end structure, designed to remain domain-agnostic thus is suited for a wide range of applications. A public poll by the Data Science Process Alliance showed 49% of respondents use CRISP-DM (Saltz, 2020). Upon first analysis, CRISP-DM seems to be a more developed and concise version of SEMMA, however SEMMA does provide the first stages to perform data understanding, but it is more superficial and general. KDD is also used among researchers as it is accurate however it has no official website or manual, while CRISP-DM is officially documented and company-oriented (Shafique & Qaiser, 2014). Using CRISP-DM means using industry best practices. Thus, it is appropriate for studying medical data as it allows the results of the model to be easily replicated and compared by others, which is a vital aspect of peer-reviewed research. The results can be quickly verified and accepted among the larger academic field, for medical data like SEER's Breast Cancer Dataset this is crucial. CRISP-DM is also a true methodology, as each phase and tasks to perform is described in detail in the documentation which also provides concrete examples in the form of case studies (Palacios et al., 2017). Best practices usage also saves time and resources as the processes are well established, understood and documented.CRISP-DM is more suitable if knowledge of the problem is insufficient, and it does not fit within typical data mining cases (Antonia & Sara, 2018). Medical data such as SEER's Breast Cancer Dataset can be complex, thus CRISP-DM can be more suitable. It has detailed tasks and activities especially for development of the business and data understanding phases, which facilitates understanding data and data transformation (Palacios et al., 2017).

2 Literature Review

According to the research by Dag et al. (2022), a Tree-Augmented Bayesian Belief Network (TAN) with Genetic Algorithm (GA) has been proposed to study and address the business goals of uncovering the intricate relationships between breast cancer-related factors and identify the most critical variables that influence cancer survival rates and compare its performance to popular and commonly used machine learning models Gradient Boosting (XGB), Artificial Neural Networks (ANN) and Random Forests (RF). A Bayesian network is most commonly used to represent a joint distribution over a random vector with nodes and arcs making up the network. Their proposed model was able to predict breast cancer survival when the data was balanced with RUS with AUPR, Accuracy, Sensitivity, and Specificity values of 0.608, 0.813, 0.751, and 0.822, respectively. According to the findings of the study, the stage of cancer is the most relevant factor determining breast cancer survival, whereas the sequence number variable (i.e., the number of reportable tumors) is the least important in determining breast cancer survival. The researchers have also noted that the Markov blanket may have been utilized as a structural learning technique, but that would have required more comprehensive patient-level data as compared to the dataset that was being used for their research.

3 Dataset Preparation

3.1 Data Cleaning and Filtering

Upon importing, it is found that the SEER Breast Cancer Dataset has rows with missing values and empty columns. While the data has been cleaned before public release, some data still needs to be filtered. The empty column with no values was removed, while 2 rows of data were missing Survival Months and Survival Status values.

The option is to either replace the missing values with mean values, or to remove them. Since only 2 rows of data were incomplete, the decision was made to remove the 2 rows, resulting in 4024 rows of data.

Filtering for columns was done by excluding empty columns during "Import Data" process. For rows with missing values, a Filter Examples operator was used to only select rows where Survival Months and Survival Status are not missing.

3.2 Dataset Structure

There are 15 attributes record the patient tumor characteristics and demographics information, with a total of 4024 rows. 3 attributes record patient's demographic data, which are Age, Race and Marital Status. 2 attributes record whether patient survived and their survival period. The remaining 10 attributes record the tumor state and other biological information. A percentage breakdown for each attribute value is given in Table 1. The size of the dataset is small at only 4024 rows, this eases model training and data preparation.

Table 1. Table of Dataset Structure.

S. No.	Attribute	Description	Type
1	Age	Patient's age (29 to 77)	Numaric
2	Sex	Gender of patient(male-0 female-1)	Nominal
3	Cp	Chest pain type	Nominal
4	Trestbps	Resting blood pressure(in mm Hg on admission to hospital ,values from 94 to 200)	Numerical
5	Chol	Serum cholesterol in mg/dl, values from 126 to 564)	Numerical
6	Fbs	Fasting blood sugar>120 mg/dl, true-1 false-0)	Nominal
7	Resting	Resting electrocardiographics result (0 to 1)	Nominal
8	Thali	Maximum heart rate achieved(71 to 202)	Numerical
9	Exang	Exercise included agina(1-yes 0-no)	Nominal
10	Oldpeak	ST depression introduced by exercise relative to rest (0 to .2)	Numerical
11	Slope	The slop of the peak exercise ST segment (0 to 1)	Nominal
12	Ca	Number of major vessels (0-3)	Numerical
13	Thal	3-normal	Nominal
14	Targets	1 or 0	Nominal

3.3 Data Exploration and Transformation

RapidMiner is used to perform basic exploratory data analysis. Shown in Fig. 1, the predicted output would be patient survival status, the dataset has 3408 alive and 616 dead, which means the data is imbalanced with a skewed distribution, with only 15% of data related to dead patients while 85% is of alive patients. Many attributes with class imbalance can dramatically skew model performance by introducing prediction bias (Leevy et al., 2018). Thus, SMOTE up sampling will be used to increase samples of the minority class, to prevent the model predicting everything as majority class (Saad Hussein et al., 2019).

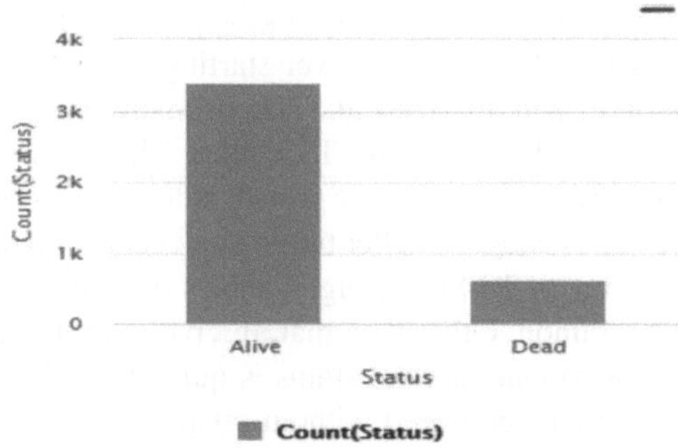

Fig. 1. Survival Status Distribution

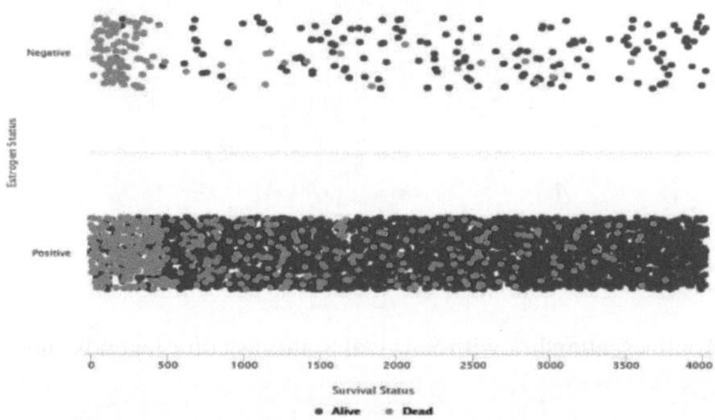

Fig. 2. Estrogen Status Scatterplot with Survival Status as colored legends

Estrogen Status scatter plot in Fig. 2 shows distribution of how majority of estrogen status values are Positive. Table 1 shows that 93.3% of Estrogen Status values are positive, meaning almost all values are positive. Attributes like this tend to lead to lower accuracy when training models, thus can be removed.

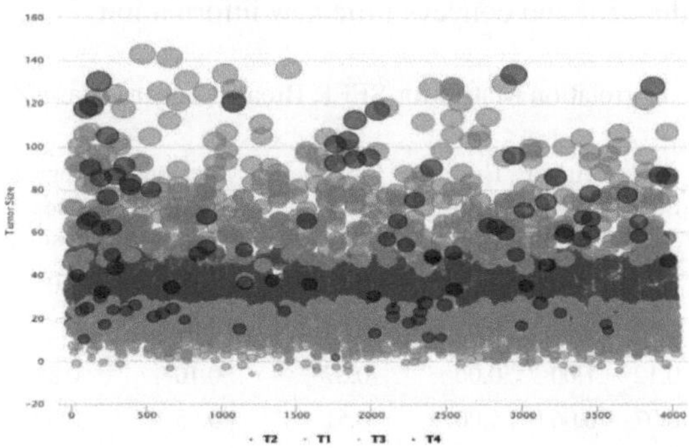

Fig. 3. Tumor Size scatterplot with T Stage as color legends and Tumor Size as size

T Stage refers to tumor size and extent of cancer spread. Figure 3 shows size ranges for T1 and T2 are mostly delineated, however starting from T3, a few of the tumor sizes are in the range of T2 and T1 sizes, and size of many of the T4 tumors are more unpredictable, ranging from T1 to beyond T3 ranges. This shows that smaller tumor sizes generally mean extent of breast cancer growth is lower, however with later stages of breast cancer such as at T3 Stage, smaller tumor sizes can still indicate serious cancer growth and even more so with T4 Stage, Fig. 4 shows that below the age of 35, cases of breast cancer are uncommon. Other than that, if a patient survived breast cancer for over 40 months, cases of patient survival status is quite high. This means after about 4 years if patient's cancer has recovered without relapse symptoms, it's very possible for a patient to survive breast cancer.

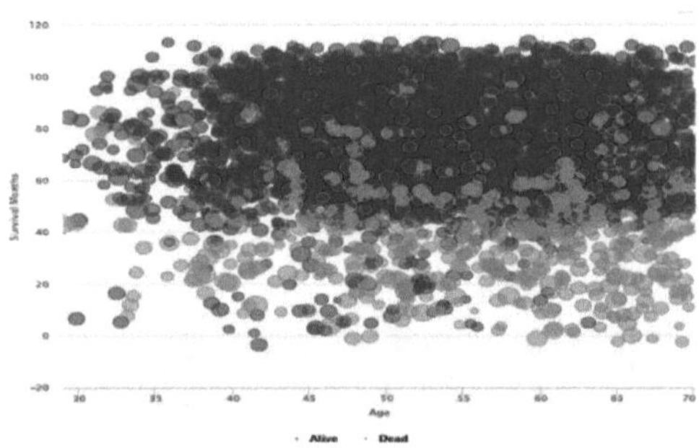

Fig. 4. Survival Months scattershot with Survival Status as color legends and Tumor Size as size.

3.4 Feature Selection

Feature selection is essential especially for larger datasets to reduce training time, by reducing the number of attributes. Even for smaller datasets, feature selection can reduce data overfitting by removing correlation between attributes. With 14 features for determining output label (Survival Status), some are likely to correlate significantly with target output, therefore feature selection is performed. With correlation between attributes (multicollinearity), information of one attribute is partially contained in another attribute, meaning data is redundant and conveys little new information.

Table 2. Correlation Matrix for SEER Breast Cancer Dataset Attributes

Attributes	Age	A Stage	Tumor Size	Estrogen Status	Progesterone Status	Regional Node Examined	Regional Node Positive	Survival Months	Survival Status
Age	1.00	–0.02	–0.08	-0.06	0.02	–0.03	0.01	–0.01	0.06
A Stage	–0.02	1.00	0.12	0.07	0.03	0.06	0.23	–0.07	0.10
Tumor Size	–0.08	0.12	1.00	0.06	0.07	0.10	0.24	–0.09	0.13
Estrogen Status	–0.06	0.07	0.06	1.00	0.51	0.05	0.09	–0.13	0.19
Progesterone Status	0.02	0.03	0.07	0.51	1.00	0.02	0.08	–0.10	0.18

(*continued*)

Table 2. (*continued*)

Attributes	Age	A Stage	Tumor Size	Estrogen Status	Progesterone Status	Regional Node Examined	Regional Node Positive	Survival Months	Survival Status
Regional Node Examined	–0.03	0.06	0.10	0.05	0.02	1.00	0.41	–0.02	0.04
Regional Node Positive	0.01	0.23	0.24	0.09	0.08	0.41	1.00	–0.14	0.26
Survival Months	0.01	0.07	0.09	–0.13	–0.10	–0.02	-0.14	1.00	-0.48
Survival Status	0.06	0.10	0.13	0.19	0.18	0.04	0.26	-0.48	1.00

Correlation matrices cannot be calculated using nominal attributes thus they are excluded from the correlation matrix in Table 2. Diagonal entries are 1, because an attribute's correlation to itself is always. Table 2 shows low correlation between most pairs of attributes. For regional nodes, it can be explained as patients who had more regional nodes examined lead to more chances of identifying positive nodes. High correlation between Survival Period and Survival Status means Survival Period is an important attribute in determining a patient's Survival State. Estrogen Status however has a correlation rate of 0.51 with Progesterone Status. This high correlation rate could result in overfitting. While accuracy of model usually doesn't get affected, quality of interpreting the independent attributes is affected. Therefore, Estrogen Status attribute is dropped, as it has slightly higher correlation with other attributes compared to Progesterone Status.

3.5 Data Transformation

Figure 5 shows all the RapidMiner operators performed on the SEER Breast Cancer Dataset, which includes preprocessing and transformations. Rows with missing values are removed, Estrogen Status attribute is dropped using Select Attributes, and Survival Status is Set Role as "label". New attributes are generated from Marital Status and Race attributes using one hot encoding. One hot encoding (also known as dummy encoding) is done for categorical attributes without ordinal relationships, where the columns are split into multiple columns. Polynomial data are converted to numeric attributes (integers) in new columns to prevent information loss of ordinal placement (Kotu & Deshpande, 2019). This is done by mapping Grade, 6th Stage, T Stage and N Stage to corresponding ordinal values, then parsing as integers. A Stage and Progesterone Status are converted to binomials. With data transformation the first step is normalization, attributes Age, Survival Months, Tumor Size, Regional Node Examined and Regional Node Positive are normalized using Z-transformation. Normalization is done because the attributes are on different scales, otherwise attributes on a lower scale might get diluted due to other attributes having larger values on a larger scale, normalization ensures all data column values lie in the same range. Z-transformation is also known as statistical normalization, it is less influenced by extreme outlier values and works by subtracting means of data and dividing by standard deviation, resulting in data distribution with zero mean and variance of one (Kotu & Deshpande, 2019).

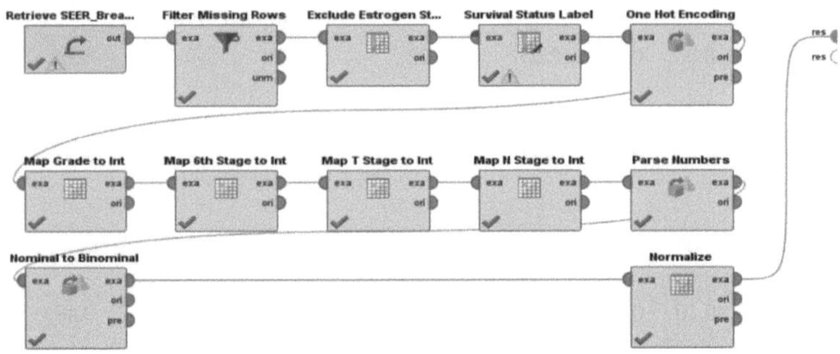

Fig. 5. RapidMiner Process Flow for Data Processing and Transformation

3.6 Type of Prediction - Gradient Boosted Tree

The type of prediction that would be used for the SEER breast cancer dataset would be the classification category. As written in the group part of the assignment, the aim of the research is to ultimately predict the breast cancer patients' mortality rate through the processed dataset which contains information such as the patients' demographics along their biological condition. The SEER breast cancer dataset also has a lot of data where the data type belongs to a boolean type where it is either positive or negative, black or white and alive or dead etc. Besides that, the mortality of humans which the group is predicting are usually put into two categories which are either alive or dead and in the SEER breast cancer dataset, the 'alive' or 'dead' status belongs to a boolean type, hence making classification the better selection for the type of prediction used in the assignment. Furthermore, various research and studies have proved that the SEER breast cancer dataset is suitable for solving classification problems. The modeling technique that has been selected and will be implemented is the Gradient Boosted Tree, also known as Gradient Boosted Decision Tree by others.

3.7 Data Mining Algorithms

The data mining algorithm which has been selected is the Gradient Boosted Tree (GBT), also known as Gradient Boosted Decision Tree (GBT). The GBT is essentially an ensemble technique which combines the use of gradient boosting and decision trees to ultimately form the GBT. The GBT is also considered as a sequential decision tree model (Rezazadeh et al., 2022). The research by Rezazadeh et al. (2022) also mentioned the fact that the GBT is widely used to perform various machine learning tasks due to its capability of providing a high accuracy and efficiency. The GBT also functions by combining multiple "weak" learner classifiers to form a strong classification model in the form of an ensemble using gradient boosting. A study by Ponnuraja et al. (2017) has conducted research to produce an accurate model which is able to classify the breast cancer patients' survival through the use of data mining techniques. The researchers used the SEER breast cancer dataset and they have identified seven important features that will be used for analysis. Pre-processing to remove missing data is also done before the decision tree algorithm is applied to test the data. Result of the algorithm shows that the mortality rate of female patients at the age of 42 to 52 is classified to be 95.1% and it is similar to the results produced by male patients. The test revealed that the highest survival rate is 96.4% while the highest death rate is 95.1%, concluding that female patients at the age of 42 to 52 and male patients at age less than 42 are classified as high-risk groups.Data pre-processing has been done to improve the performance of the models. Results of the study have shown that the GLM model displayed an accuracy of 99.4% followed by GBDT with 98.7% and finally LR with 98.2%. Despite the GLM model having the highest accuracy, the researchers have stated that the GBDT model is more measurable, effective, strong and flexible.

3.8 Model Evaluation, Analysis and Recommendations

Based on Fig. 7, the GBT model achieved an accuracy of 87.80% with the class recall producing a rate of 92.58% for alive patients and 61.63% for dead patients. As for the class precision, the model displayed a result of 92.99% for alive patients and 59.90% for dead patients. However, despite the model producing an accuracy of 87.80%, the credibility of the accuracy is judged due to the class recall and class precision for the dead class being very low when compared to the alive class. This low performance may be due to the alive patients data having much more number than the dead patients data. The alive patients data have a total number of 3408 while the dead patients data have a total number of 616, showing a huge contrast. Figures 7.2 to 7.4 shows the process flow of the GBT model (Figs. 8, 9, 10, 11, 12, 13, 14 and 15).

accuracy: 87.80% +/- 1.58% (micro average: 87.80%)

	true Alive	true Dead	class precision
pred. Alive	3155	238	92.99%
pred. Dead	253	378	59.90%
class recall	92.58%	61.36%	

Fig. 6. Performance of base GBT model

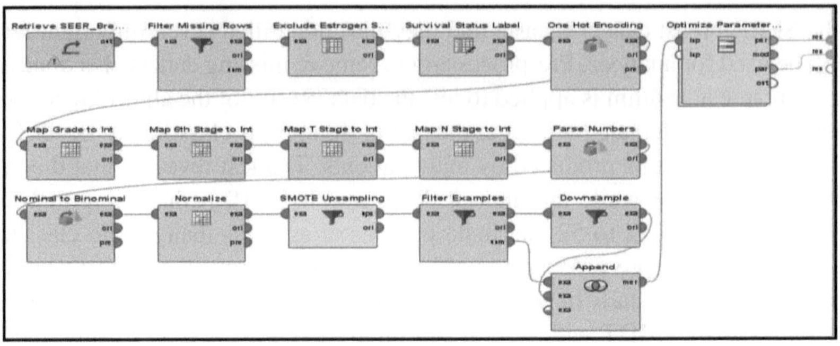

Fig. 7. GBT Model Process Flow with Up sampling and Down sampling

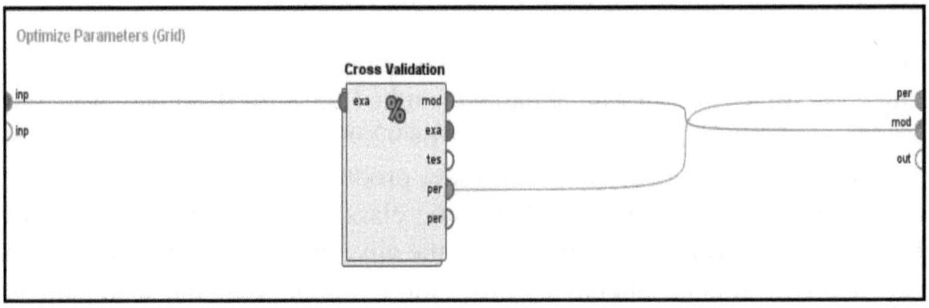

Fig. 8. Inside Optimize Parameter Function

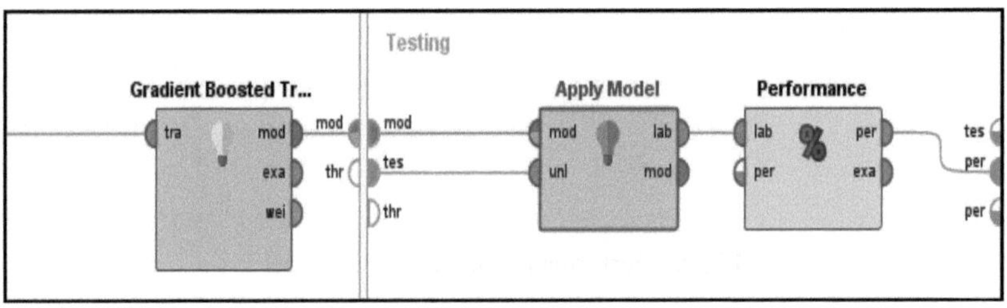

Fig. 9. Inside Cross Validation Function

accuracy: 83.02% +/- 2.16% (micro average: 83.02%)

	true Alive	true Dead	class precision
pred. Alive	2209	391	84.96%
pred. Dead	291	1125	79.45%
class recall	88.36%	74.21%	

Fig. 10. Performance of GBT after up sampling and down sampling

To solve the previous issue, the upsampling and downsampling process has been added as seen in Fig. 7.2. After going through the upsampling process to increase the number of patients who are dead and the downsampling to decrease the number of patients who are alive so that the data can be more balanced and optimized, the GBT model achieved an accuracy of 83.02% with a class recall of 84.96% for alive patients and 79.45% for dead patients in Fig. 7.5. As for the class recall, the model achieved 88.36% for alive patients and 74.21% for dead patients respectively. This shows a slight improvement to the class recall and precision for dead class, but it is noticeable that the class recall and precision for the alive class has been reduced due to the downsampling of the alive patients data.

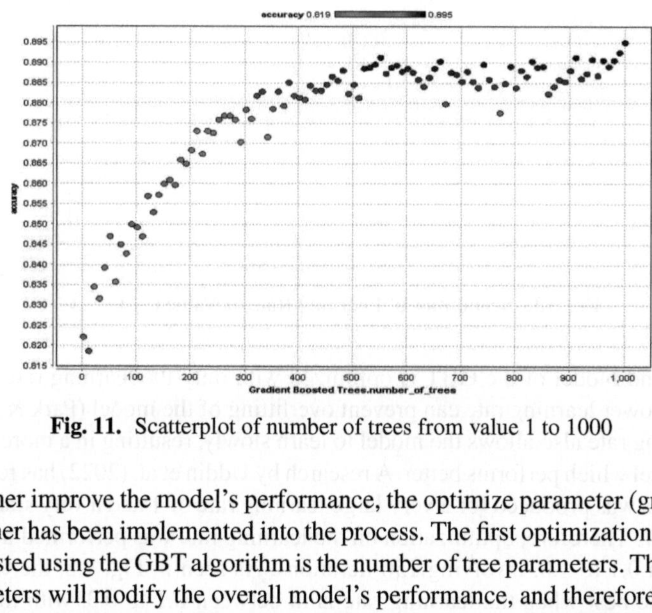

Fig. 11. Scatterplot of number of trees from value 1 to 1000

To further improve the model's performance, the optimize parameter (grid) function in rapidminer has been implemented into the process. The first optimization method that has been tested using the GBT algorithm is the number of tree parameters. The number of tree parameters will modify the overall model's performance, and therefore it is chosen as one of the parameters to be optimized. However, it was stated that the number of trees should be carefully selected as having too many trees can lead to the emergence of too many weak learners in the model, which would then result in overfitting (Gaurav, 2021). Therefore, the number of tree parameter have been tuned to a value of 1 to 1000 with 100 iterations involved. As seen in Fig. 6 , the accuracy of the model increases with the increase in the number of trees, with the range of 800 to 1000 trees producing the better accuracy.

accuracy: 89.52% +/- 1.27% (micro average: 89.52%)

	true Alive	true Dead	class precision
pred. Alive	2299	220	91.27%
pred. Dead	201	1296	86.57%
class recall	91.96%	85.49%	

Fig. 12. GBT performance with optimized number of trees

In Fig. 7.7, the GBT model optimized with only the number of trees has produced an accuracy of 89.52% with a class precision of 91.27% for alive patients and 86.57% for dead patients. As for class recall, the model produced 91.96% for alive patients and 85.49% for dead patients. An improvement to the class recall and class precision for both the dead and alive class can be observed. For instance, the class recall for the dead class have improved from 74.21% from the previous model to 85.49% with the current model. A similar improvement can be seen in the alive class as well.

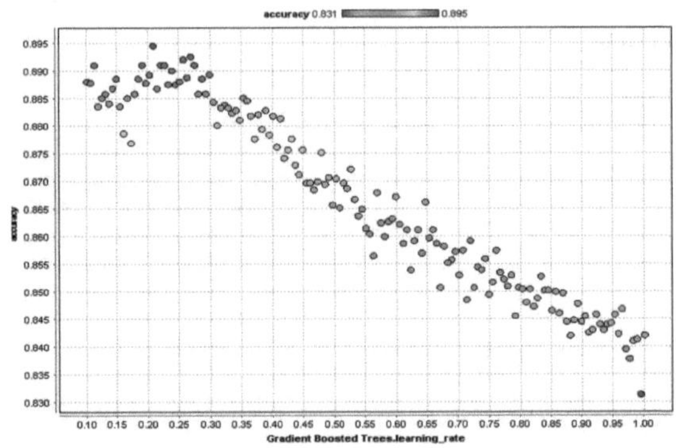

Fig. 13. Scatterplot of Learning Rate of Value 0.1 to 1.0

The second model of the GBT is optimized with only the learning rate parameter. Typically, a lower learning rate can prevent overfitting of the model (Park & Ho, 2018). A low learning rate also allows the model to learn slowly, resulting in a more robust and efficient model which performs better. A research by Uddin et al. (2022) has revealed that very high loss was produced for a very large learning rate of 1 and a very small learning rate of 0.0001. Therefore, optimization on the learning rate was performed using values ranging from 0.1 to 1.0, involving 150 iterations. As seen in Fig. 7.8, the GBT model performed the best when the learning rate is in between 0.1 to 0.3, with the accuracy gradually decreasing with the increase in learning rate. This is because a high learning rate may result in the model converging too quickly on a suboptimal solution, which causes poor performance despite the model learning quickly. However, the low learning rate results in a long training time and it is extremely computational heavy despite its higher accuracy.

ParameterSet

```
Parameter set:

Performance:
PerformanceVector [
-----accuracy: 89.47% +/- 1.49% (micro average: 89.47%)
ConfusionMatrix:
True:   Alive   Dead
Alive:  2328    251
Dead:   172     1265
]
Gradient Boosted Trees.learning_rate    = 0.20800000000000002
```

accuracy: 89.47% +/- 1.49% (micro average: 89.47%)

	true Alive	true Dead	class precision
pred. Alive	2328	251	90.27%
pred. Dead	172	1265	88.03%
class recall	93.12%	83.44%	

Fig. 14. Performance of GBT with optimized learning rate

As seen in Fig. 7.9, the GBT model performed the best with a learning rate of 0.208. This proved that a lower learning rate produces a better model at the cost of training time It managed to achieve an accuracy of 89.47% with the class recall being 93.12% and 83.44 for alive patients and dead patients respectively. The class precision produced a 90.27% and 88.03% for alive patients and dead patients respectively as well. This is also an improvement to the model with only up sampling and down sampling without any optimization. For instance, the class recall for the dead class has improved from 74.21% to 83.44% while the class recall for the alive class improved from 88.36% to 93.12%.

ParameterSet

```
Parameter set:

Performance:
PerformanceVector [
-----accuracy: 89.47% +/- 1.31% (micro average: 89.47%)
ConfusionMatrix:
True:    Alive   Dead
Alive:   2298    221
Dead:    202     1295
]
Gradient Boosted Trees.number_of_trees   = 80
Gradient Boosted Trees.learning_rate     = 0.28072
```

accuracy: 89.47% +/- 1.31% (micro average: 89.47%)

	true Alive	true Dead	class precision
pred. Alive	2298	221	91.23%
pred. Dead	202	1295	86.51%
class recall	91.92%	85.42%	

Fig. 15. Performance of GBT with combined optimization of learning rate and number of tree

When the combined optimization of number of trees and learning rate was performed with the learning rate ranging from 0.01 to 0.3 and the number of trees ranging from 1 to 100 having 50 iterations each, the result can be observed in Fig. 7.10. The combined optimization on the GBT model produced an accuracy of 89.47% with the class recall showing 91.92% and 85.42% for alive and dead patients respectively. The class precision on the other hand showed a result of 91.23% and 86.51% for alive and dead patients respectively.

Table 3. Performance comparison on the model variations of the GBT

	Model 1 (Base Model)	Model 2 (Upsampling + Downsampling)	Model 3 (Upsampling + Downsampling + Optimized number of tree)	Model 4 (Upsampling + Downsampling + Optimized learning rate)	Model 5 (Upsampling + Downsampling + Optimized number of tree + optimized learning rate)
Accuracy	87.80%	83.02%	89.52%	89.47%	89.47%
Precision (alive)	92.99%	84.96%	91.27%	90.27%	91.23%
Precision (dead)	59.90%	79.45%	86.57%	88.03%	86.51%
Recall (alive)	92.58%	88.36%	91.96%	93.12%	91.92%
Recall (dead)	61.36%	74.21%	85.49%	83.44%	85.42%

The comparison on all the variations of the GBT model is shown in Table 3. As can be observed, the performance of the GBT model improved with more iterations. The precision of the dead class has improved from 59.90% in the base model to 88.03% in model 4. The recall rate has also improved from 61.36% to 85.49% in model 3. Overall, an improvement can be seen for both the precision and recall of the dead class over the other 4 models. This is likely due to the help of the upsampling and downsampling process, which helps to balance the dataset. The optimization of two separate parameters also greatly improves the overall performance of the GBT model in general.

However, in terms of consideration for the computational cost and time required for training of model, model 3 or model 4 would be the optional choice as it takes less time for training while at the same time, they produce a good overall performance, with improvements on the base mode as well as the model with upsampling and downsampling performed. Despite model 5 having slightly better results, it takes a long time of training due to the use of a combination of two parameters. When compared with Fitrah Umami & Sarno (2020) and Ponnuraja et al. (2017), their model has a higher accuracy, but a full comparison cannot be performed as both of their experiments do not include precision and recall into their final results.

4 Conclusion

With the comparisons and evidence presented in the above, it is evident that a model to accurately predict patient mortality based on patient demographics and their biological condition and information using the factors leading to breast cancer mortality is created. Gradient Boosted Trees is the ideal algorithm for prediction of patient mortality on the SEER Breast Cancer Dataset, future work can be performed on other datasets to determine the performance and suitability of the model. Other algorithms, Random Forest, Decision Tree, and Generalized Linear Model were also fitted and compared in this study, with justifications provided for the selection of models, along with interpretations of their results. The results of this study can helpfully educate the public on cancer awareness and help further research into breast cancer research.

References

Al-Salihy, N.K., Ibrikci, T.: Classifying breast cancer by using decision tree algorithms. In: Proceedings of the 6th International Conference on Software and Computer Applications - ICSCA 2017, 6(1), 144–148 (2017). https://doi.org/10.1145/3056662.3056716

Anisha, P.R., Kishor Kumar Reddy, C., Apoorva, K., Meghana Mangipudi, C.: Early diagnosis of breast cancer prediction using random forest classifier. IOP Conf. Ser. Mater. Sci. Eng. **1116**(1), 012187 (2021). https://doi.org/10.1088/1757-899x/1116/1/012187

Antonia, D., Sara, R.: Evaluating Frameworks for Implementing Machine Learning in Signal Processing: a Comparative Study of CRISP-DM, SEMMA and KDD [Thesis] (2018). http://urn.kb.se/resolve?urn=urn%3Anbn%3Ase%3Akth%3Adiva-235408

Breast Cancer Research Foundation: Breast Cancer Statistics And Resources. Breast Cancer Research Foundation (2022). https://www.bcrf.org/breast-cancer-statistics-and-resources/

Dag, A.Z., Akcam, Z., Kibis, E., Simsek, S., Delen, D.: A probabilistic data analytics methodology based on Bayesian Belief network for predicting and understanding breast cancer survival. Knowl.-Based Syst. **242**, 108407 (2022). https://doi.org/10.1016/j.knosys.2022.108407

De Rosa, R., Cesa-Bianchi, N.: Confidence decision trees via online and active learning for streaming data. J. Artif. Intell. Res. **60**(1), 1031–1055 (2017). https://doi.org/10.1613/jair.5440

Diami, K., Hazzazi, N.: Breast Cancer Dataset Analytics. In: Big Data Analytics: 9th International Conference, pp. 13–20 (2020). http://www.thinkmind.org/index.php?view=article&articleid=data_analytics_2020_2_10_60008

Fitrah Umami, R., Sarno, R.: Analysis of classification algorithm for wisconsin diagnosis breast cancer data study. In: 2020 International Seminar on Application for Technology of Information and Communication (ISemantic) (2020). https://doi.org/10.1109/isemantic50169.2020.9234295

García-Munoz, S., Laird, C.D., Realff, M.J.: Proceedings of the 9th International Conference on Foundations of Computer-Aided Process Design. Elsevier (2019)

Gaurav: An Introduction to Gradient Boosting Decision Trees. Machine Learning Plus, 12 June 2021. https://www.machinelearningplus.com/machine-learning/an-introduction-to-gradient-boosting-decision-trees/#Gradient-Boosting-Decision-Trees

Kangralkar, S.: Regularization and Cross-Validation — How to choose the penalty value (lambda). Analytics Vidhya 26 August 2021. https://medium.com/analytics-vidhya/regularization-and-cross-validation-how-to-choose-the-penalty-value-lambda-1217fa4351e5

Kim, H.J., Kim, S., Freedman, R.A., Partridge, A.H.: The impact of young age at diagnosis (age <40 years) on prognosis varies by breast cancer subtype: a U.S SEER database analysis. Breast **61**, 77–83 (2022). https://doi.org/10.1016/j.breast.2021.12.006

Epileptical Seizure Classification from Electroencephalography Signals and 2-D Images: A Comprehensive Approach

M. Anita[1](✉), A. Meena Kowshalya[2], K. Johny Elma[1], A. Muthuram[3], M. A. Y. Peer Mohamed Appa[4], and P. Sivasakthi[1]

[1] Department of Computing Technologies, School of Computing, SRM Institute of Science and Technology, Kattankulathur, Chennai, India
anita.21820@gmail.com

[2] Department of CSE, Government College of Technology, Coimbatore, India
meenakowsalya@gct.ac.in

[3] Department of Aerospace Engineering, SRM Institute of Science and Technology, Kattankulathur, India
muthuraa@srmist.edu.in

[4] Department of CSE School of Computing, Vel Tech Rangarajan Dr. Sangunthala R&D Institute of Science and Technology, Morai, India
peer.appa@gmail.com

Abstract. Epilepsy, a chronic neurological disorder that impacts countless individuals globally, is characterized by recurring seizures. Accurate and timely classification of different seizure types is crucial for effective medical intervention and patient care. By employing a deep stacking approach, this paper aims to enhance the accuracy and precision of seizure classification, which can significantly improve treatment strategies and outcomes for individuals with epilepsy. The paper aims to enhance the effectiveness of EEG signals, the primary diagnostic tool for epilepsy. In addition to analyzing EEG data, this paper incorporates images derived from these signals, providing a multi-modal perspective that captures the diverse aspects of the underlying condition.

Keywords: Epilepsy · Seizure Classification · EEG Signals · CNN · LSTM · HVD · CWT

1 Introduction

Seizures, unpredictable bursts of electrical activity in the brain, are a pressing worry for countless people around the globe. These neurological occurrences can greatly impact a person's day-to-day, jeopardizing their safety, mental health, and general sense of wellness. Identifying and categorizing seizures accurately is essential in the realm of neurology, and has a direct impact on how patients are treated, their care, and their long-term outlook. The process of seizure classification involves the careful categorization of

R. Geetha et al. (Eds.): AAIMB 2024, CCIS 2653, pp. 237–256, 2026.
https://doi.org/10.1007/978-3-032-13026-6_17

seizures into specific types, taking into consideration characteristics such as their clinical manifestations, EEG readings, and other diagnostic markers. This differentiation is crucial for healthcare professionals, as it allows for customized treatment plans, accurate prediction of recurrence probabilities, and overall improved patient care. Furthermore, seizure classification plays a vital role in the advancement of better therapies, ultimately leading to a better quality of life for those suffering from epilepsy. The significance of accurately classifying seizures goes beyond the medical field. In fact, individuals with epilepsy depend on the precise categorization of their seizures to make informed decisions about their conduct of life. Moreover, progress in seizure classification plays a crucial role in the creation of medical devices like neuro-stimulators and seizure detection systems, providing timely interventions and ensuring greater safety for those with epilepsy. Seizure classification is an essential link between the medical community, individuals living with epilepsy, and society as a whole. Its role is crucial in improving healthcare, ensuring safety, and enhancing the quality of life for people impacted by seizures.

The purpose of our work comprises of 4 major goals. Our primary goal is to design a powerful categorization outline from a cutting-edge methodology. Our focus will be on developing a system that is specifically and efficiently analyzes both EEG signals and images, leading to the accurate classification of various seizure types. To accomplish this, we intend to employ variety of advanced methods namely the t-SNE (t-distributed Stochastic Neighbor Embedding), Long Short-Term Memory (LSTM), CWT –Continuous Wavelet Transform, HVD - Hilbert Vibration Decomposition, and CNN (Convolutional Neural Network). By combining these components to create a flexible and high-accuracy model for accurately distinguishing between types of seizures. Our second goal, is to optimize our classification model's performance by aiming for the best possible levels of specificity, sensitivity, and accuracy. This will require a meticulous process of fine-tuning model parameters, carefully selecting algorithms that best suit the task, and implementing strategies to tackle imbalanced datasets. Our third goal is to test and assess the efficacy of the classification model we have built. This involves evaluating its performance on various datasets that represent a wide range of patient populations and data sources. We employ comprehensive assessment tools to fully grasp the model's advantages and disadvantages as well as its suitability for implementation in practical situations. Our fourth goal focuses on making seizure classification systems practical for use. To make it easily accessible and understandable to healthcare professionals, developed an intuitive user interface. Within the field of neurology, categorizing different forms of seizures is an important yet difficult undertaking. Epilepsy presents complex problems for medical practitioners due to its characteristic of repeated and unexpected episodes. Correct classification of seizure types is critical because it affects diagnosis, treatment decisions, and, in the end, patient outcomes. Our approach uses a cutting-edge deep learning technology to tackle this complex task by utilizing the analytical power of EEG data and supporting visuals. The goal of this novel method is to offer a comprehensive and sophisticated categorization of seizure types. It is impossible to overestimate the importance of Electroencephalography (EEG) signals in diagnosing and treating epilepsy as they record the electrical activity of the brain. However, a significant barrier to accurate categorization is the varied and subtle differences in EEG patterns throughout seizure types. Our methodology aims to overcome this difficulty by creating a thorough and precise categorization system. This methodology offers a sophisticated knowledge of seizure types that goes beyond traditional methods, which is intended to be beneficial to both medical professionals and people with epilepsy. This paper has the potential to completely transform how epilepsy is identified and cured. When the intricacies of seizure type categorization are effectively tackled, patient outcomes can be significantly enhanced and overall quality of life can be raised. Our initiative intends to provide more efficient and individualized care for individuals affected by epilepsy by bridging the gap between the complex neurobiological subtleties of seizure types and the practical demands of healthcare practitioners.

2 Major Contributions

EEG signals, providing valuable insights into brain activity, are renowned for their sophistication and complexity. A highly advanced approach is utilized in the existing project to separate the EEG signals into various subcomponents, each representing a unique aspect of the signal. One of the noteworthy methods utilized in the decomposition process of the existing system is the Hilbert Vibration Decomposition (HVD). HVD not only break down the EEG signals into their individual parts but also preserve critical phase information. The first three specific subcomponents are for further completion of the classification task due to its characteristics. The continuous wavelet transform technique translates the subcomponents into 2D images to provide a structured and spatial representation of the EEG data. Classification is accomplished through a hybrid deep stacking pipeline, which integrates two intelligent neural networks, which are CNN and the LSTM. This combination resulted in precise classification of EEG signals into various categories, including different types of seizures [3–6]. We employ HVD to break down signals, identify significant, high-energy subcomponents, convert them into two-dimensional images, and employs a powerful fusion of CNN and LSTM to achieve precise classification of seizure types.

3 Literature Survey

The author in [1] accurately differentiated seizures into three categories, by utilizing a combination of Mel frequency cepstral coefficient, independent component analysis and Hjorth descriptor. Their approach, yielded an impressive accuracy of 91% by employing a SVM classifier. Similarly, Wijayanto [2] proposed a unique technique using empirical mode decomposition paired with SVM classification to effectively distinguish into seizures of four categories achieving a classification accuracy of up to 95%. In another study by Kassahun, [3], the authors utilized ontology to classify two types of seizures along with genetics, demonstrating a similar level of thoroughness in their methodology. Additionally, the success of these methods is heavily reliant on the meticulous selection of relevant hand-crafted features, posing significant challenges, especially in distinguishing small discrepancies between different types of seizures [4–6, 12]. In response to this obstacle, the implementation of a DL-based approach that eliminates the need for manual feature selection has shown promise [4, 5]. Moreover, in the realm of image-based classification, including biomedical signals, the efficacy of Deep Learning-based techniques has already been demonstrated [7–10]. However, the effectiveness of using a Deep Learning method for specific uses heavily hinges - the availability of a substantial and diverse dataset for automatic feature discovery [7–9, 12].

Several studies have explored DS-based pipelines for distinguishing between various types of seizures. For instance, Roy [13] have utilized a basic CNN and achieved 72.2% as an impressive F1-score in distinguishing eight distinct seizure types. However, these studies have not yet tackled the critical issue of developing an effectivetechnique that generates images as input.

The authors in [6] expertly utilized four diverse CNN models - spanning from basic to advanced, including AlexNet, VGG19, VGG16in order to effectively categorize eight distinct seizure types. Through the implementation of a vertically stacked 2D spectrogram of unprocessed input EEG data, their efforts resulted in astonishing accuracies of 84.06%, 81.14%, 79.71%, and 76.81%. Authors in [5] also exemplified innovation in their research by incorporating sophisticated techniques such as the saliency-encoded spectrogram. Additionally, they harnessed the powerful SeizureNet algorithm - a fusion of multiple CNN building blocks - to surpass their previous findings and achieve further significant enhancements.

Although their image generation mechanism is highly effective, it is worth considering that a Deep Stacking (DS) pipeline comprised solely of convolutional neural network (CNN) blocks may not be the most optimal choice. While CNNs excel at extracting spatial features, temporal information is crucial for accurately detecting and differentiating between different types of seizures. To this end, Ahmedt et al. [13] utilized a stacked auto-encoder, RNN, CNN and recurrent CNN in their architecture in order classify eight seizure types, achieving an impressive 94.0% weighted F1-score. Similarly, Liu et al. [4] used a bilinear hybrid DL approach that utilized 2D depiction of the unfiltered input data for the classification of seizures into eight different types. They achieved their results through a fusion of a STFT—short-time Fourier transform and a deep learning architecture.

Emami et al.[14] explored the effectiveness of using different durations of EEG components (ranging from 0.5s to 10s) in accurately classifying epileptic seizures. Similarly, Gao [9], investigated the role of 4s frames in distinguishing between seizure episodes [9]. Additionally, authors [12, 15] examined the impact of dividing EEG signals into segments ranging from 5.9s and 10s to classify seizure activity and identify different seizure types among other related studies. The new study uses the HVD approach on EEG fragments that were previously recovered from larger recordings, building on this earlier work. Furthermore, the 1D signals are decomposed and converted into 2D images for use in a deep learning-based classification approach. This innovative method shows promise in accurately identifying and categorizing epileptic seizures.

Emami [14] captured 2D images by directly capturing snippets of EEG segments on specific durations. While [9] utilized power spectrum density energy diagrams to generate 2D images, the low-energy frequency accuracy of EEG may limit their efficiency. In contrast, Shankar [12, 15] introduced time series techniques such as gramian angular field and recurrence plot to construct 2D images for their deep learning-based seizure classification. However, these methods disregard the importance of spatial features.

Other techniques for generating 2D images includes using the TFR—Time and Frequency Representation for one dimensional input signals. This method is particularly effective as it retains the critical frequency and time data [11, 13]. Among the various TFR methods, the CWT has displayed a significant success in seizure analysis due to its ability to capture rapid frequency changes with high resolution. As such, the current study utilizes TFR with CWT to transform 1D EEG signals into 2D images.

4 Proposed System

Our proposed methodology to analyze and interpret electroencephalogram (EEG) signals. Our primary goal is to uncover underlying patterns hidden within the data. To achieve this, we rely on the powerful Hilbert Vibration Decomposition (HVD) technique, which accurately deconstructs the EEG signals while preserving crucial phase details. To begin, our system applies HVD to break down the signals. Then, it identifies the most energetic subcomponents. These selected subcomponents are then further examined through CWT, which transforms the selected subcomponents into two-dimensional images. This process provides a deeper and more comprehensive analysis is hybrid deep stacking. This integration enables the simultaneous capture of the temporal as well as spatial properties present in the data collected from the EEG. This improves the system's capability to identify complex patterns and correlations in the data. For dimensionality reduction, we used the t-SNE method to further optimize the system's performance. By optimizing the EEG data format, the analysis got streamlined and effective. Our methodology combines deep learning framework with sophisticated signal processing techniques to provide a solid foundation for deriving useful insights from EEG signals. The framework and working of the recommended system is depicted in Fig. 1.

4.1 Dataset

Temple University Hospital provided the EEG dataset. TUH v1.5.2 is the version of the data set. This dataset provides a broad and comprehensive set of data for our research since it includes EEG signals from five different seizure types with different sampling frequencies and montages. Table 1 shows the five kinds of seizure and its related durations. However, EEG data were not evenly distributed between seizure categories. Researchers concentrated on 153 seizure sessions in particular to guarantee the validity of our findings. This guarantees a balanced representation of all five seizure types and prevents an imbalance in the data from skewing our findings.

The EEG data was recorded using the Av hardware setup. The data has a resolution of 16 bits and a sampling frequency of 250 Hz. The collection includes EEG signals from 23 common channels, each of which corresponds to a specific electrode placement. The channels used are: FP1-F7, P4-O2, FP2-F8, F8-T8, F7-T7, T7-P7, P7-O1, FP1- F3, F3-C3, T7-FT9, FT9-FT10, FT10-T8, C3-P3, P3-O1, FP2-F4, F4-C4, C4-P4, T8-P8-0, P8-O2, FZ-CZ, CZ-PZ, P7-T7 and T8-P8-1. These channels, corresponds to specific electrode placements on the scalp. These electrodes are strategically positioned to capture electrical impulses produced by the brain. Each channel represents the electrical responses between two designated electrode locations. This dataset's comprehensiveness and the careful selection of seizure episodes guarantee that our study is based on

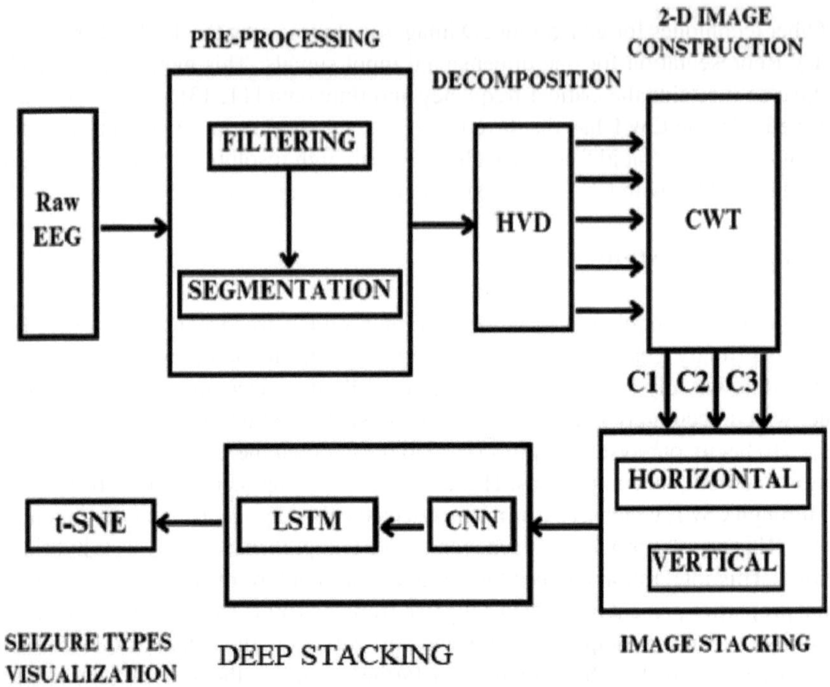

Fig. 1. System architecture of the recommended system

Table 1. Description of TUH v1.5.2 dataset.

SEIZURE TYPE	DURATION (s)
Complex partial seizure (CPS)	1904.09
Focal non-specific seizure (FNS)	1829.19
Generalized non-specific seizure (GNS)	2060.84
Simple partial seizure (SPS)	1329.5
Tonic-clone seizure (TCS)	517.17
Seizure free (SF)	2000

a representative and varied sample, which improves the validity and generalizability of our experimental results.

4.2 Preprocessing

The first step in preparing the recorded EEG signals is to remove noise and artifacts [7, 8]. We employed a Butterworth band-pass filter (Fifth order) with a frequency ranging from 0.5Hz to 50Hz [12]. This carefully selected range greatly enhanced our ability to detect the intricate patterns present in epileptic seizures. The bandpass filter played a pivotal role in the signal processing process, effectively removing irrelevant high frequencies and allowing us to focus solely on the vital frequency band[4–6, 11, 12]. We were able to isolate the distinctive features of epileptic seizure patterns with remarkable accuracy and precision by successfully removing unnecessary high frequencies from the filtering process[12]. This crucial step enabled us to create a refined and concentrated dataset, laying the foundation for further analysis. In order to enable classification using deep learning, the produced signals were split into 10-second intervals that had a 50% overlap [14]. In order to preserve the data integrity and provide a continuous and comprehensive collection of samples for analysis, this segmentation approach was used. Because the overlapping segments capture possible transitions and changes within the data, they aid in a more detailed understanding of the temporal dynamics in the EEG signals [7, 8, 14]. The Hilbert vibrant Decomposition (HVD) method is utilized to enhance the accuracy of the segmented signals of EEG. Every segmented signal was subjected to HVD, enabling a thorough decomposition that maintains crucial phase information. By ensuring that the ensuing deep learning models are trained on high-quality, representative data, this stage helps to provide a more nuanced analysis and raises the system's accuracy and reliability in classifying epileptic episodes

4.3 HVD-Hilbert Vibration Decomposition

The next step in our method is the complex decomposition of filtered electroencephalogram (EEG) data into many smaller components. HVD is a signal processing technique that is distinguished by its special qualities. It is best suited that can be used for variable data like EEG signal. HVD varies the IA–instantaneous amplitude and IF–instantaneous frequency gradually. This procedure fulfils an important function by allowing researchers to separate and extract discrete elements or patterns from the EEG data. The analogy of breaking the signal down into its constituent parts is appropriate since it reflects the goal of obtaining a more complex understanding of the EEG signal's underlying structure. Filtering the IFs in each cycle, separates the slowly varying synchronous components. In addition to its ability to break down the signal, HVD is especially valuable because of its ability to maintain important phase information. This phase data is essential to the analysis since it provides information on the timing and alignment of different frequency components in the EEG signal. With HVD, our approach recognizes the importance of preserving the temporal correlations and synchronization of these components in addition to revealing their individual components. However, the HVD method is used under two main primary criteria: (i) a minimum of a single quasi-harmonic with numerous full cycles is superposed to create the underlying signal and ii) the frequency and envelope of each vibrating components are distinct from one another.

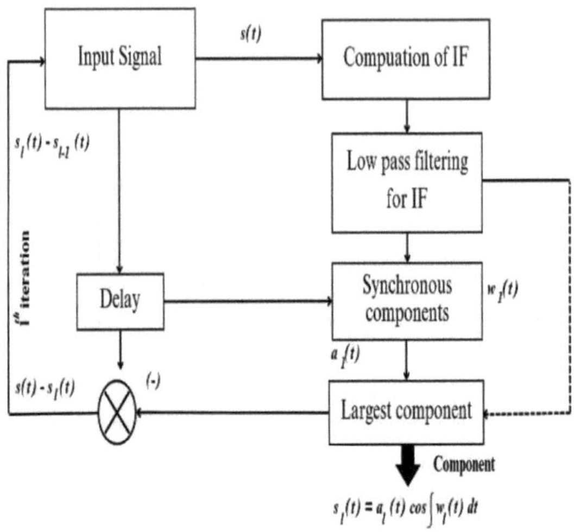

Fig. 2. Systematic working of HVD

The ECG signal does, in fact, meet these two main requirements since displays numerous full-cycle quasi-harmonic components with varying IFs. Undoubtedly, it has already been effectively used to analyze EEG signals. While the HVD is operating, the IF estimate as well as synchronized detection operations. Because of the synchronous detection, it is finally possible to keep the elements with slight fluctuations in amplitude, are overpowered by other major elements. The IFs can be utilized to estimate it. As a result, both the IA as well as IF associated with an EEG are calculated first the IF is subsequently filtered using a low-pass filter (LPF) with an exceptionally low threshold frequency. The synchronous component is produced by the filtered IF and IA. The biggest amplitude component is extracted from its source signal, followed by the utilization of the remaining portion in the further process. Figure 2. Illustrates the many phases of the HVD process.

A composite signal, $s(t)$, is defined by Eq. (1). It is made up of amplitude subcomponent $a(t)$, and l subcomponents of slowly varying frequencies:

$$s(t) = \sum\nolimits_{l} a_l(t) coscos\left(\int \omega_l(t)dt\right) \tag{1}$$

Next, the IF may be estimated by utilizing the computational notation $A(t)$ of $s(t)$:

$$A(t) = s(t) + j\tilde{s}(t) \tag{2}$$

Using (3), we can assess the integral's Cauchy principal value (P.V.) and s̃(t) are represented as the Hilbert transform (HT) of $s(t)$.

$$\tilde{s}(t) = HT[s(t)] = \frac{1}{\Pi} P \cdot V \int_{-\infty}^{\infty} \frac{s(\tau)}{t - \tau} d\tau \tag{3}$$

Then, IF can be found using (4):

$$\omega(t) = \frac{d}{dt}\varphi(t) = \frac{d}{dt}\left(\frac{\tilde{s}(t)}{s(t)}\right) \tag{4}$$

where, respectively, $\omega(t)$ and $\varphi(t)$ stand for the instantaneous phase (IP) and the IF. Afterwards, synchronous demodulation has been used to identify the greatest envelope subcomponent. In fact, synchronous detection makes it possible to identify even minute changes in $s(t)$.

$$s_l{=}_r(t) = \frac{1}{2}a_l(t)[coscos(\phi_l(t)) + coscos(\int (\omega_l(t) + \omega_r(t)dt + \phi_l(t))]$$

$$\tilde{s}_l{=}_r(t) = \frac{1}{2}a_l(t)[sinsin(\phi_l(t)) - sinsin(\int (\omega_l(t) + \omega_r(t)dt + \phi_l(t))] \tag{5}$$

Using Eq. (5), one can now determine the HT phase ($\tilde{s}_l{=}_r(t)$ and in-phase $s_l{=}_r(t)$ in the r-referenc5e of the l th component.In this case, the IF of the greatest factor in the r-reference is represented by $\omega_r(t)$, and the IP, IF, and IA from the l th components are denoted by $\phi_l(t)$, $\omega_l(t)$ and $a_l(t)$.

The subcomponent with the highest energy, $s_1(t)$ is extracted from the original compositions,$s(t)$. The lower energy signal, $s_{l-1} = s_l(t) - s_1(t)$, can be broken down in the ensuing iterations. The energy of the decomposed subcomponents eventually decreases in order; that is, the energy of the first subcomponent is greater than that of the next, and so forth. As a result, the EEG signals are disassembled into several smaller parts, which were then taken into consideration for the creation of 2D images. In addition, of all the subcomponents that are deconstructed using HVD, the first three together maintain approximately 80 percent of the energy of the primary signal; for this reason, they have been given attention for additional processing. Figure 3. Depicts the plotting of all the five subcomponents (sub_0 to sub_4) obtained after performing HVD algorithm.

Additionally, to ensure that the subcomponents are appropriate, the correlation (Cc) between them and the main signal has been examined in Table 2.

Table 2. Correlation (Cc)and Energy percentage of all five subcomponents.

SUBCOMPONENTS	ENERGY (%)	C_c
C1	67.25	0.854
C2	11.35	0.386
C3	2.74	0.178
C4	0.715	0.120
C5	0.695	0.083

4.4 2D Image Generation by CWT

Upon identification of the three high-energy subcomponents via Hilbert Vibration Decomposition (HVD), the use of continuous wavelet transform (CWT) is a crucial step towards optimising them for deep stacking analysis. The chosen subcomponents are transformed into 2D representations, which is useful for more efficient analysis and classification.

CWT helps in dividing input signal into parent wavelet. These are microscopic oscillations within an extremely small period. Excellent time and frequency translation is achieved by scaling and moving the parent wavelet across the period plane of the intended signals to form a time-frequency pattern. It has been discovered that the CWT is incredibly effective and appropriate for modeling nonlinear and irregular signals, like

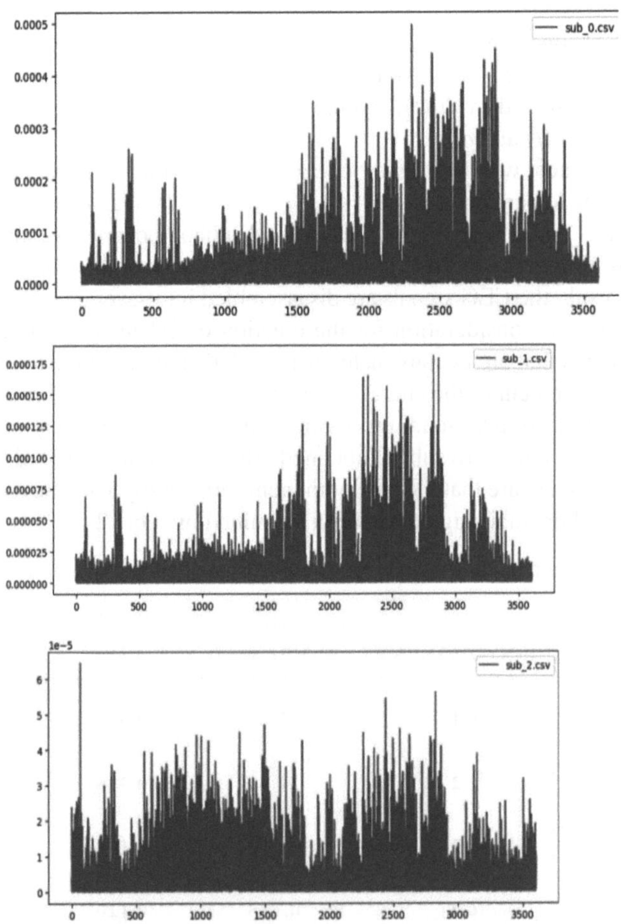

Fig. 3. HVD plotting of all subcomponents.

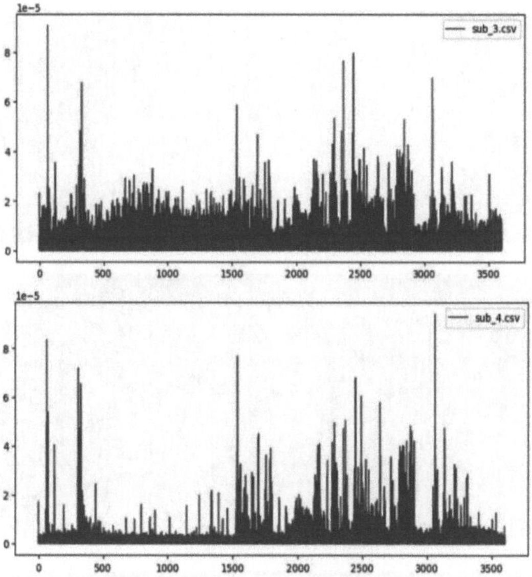

Fig. 3. (*continued*)

EEG [7, 8], Mathematically the CWT of x(t)signal can be determined by (6):

$$CWT_s(d, \Delta) = \frac{1}{\sqrt{d}} \int x(t)\psi\left(\frac{t-\Delta}{d}\right)dt \quad (6)$$

where the symbols ψ and CWT(d,Δ)stand for the base wavelet function and wavelet coefficient, respectively. Δ shifts theψ(t), and d contracts or dilates it. When d is less than 1, or contraction, a high temporal resolution is provided by ψ(t), which is useful for measuring brief activities; when d is more than 1, or dilatation,ψ(t) yields a high spectrum resolution. The Morlet wavelet was used for this work because it is effective at evaluating complicated signals, such as EEGs, and it has shown respectable results in a number of EEG-based research [6, 13]. It can accurately record short spike spines that alternate and repeat, together with a start and finish time. Since t represents the sample instants, Eq. (7) provides a mathematical definition of the Morlet wavelet function.

$$\psi(t) = e^{-t\frac{2}{2}}coscos(5t) \quad (7)$$

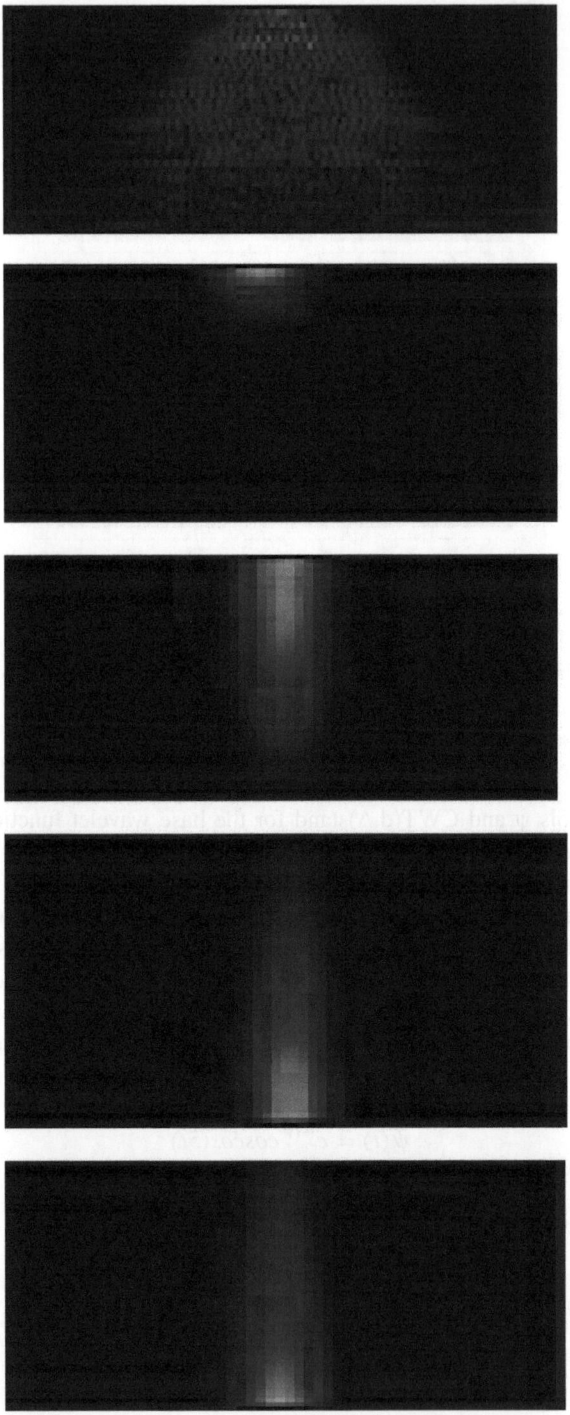

Fig. 4. CWT Image Generation for all subcomponents

The Fig. 4 shows each of the five subcomponents' two-dimensional pictures. The complex information contained in the subcomponents and the analytical capability of deep learning are connected by the continuous wavelet transform. Deep learning models are presented with a structured format by graphically organising the data through the conversion of subcomponents into 2D representations. This conversion is important because it improves the data's fit for deep learning algorithms while preserving the critical information included in the subcomponents. Deep learning harnesses immense power in unraveling the intricate spatial ties, complex patterns, and unique attributes within EEG data. These subtle nuances may be obscured in the raw signal, but when transformed into 2D representations, they become clearer. The deep learning model's efficacy in processing this organized and enhanced data enables it to extract higher-level features, leading to a deeper comprehension of the underlying neural function. Essentially, the 20 continuous wavelet transform serves as a crucial precursor, paving the path for the deep learning model to expose hidden intricacies within the EEG data.

4.5 Image Stacking

One of the most essential phases in using DL-based classifiers successfully is input data preparation [7, 8, 10]. The subcomponents are stacked in both vertically and horizontally to produce one image as displayed in Figs. 5 and 6.One image has been created by stacking the various images that each subcomponent generates in a horizontal and vertical manner. For uniformity, all photographs have been resized to 32 by 32.Ultimately, in order to determine the optimal image stacking technique, the hybrid Deep Stacking pipeline receives the images obtained through stacking individually.

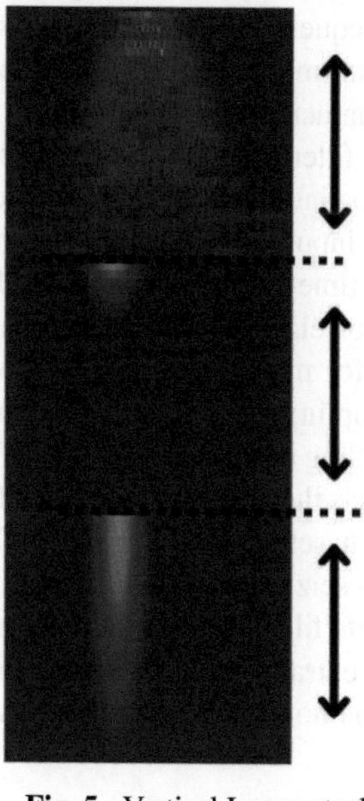

Fig. 5. Vertical Image stacking

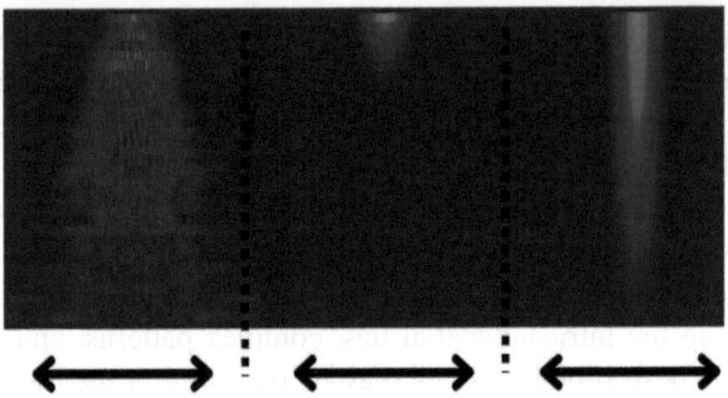

Fig. 6. Horizontal Image stacking

4.6 Deep Stacking Pipeline

Two popular models—LSTM and CNN—make up the suggested hybrid DS pipeline. While the LSTM proficiently distinguishes time-related events, the CNN excels at extracting spatial features. This distinction can be crucial in the analysis of epileptic seizures [8]. The combination of the two models is highly effective in seizure analysis. Two combinations have been made out of them: parallel and serial. It has been discovered that the serial combination CNN followed by LSTM is more effective and have been used in this work. Additionally, 20% of the training samples were taken into account while validating the model training.

Convolutional Neural Network (CNN)
CNN's framework is designed to identify seizures by effectively extracting and processing spatial patterns in the frequency as well as time spectrum of the input data. Convolution, pooling, and batch normalization are the three hidden layers of CNN that automatically extract pertinent characteristics from the raw input data [9, 10, 12]. Convolutional layers are essential for filter-based input EEG picture scanning. The purpose of these filters is to recognise spatial patterns that point to seizure activity. Through convolving these filters over the input pictures, the network learns to identify important elements in the EEG signals' time-frequency domain. This makes it possible for the model to identify minute spatial correlations and patterns that might be signs of seizures.

Pooling layers are essential for minimizing data size while preserving important features. This involves down sampling the output from convolutional layers, selecting relevant information and disregarding unnecessary details. This not only improves computational efficiency but also directs the network's attention towards the most significant aspects of the input. A labelled dataset is used to train the CNN, designating each EEG fragment as either a non-seizure or seizure condition. This process uses backpropagation, enabling the network to modify its filter and neuron weights. The inputs are convolutioned with a predefined kernel to extract structural information, which is then processed by an activation function which is non-linear. The result is calculated mathematically using (8):

$$C_p^m = f\left(\sum_{q=1}^{N_m-1} conv2D\left(w_{q,p}^m, F_q^{m-1}\right) + b_p^m\right) \tag{8}$$

Where conv 2D represents the 2D convolution operation and C_p^mrefers to the convolution output of the pth node in the mthstratum. The pthnode and the total nodes in the (m-1)thstratum are denoted by F_q^(m-1)andN_m-1, respectively. The trainable kernel's weight is denoted by w_(q,p)^m. The convolution result was then followed by an activation that was non-linear(f(y) = max(0,y)). The normalization layer of the batch and the layer with the greatest pooling then receive the output from the convolution layer. By creating a smooth and recognizable collection of various unique features, maximum pooling stratum lowers the model's spatial size, feature variance, and complexity. [9, 12]. Additionally, batch normalization helps reduce the problem of internal covariance shift [12, 14]. As a result, the CNN continually improves its ability to identify spatial patterns that indicate seizures. The CNN's fully connected layers offer the final categorization output. These layers use the high-level characteristics acquired from both the pooling layer and convolutional stratum in order to evaluate if input EEG segment is either from a seizure state or from a non-seizure state. Because of its extensive architecture, the CNN is an effective tool for automated seizure identification in neurological monitoring applications. It can learn and recognise spatial patterns in EEG signals on its own.

Long Short-Term Memory (LSTM)

LSTM is essential to seizure categorization models because they can store information across long sequences. Because of their unique architecture—which includes gating mechanisms and memory cells—LSTMs are able to recognise and comprehend complex patterns linked to seizures. During the learning process, the memory cells in long short-term memory (LSTMs) are educated to adjust their characteristics, such weights and biases. To properly capture the temporal patterns that define seizures, this modification is essential. LSTMs employ the backpropagation through time (BPTT) technique to address the issue related to vanishing gradient. It is a prevalent difficulty in lengthy sequences, in contrast to conventional RNN. BPTT takes the full sequence into account when updating the weights, which helps LSTMs identify long-term dependencies from the input data that is sequential in nature. The handling of long-term dependence by LSTMs is very important when it comes to seizure identification. Since seizure patterns can span several time periods, the model must identify connections and dependencies over a long period of time.

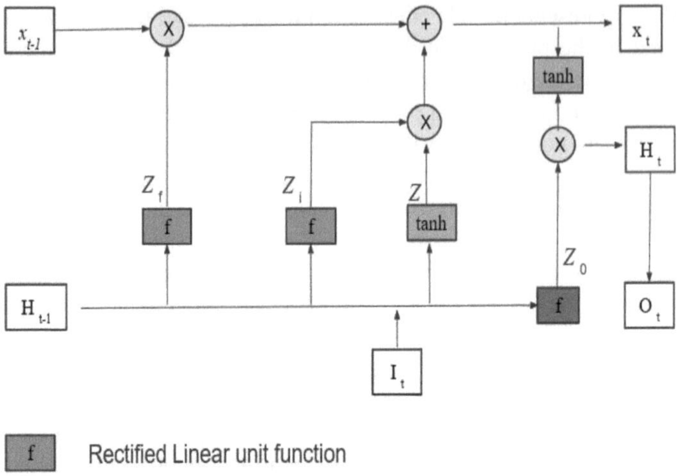

Fig. 7. The internal operation of the LSTM framework.

From the Fig. 7. The cell gate (z) in an LSTM is in charge of retaining information over time. The z_f stands for the forget gate, z_i is the input gate, and zo, this controls how much data in the cell is used to calculate the result using the weight metrics, W. The x_t, H_t, and O_t are are three outputs that show the current output, hidden state, and cell state, respectively, the H_{t-1}, x_{t-1} and It designate the prior hidden state, cell state, and current input.

T-DISTRIBUTION STOCHASTIC NEIGHBOR EMBEDDING (t-SNE)

One powerful method that is often used to make complicated data visualization easier is t-SNE.When high-dimensional features are difficult to grasp, this algorithm is very helpful. CNN extracts features extensively, however these features that are produced are frequently located in a space with a high number of dimensions, which makes it challenging to understand the patterns that they contain. This is exactly the situation in which t-SNE shines, providing a solution by mapping these complex properties onto a space with fewer dimensions, typically, usually in 2D or 3D. The efficacy of t-SNE is based on its capacity to maintain the correlations among data points while reducing dimensionality. In other words, data points which are closer to one each other will remain close in the lower-dimensional space. The maintenance of the relative positioning of data points, which indicates similarities or differences, is dependent upon the preservation of local structure, which is essential. t-SNE makes it easier to find unique clusters in the data by lowering dimensionality while maintaining local structure.

Visualising and comprehending patterns or groupings that might be hidden in the higher-dimensional feature space is greatly aided by this. Through the identification of significant correlations and the discovery of clusters that may represent certain patterns or categories, researchers and practitioners can obtain understanding of the data's fundamental structure.t-SNE is an essential tool for improving interpretability and extracting insights from high-dimensional feature spaces. This helps to make complex datasets easier to grasp, particularly when it comes to CNN-derived features.

In this case, LSTMs perform exceptionally well because they are able to capture the complex temporal dynamics related to seizures. The hidden states of the LSTM generate abstract representations of the input sequence. During classification, these concealed states provide an essential basis for differentiating between seizure and non-seizure states by encapsulating contextual information and learning properties. The LSTM is a reliable and efficient tool for modelling and classifying temporal patterns in EEG data. It is especially well-suited for the complex task of seizure identification because of its innate ability to retain and use information over extended sequences and resistance to the vanishing gradient problem.

5 Results

The metrics namely F1-score, accuracy, specificity, and sensitivity are used for evaluation purpose for both the datasets. When all data sets are not proportionally available, the weighted F1-score becomes extremely essential [5, 7, 8]. It also provides a more accurate assessment of misclassified instances. The performance was evaluated using weighted F1-score, sensitivity, specificity, and accuracy.

$$A_{cc} = \frac{TP + TN}{TP + TN + FP + FN} \tag{9}$$

$$S_e = \frac{TP}{TP + FN} \tag{10}$$

$$S_p = \frac{TN}{TN + FP} \tag{11}$$

$$F1 = \frac{2TP}{2TP + FP + FN} \tag{12}$$

where FN stands for false negative and FP for false positive. TN and TP stand for the true positive and negative, respectively. Additionally, the receiver operating characteristics (ROC) analysis has been used to assess the resilience and systematic capabilities of the proposed model [7, 8, 10].

The suggested approach produced a range of performance indicators, including as weighted F1-score, accuracy, sensitivity, and specificity, as demonstrated in Fig. 9. The The vertical axis shows the relevant proportion of performance measures, while the horizontal axis shows several sets of input images.

The Table 3 elucidates an analysis on similar works by comparing various factors.

Table. 3. Analysis Table.

WORKS	INPUT TO CLASSIFY		CLASSIFIER	ST	PM (%)	
	TYPES	IGM			Acc	F1
Liu et al.	2D IMAGES	STFT	CNN	8	-	95.5
			RNN		-	95.8
			HYBRID		-	97.4
Asif, et al.	IMAGES	SES	CNN	8	-	94
Sriraam, et al.	2D IMAGES	STFT	CNN	8	84.1	-
Ahmedt, et al.	EEG	FFT	P-NMN	7	-	94.5
Wijayanto, et al.	SF	-	SVM	4	95	-
Kassahun et al.	MD	-	GB-ML	2	-	77.8
Roy et al.	EEG	FFT	k-NN	8	-	90.1
			XGB		-	86.6
			CNN		-	72.2
Shankar el al.	2D IMAGES	GAF	CNN	5	84.2	84
This Work	2D IMAGES	HVD, CWT	CNN, LSTM	6	99.09	99.01

The Fig. 8 Illustrates the t-SNE plotting using symbols of various colors which correspond to the numerous seizure types that have been identified.

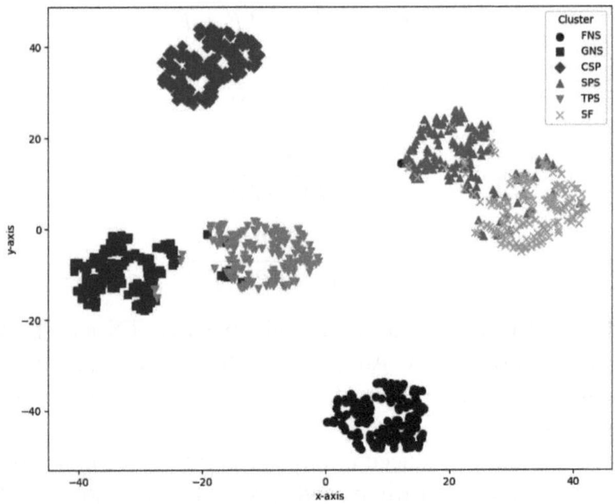

Fig. 8. t-SNE plotting for seizure types.

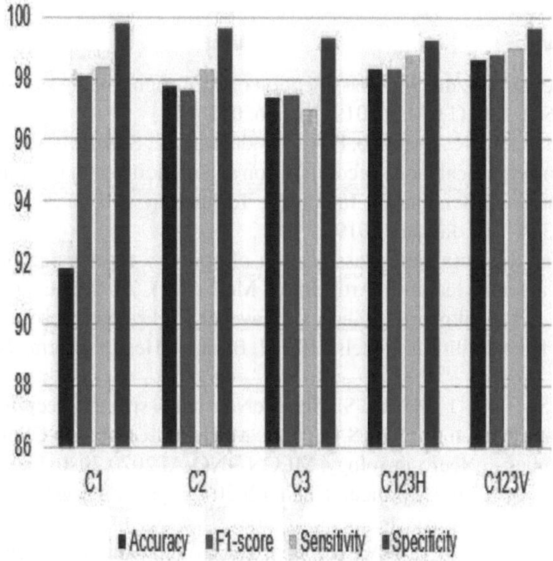

Fig. 9. Performance Metrics

6 Conclusion

Utilize the researcher work aims to detect seizure using the Hilbert Vibration Decomposition (HVD) technique to carefully dissect electroencephalogram (EEG) segments into smaller constituents. In order to improve the analysis even further, the first three subcomponents that are produced from HVD are used to create 2D pictures using the CWT. CNN and LSTM networks are combined into a single representation for a hybrid Deep Stacking (DS) pipeline, which allows for extensive feature extraction and categorization. Utilizing the Temple University EEG dataset (TUH), the experimental validation produces impressive outcomes. With an exceptional classification accuracy of 99.09%, the suggested technique gets a weighted F1-score of 99.01%. The application of t-SNE increases trust in feature extraction procedure's thoroughness even further. The suggested strategy demonstrates exceptional classification performance in EEG signal processing, outperforming current approaches in a comparative comparison. By using deep stacking to improve and automate seizure type categorization, this study fills a major need in the medical industry. The ramifications are significant, resulting in improvements in patient care via early detection, individualized treatment plans, and a better understanding of seizure diseases. This research makes a substantial contribution to the area of neurology by successfully categorizing seizure types with such high accuracy. In addition to streamlining medical procedures, the hybrid DS pipeline's accurate and automatic categorization gives medical staff members insightful information that helps them provide patients with individualized treatment. In the end, this study might revolutionize the field of diagnosing and treating seizure disorders, highlighting the critical role that deep stacking plays in improving medical practices.

References

1. Saputro, R.D., et al.: Seizure type classification on EEG signal using support vector machine. J. Phys. Conf. Ser. **1201**(1), May 2019. Art. no. 012065
2. Wijayanto, I., Hartanto, R., Nugroho, H.A., Winduratna, B.: Seizure type detection in epileptic EEG signal using empirical mode decomposition and support vector machine. In: Proceedings of IEEE International Seminar on Intelligent Technology and Its Applications, Surabaya, Indonesia, pp. 314–319, January 2019
3. Kassahun, Y., et al.: Automatic classification of epilepsy types using ontology-based and genetics-based machine learning. Artif. Intell. Med. **61**(2), 79–88, March 2014
4. Liu, T., Truong, N.D., Nikpour, A., Zhou, L., Kavehei, O.: Epileptic seizure classification with symmetric and hybrid bilinear models. IEEE J. Biomed. Health Inform. **24**(10), 2844–2851, October 2020
5. Asif, U., Roy, S., Tang, J., Harrer, S.: SeizureNet: multi-spectral deep feature learning for seizure type classification. In: Kia, S.M., et al. Machine Learning in Clinical Neuroimaging and Radiogenomics in Neuro-oncology. MLCN RNO-AI 2020 2020. Lecture Notes in Computer Science(), vol. 12449. Springer, Cham (2020). https://doi.org/10.1007/978-3-030-66843-3_8
6. Sriraam, N., Temel, Y., Rao, S.V., Kubben, P.L.: A convolutional neural network-based framework for classification of seizure types. In: Proceedings of IEEE 41st Annual International Conference on Engineering in Medicine and Biology Society, Berlin, Germany, pp. 2547–2550, July 2019
7. Shoeibi, A., et al.: Epileptic seizures detection using deep learning techniques: a review. Int. J. Environ. Res. Public Health **18**(11), May 2021. Art. no. 5780
8. Roy, Y., et al.: Deep learning-based electroencephalography analysis: a systematic review. J. Neural Eng. **16**(5), August 2019. Art. no. 051001
9. Gao, Y., Gao, B., Chen, Q., Liu, J., Zhang, Y.: Deep convolutional neural network-based epileptic electroencephalogram signal classification. Front. Neurol. **11**, 375–385, May 2020
10. Cho, K.O., Jang, H.J.: Comparison of different input modalities and network structures for deep learning-based seizure detection. Sci. Rep. **10**(1), 1–11, January 2020
11. Roy, S., Asif, U., Tang, J., Harrer, S.: Seizure type classification using EEG signals and machine learning: Setting a benchmark. In: Proceedings of IEEE Signal Processing in Medicine and Biology Symposium, PA, USA, pp. 1–6, December 2020
12. Shankar, A., Dandapat, S., Barma, S.: Seizure type classification using EEG based on Gramian angular field transformation and deep learning. In: Proceedings of IEEE 43rd Annual International Conference of the IEEE Engineering in Medicine and Biology Society, Mexico, pp. 3340–3343, November 2021
13. Ahmedt-Aristizabal, D., Fernando, T., Denman, S., Petersson, L., Aburn, M.J., Fookes, C.: Neural memory networks for seizure type classification. In: Proceedings of IEEE 42nd Annual International Conference of the IEEE Engineering in Medicine and Biology Society, Montreal, QC, Canada, pp. 569–575, January 2020
14. Emami, A., Kunii, N., Matsuo, T., Shinozaki, T., Kawai, K., Takahashi, H.: Seizure detection by convolutional neural network-based analysis of scalp electroencephalography plot images. NeuroImage Clin. **22**, January 2019. Art. no. 101684
15. Anita, M., Meena Kowshalya, A.: Automatic epileptic seizure detection using MSA-DCNN and LSTM techniques with EEG signals. Expert Syst. App **238**, 121727 (2024)

Evaluating Machine Learning Algorithms for Breast Cancer Detection: Logistic Regression, Naive Bayes, and Decision Tree Models

K. E. Narayana[1](✉), S. Ravikumar[2], S. Muruganandam[3], R. Pradeep[4], and M. Sindhuja[5]

[1] Department of Information Technology, Rajalakshmi Engineering College, Chennai, India
narayana.ke@rajalakshmi.edu.in

[2] Department of Computer Science and Engineering, Vel Tech Rangarajan and Dr. Sangunthala R&D Institute of Science and Technology, Chennai, India

[3] Department of Computer Science and Business Systems, Panimalar Engineeirng College, Chennai, India

[4] Department of Computer Science and Engineering (Cyber Security), Saveetha Engineering College, Chennai, India

[5] Department of Computer Science and Engineering, Acharya University, Karakul, Uzbekistan

Abstract. Breast cancer is one of the most prevalent cancers in women worldwide; thus, it's important to diagnose the disease quickly and accurately so patients have the best chance of survival. Using the Wisconsin Diagnostic Breast Cancer (WDBC) dataset, our experiment aims to see how well three machine learning algorithms—very LR is a statistical model used for binary classification, naïve Bayes (NB) probabilistic classifier based on Bayes theorem, assuming independence between predictors, and decision tree (DT) A flowchart-like tree structure where internal nodes represent decisions based on features, and leaves represent outcomes. do at classifying breast cancer patients. 30 numerical traits extracted from FNA images describe each of the 569 cases in Cottontail-DB. At 97.62% precision, 96.47% accuracy, 96.47% F1 score, and 99.74% ROC AUC, LR outperforms NB and DT. NB, on the other hand worked relatively better (mean RA of 99%) in general but particularly gave worst service for ROC AUC(98.76%), an improvement from DT which had low accuracy and few performance measures so score comparison could be done fairly easily. These findings reveal the robustness and reliability of LR for BC screening. The use of appropriate machine-learning models is vital for increasing the precision of diagnosis and aiding in clinical decision-making, underline the investigators.

Keywords: Diagnosis of Breast Cancer · Logistic Regression · Bayesian · Machine Learning · Medical Diagnostics · WDBC

R. Geetha et al. (Eds.): AAIMB 2024, CCIS 2653, pp. 257–266, 2026.
https://doi.org/10.1007/978-3-032-13026-6_18

1 Introduction

According to the data, breast cancer is one of the main killers of females worldwide [1]. The World Health Organization (WHO) reports that this type of cancer affects approximately 25% of all cases and is the most common among women [2]. According to the American Cancer Society [3], approximately one-eighth of a woman's female relatives will receive a breast cancer diagnosis during her lifetime. In light of these numbers, it is quite clear that we need effective treatment interventions and robust early detection technologies to drastically cut mortality rates and improve patient outcomes.

The earlier breast cancer is detected the better, because it can improve your chances of a successful therapeutic course and survival [4]. Besides, some conventional diagnostic methods are being adopted for the diagnosis of breast cancer such as Mammography, ultrasound and a biopsy [5]. Unfortunately, these approaches are invasive and time consuming, may give false positives or negatives. Machine learning, as technology is strides forward, grows even more popular as a modality to facilitate faster and accurate breast cancer diagnosis.

To put it simply, machine learning (ML) is a subfield of AI that deals with systems that can learn from data [6]. ML algorithms can analyze medical diagnostic big data sets to identify patterns and potentially predict the emergence of diseases. In situations such as cancer, this is a vital capability since the early and accurate recognition of cancers can be lifesaving. With machine learning algorithms feature rich data may be worked on, and exclusive insights as well that human analysts might miss otherwise [7]. As an example, a machine learning algorithm can be trained to distinguish between benign and malignant tumors in the field of breast cancer diagnosis using patient demographics, medical history, and imaging results. Less harsh procedures may be necessary if these predictions help doctors make better-informed decisions.

The purpose of this research was to evaluate three well-known ML classification algorithms for use in breast cancer diagnoses. Finding the best approach for detecting breast tumors in terms of ROC, AUC, F1 scores, recall, precision, and accuracy is the main goal of this research. It accomplishes this by testing several methods on the WDBC dataset [8].

Logistic Regression (LR): Used for binary classification problems and is a statistical model most of the time. LR computes the probability that a particular point belongs to any class. Its interpretability and simplicity make it a popular choice among medical applications.

*Naive Bayes (NB):*A Bayesian classifier that assumes strong feature independence and uses the Bayes theorem for probabilistic classification. One of the most recent applications of Naive Bayes (NB) is in real-world problems such as medical diagnosis, despite its simplicity..

Decision Tree (DT): A tree-shaped model derived in such a way that its end-nodes are always the values of some variable terminal node based on input data by making decision at each level. Internal node - represents a feature. Branch - represents a decision Leaf Node: that represent the outcome DTs: Decision Trees that can take both numeric and categorical data, are interpretable.

The increased awareness and incidence rate of breast cancer along with the need for early detection, seed the requirement to developed accurate diagnosing tools. In the

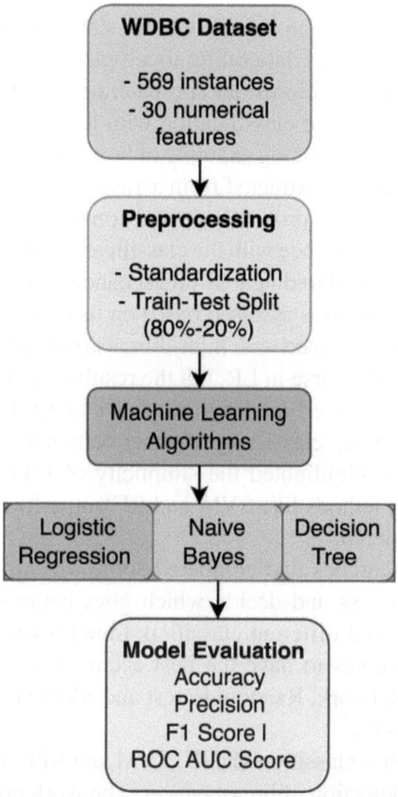

Fig. 1. Architecture diagram for the work flow on Breast Cancer detection

same breath, machine learning algorithms are highly valuable because they can process complex dataset and provide insights that is expected. In this paper, we contribute to ongoing efforts in this area by comparing the performance of three common classification algorithms LR,NB and DT on WDBC dataset. These findings of this work can inform the development of more sensitive diagnostic models that will ultimately improve breast cancer patient outcomes.

2 Related Works

When it comes to cancers affecting women, breast cancer is both the most prevalent and one of the main killers on a global scale. Early disease detection and advancement in effective medicines substantially improve the odds of successful therapy. Medical data-driven ML algorithms have demonstrated promise in early breast cancer detection for differentiating between benign and malignant tumors. However, the main challenge is still to select appropriate classifier. ML algorithms have been used in several medical datasets research, and the results are quite promising. Against the above backdrop, this work studies many machine learning classifiers like DT, LR and NB for predicting breast cancer in light of literature review.

You can use logistic regression, a well-known statistical method, to solve binary classification problems. It utilizes data fitting to a logistic function to forecast the likelihood of a binary result [9] developed an approach to classify breast cancer utilizing maximum perimeter and a texture classification with logistic regression, yielding 95% true positive rate. This is an illustrating example of how LR can be used to predict breast cancer using numerical features extracted from x-rays.

The Naive Bayes classifier is probabilistic and considers its features to be independent; this model works in accordance with the classification rule given by bayes theorem. The good old Naive Bayes used to diagnose breast cancer was mentioned in [10], and it worked quite well. The model was assessed based on its correctness, and more test data was used to analyze its performance with high-dimensional input variables. Needless to say, NB did perform slightly worse in LR, but the results are still very promising.

Decision Tree models are tree-based algorithms which splits the nodes into branches in a way such that it will end up classifying whether certain properties and data features have been met. This work highlighted the simplicity of DTs to interpret and utilize. However, nonparametric methods like SVM and RF outperformed DTs with respect to most of the criteria considered.

There are a lot of researches that compared some Machine Learning functions to model the acquisition process and decide which goes better in detecting breast cancer. One work that compared different classifiers found SVM, which did not use fast correlations-based Streamlines to have the best accuracy of 97%. For example, [11] compared the Bayesian Network, Random Forest and SVM analysis approach showing that BN achieves best results.

In [12] employed several classifiers (KNN, SVM and RFE) to automatically classify digital data/facts for the detection of breast cancer. The work provided a comprehensive analysis of each classifier along with its pros and cons, importance of feature selection to the classification performance accuracy. Recent machine-learning advances have led to more sophisticated models and methods for the breast cancer identification. Many DL methods, such as CNNANN have shown great success for image based cancer diagnosis, also reached very high sensitivity rates using ANNs and CNNs based DL technique [13]. To improve model efficiency, their work employed data augmentation and regularization methods; they achieved accuracy of 82.68% by ANN and 86.24% using CNN.

The work by Singh and Kumar focused on histological image analysis using a cubic support vector machine (SVM) with features extracted from the histology itself to diagnose breast cancer. Our approach benefited from complex image structures. To improve detection accuracy, [14] proposed a new CNN architecture for breast cancer classification with 2D and 3D mammograms fusion. Similar to this an algorithm named BDR-CNN-GCN proposed and that improves identification of malignant lesions on breast mammograms combining CNNs with GCN.

Most healthcare datasets provide a class imbalance that could bias the results of classification. The breast cancer HPI dataset to evaluate different over- or under-sampling strategies in order have a more comprehensive investigation on how to reduce class imbalance when it comes of CNN classifier accuracy. Synthetic over-sampling improved the performance of CNN-based classifiers as their research demonstrated, providing higher balanced accuracy, sensitivity and specificity with also an increase in F1-score.

In [15] work obtained high accuracy and AUC scores on a CNN model for diagnosing Invasive Ductal Carcinoma(IDC). This work proved the utility of CNN network for automatic diagnosis to IDC with strong emphasis on overcoming class imbalances within medical data sets.

Logistic Regression, Predictive Power: LR is excellent for binary classification problems (e.g., disease vs. no disease). Probabilistic Output: It provides probability scores, which help assess the *risk* of conditions (e.g., 80% chance of heart disease). Simplicity and Interpretability: Coefficients are easy to interpret, making it transparent for doctors to understand what factors contribute to predictions. Naive Bayes (NB),Fast and Efficient: NB performs well with high-dimensional data, common in medical records and text (e.g., patient notes). Works Well with Small Data: Often, medical datasets are small due to privacy or rarity of diseases; NB is effective in such cases. Handles Missing Data: With appropriate handling, NB can be robust even when some patient features are missing. Decision Trees (DT), Highly Interpretable: DTs provide visual flowcharts of decisions, making them ideal for clinical decision support. Mimic Human Reasoning: They follow a logical step-by-step rule-based system, similar to how clinicians think. Handles Mixed Data Types: Can work with both numerical (blood pressure) and categorical (smoker/non-smoker) features.

3 Proposed Work

One of the most well-known benchmark datasets in machine learning and medical diagnostics, is called as WDBC dataset. This dataset is used in breast mass classification such as benign(non-cancerous) or malignant(cancerous). The WDBC dataset has 569 cases, where each is an characterized by a single breast cancer case. The data set is composed of 569 cases with a total of thirty numerical features, describing the digitized images as shown in Table I. We have derived these characteristics from digital images of breast mass biopsies taken by fine needle aspirate (FNA). Fine needle aspirations (FNAs) are performed by inserting a fine, hollowed out-needle into an area that is suspected to have a breast lump and aspirating cells for examination. Once the cells are extracted, they go through an examination and a photography process for all other diagnostic procedures [16–18].

Pre-processing: There can be no machine learning techniques applied to the WDBC dataset without first completing this stage. This procedure ensures the correct preparation of data for training and evaluating models through data separation and standardization.

Standardization: It is a useful technique that allows all characteristics to be brought to the same scale, if they have different ranges or units. With bigger range features, this can dominate the learning process and might tilt model towards bias.

$$x' = x - \mu/\sigma$$

where x is the initial feature value, μ is the mean of features σ is standard deviation and x' represent standardized value. The result is that the features look like they have been scaled to a mean of 0 and standard deviation of 1.

Split Dataset: We divided the dataset into training and testing sets to validate our model's effectiveness on fresh data. It requires this in order to understand how well the model generalizes with unseen, out of distribution data points. The 80% was used to train the model and 20% of data was tested with it, as input. A large percentage kept for validation means that we have a good amount of data to train the model on.

Machine Learning Systems: Three well-known machine learning algorithms, namely DT, NB and LR are employed in this work. Each algorithm has a unique approach and a based on some math for categorization.

Logistics Regression: The logistic regression model is commonly employed for binary classification problems. The logistic function is used to estimate the likelihood of a sample being either benign or cancerous based on its attributes.

$$P(y = 1|X) = 1/1 + e^{-(\beta 0+\beta 1\mathrm{X1}+\beta 2\mathrm{X2}+\cdots+\beta n\mathrm{Xn})}$$

In this case, the probability that instance is malignant is denoted by $(y = 1\ |X)$ P(y = 1|X). Denote the intercept by β_0 while write coefficients to characteristics as $X_1, X_2, \ldots, X_n$.

Objective in logistic regression is log-loss, Also known as binary cross-entropy:

$$Log - \mathrm{Loss} = -\frac{1}{m}\sum_{i=1}^{m} [y_i \log(\mathrm{h}\theta(\mathrm{xi})) + (1 - \mathrm{yi})\log(1 - \mathrm{h}\theta(\mathrm{xi}))]$$

Where yi is the true label, m is its frequency till now and $h(xi)$ being predicted probability.

Naive Bayes: Probabilistic variant of linear RNN and naive Bayes classifier based on the last post. Independence of Features- The features are mutually assumed to be independent, considering a class label there is no dependency between existence and absence of one feature by other.

$$\mathrm{P(C\,k|X)} = \frac{\mathrm{P(Ck)}\Pi_{i=1}^{n}\mathrm{P(xi|CK)}}{\mathrm{P(X)}}$$

Here, P(Ck|X) is the posterior probability of class Ck given feature vector X, P(Ck) is the prior probability of class Ck, P(Xi|Ck) is the likelihood of feature Xi given class Ck , and P(X) is the evidence or marginal likelihood of the feature vector.

Constrains: *Independence:* Features are Independent because this is the class title.

Gaussian distribution: The probability of the features is assumed to be a Gaussian distribution for continuous data.

Decision Tree: A decision tree is a non-linear model that makes decisions about the data as it moves from left to right, shaped like what looks similar to branches on a tree. They split the data at nodes according to feature values so as to maximize the separation between classes.

Dividing Criterion: splits on decision trees were made using information gain or the Gini impurity.

Gini Impurity = $1 - \sum_{i=1}^{C}(Pi)^2$

Here, C indicates the classes number while Pi is class probability at some node.

Knowledge Gain:

Profit in Information = Entropy(parent) -$\sum$

Information Gain = Entropy(parent) $- \sum_i \frac{n_i}{n} Entropy(child_i)$

where, n is the count of total samples at parent node and n is the number of samples that has been classified as classi.

Tree Pruning: Decision Trees can cut off to prevent from overfitting. Pruning is the act of cutting out non-important branches from a tree in order to simplify it and make it more generalizable.

Metrics for Model Evaluation: The efficiency of the Decision Tree, Naive Bayes and Logistic Regression models was evaluated using a number of key indicators: prior to processing. Preparing the WDBC dataset for machine learning involves a few different steps, one of which is pre-processing. This step, also known as data pre-processing includes separating and normalizing the data so that it is structured properly for building of models or evaluation. As the features have different ranges and units, to bring all features on a same scale you need standardization. Without standardization larger range features could be the dominant feature of creating the model which would lead to bias.

Metrics for Model Evaluation: The efficacy of the Decision Tree, Naive Bayes and Logistic Regression models were shown in Table 1.

Table 1. Performance comparison between LR, NB and DT

Model	Accuracy	Precision	Recall	F1 Score	ROC AUC
Logistic Regression (LR)	0.9737	0.9762	0.9535	0.9647	0.9974
Naive Bayes (NB)	0.9649	0.9756	0.9302	0.9524	0.9974
Decision Tree (DT)	0.9474	0.9302	0.9302	0.9302	0.944

We only used NB and DT as machine learning models to compare with LR. All performance indicators, including accuracy, precision, F1 score, and ROC AUC score, showed that logistic regression performed better than the other models. Logistic regression resulted in the highest accuracy (97.37%) The LR showed good performance (e.g., of most breast cancer cases, correctly assigned 1 [=True Positive] for malignant or benign). LR also had by far and away the highest precision, at 97.62%, where this is how many of your predictions that actually are real positive (so % true pos / (% content total - false neg)). This demonstrates that LR is reliable in reducing the rate of false positives, which could have serious consequences for patients receiving unnecessary treatments or anxiety about having a problem. Figure 1 shows the graphical representation of performance comparison (FIg. 2).

LR also sent an F1 score of 96.47%. The F1 score is the best mix of recall and accuracy, tells us how good generally LR does on positive instances not beloved to have bias towards FP/ FN. In addition, LR is shown to be a good classifier with ROC AUC score of 99.74% which further supports the reliability and suitability of using it for breast cancer diagnosis.

Naive Byes also performed a good job, especially in terms of its ROC AUC score: 99.74%% (The same as LR) This means that NB does as well on the cancerous examples versus benign ones. But, the Total Accuracy and Precision of NB is slightly lesser than LR i.e., 96.49% accuracy and 97.56 precision respectively NB's F1 score of 95.24% slightly

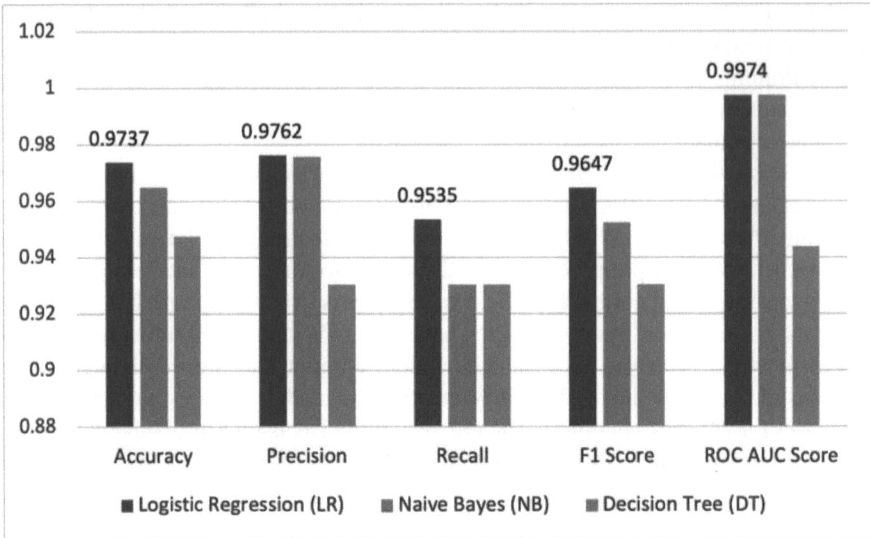

Fig. 2. Performance comparison plots for LR, NB and DT

below LR, although it was still an impressive demonstration for this categorization assignment. Despite these slight differences, NB is a high performing method especially considering it uses the Bayes theorem and has probabilistic methodology. The impressive results coming from NB demonstrate how well it works, especially in applications where attribute independence is assumed or largely unimportant.

Decision Tree is less black box yet more human readable and easier to interpret than LR or NB, despite being the worst performer regardless of metric. These results are less accurate than both LR and NB, despite having a much lower accuracy of 94.74% due to DT Although the decrease in precision indicates that DT may not be able to correctly diagnose instances of breast cancer as regularly. However, with 93.02% accuracy and recall, it had an F1 score of –93.02%. These measures have more balanced but mostly less good values when compared with those of the other ways.ROC AUC score (94.4%). Despite being the slowest, DT is useful for insight into data structure in addition to providing crisp categorization rules based on its explicit decision-making process.

The comparison work displays in no uncertain terms that among the three models Logistic Regression is, by far, the most effective algorithm for classification of instances of breast cancer. The high ROC AUC score, F1 of this model implies that its durability and reliability when it comes to medical diagnostics. While Naive Bayes also has fine performance, especially for ROC AUC column-wise because we care them most nowadays, it is lower than LR in the other measures. Although it is the simplest from a human perspective, The decision tree consistently performed worse displaying us again how more accurate models need to be complex (Fig. 3).

Confusion matrix is a simple table used to measure how well a classification model is performing. It compares the predictions made by the model with the actual results and shows where the model was right or wrong. A confusion matrix consists of four

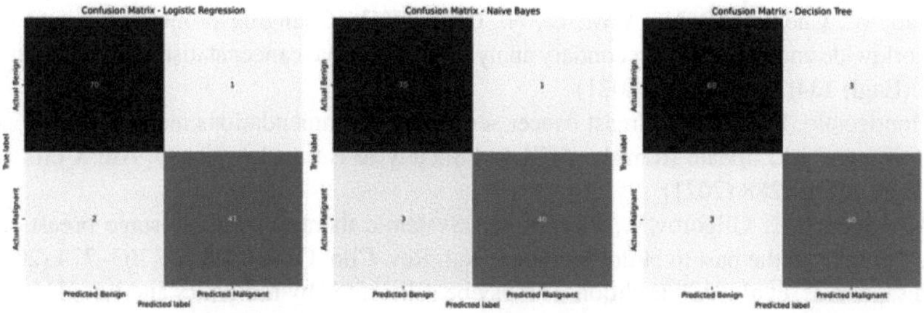

Fig. 3. Confusion matrix for the Machine Learning Models LR, NB, DT

main components: True Positives (TP), False Positives (FP), True Negatives (TN), and False Negatives (FN). These values help calculate key performance metrics such as accuracy, precision, recall, and F1-score, which provide a deeper understanding of a model's strengths and weaknesses.

These confusion matrices result in an analytical performance analysis of each model. LR matrix showing 70 true positives (TP) and high specificity, with marginal false positive rate of only one while also having minimum number of occurrences in the classification problem. NB, again shows strong performance with true positive and false negative counts very close to what LR was able achieve - but one additional False Negative. DT performs well, but with a little bit of errors. LR had a best predictive accuracy and reliability in identifying breast cancer cases compared to NB and DT.

4 Conclusion

This study examined the efficacy of three machine learning algorithms—LR, NB, and DT—in the diagnosis of breast cancer using the WDBC dataset. The main metrics used for comparison were area under the curve (AUC), first-order (F1), accuracy, precision, and recall scores. LR outperformed the other models on multiple metrics. These metrics included precision (97.62%), accuracy (97.37%), F1 score (96.47%), and ROC AUC analysis (99.74%).Because of its success in distinguishing between benign and malignant breast cancers, LR may have practical applications in clinical diagnostics. The NB method also yielded good results, showing high capacity of discrimination (AUC 99.74%). Its total efficiency figures had been somewhat worse than LR's, however. Decision Tree being the most explainable of all 3 models, simultaneously performed worst in each evaluation metric. The results indicated that for the classification of breast cancer, model selection is very important and LR was identified as the most reliable (stable) algorithm. Future research should explore the incorporation of additional machine learning methods and utilizing larger, more varied datasets to enhance diagnostic accuracy as well as reliability.

References

1. Huang, J., et al.: Global incidence and mortality of breast cancer: a trend analysis. Aging (Albany NY) **13**(4), 5748–5803 (2021)

2. Cao, W., Chen, H.D., Yu, Y.W., Li, N., Chen, W.Q.: Changing profiles of cancer burden worldwide and in China: a secondary analysis of the global cancer statistics 2020. Chin. Med. J. (Engl) **134**(7), 783–791 (2021)
3. Monticciolo, D.L., et al.: Breast cancer screening recommendations inclusive of all women at average risk: update from the ACR and Society of Breast Imaging. J. Am. Coll. Radiol. **18**(9), 1280–1288 (2021)
4. Agostinetto, E., Gligorov, J., Piccart, M.: Systemic therapy for early-stage breast cancer: learning from the past to build the future. Nat. Rev. Clin. Oncol. **19**(12), 763–774 (2022)
5. Pesapane, F., et al.: Will traditional biopsy be substituted by radiomics and liquid biopsy for breast cancer diagnosis and characterisation? Med. Oncol. **37**, 1–18 (2020)
6. Carmel Mary Belinda, M.J., Kannan, E., Alex David, S., Thangarasu, G., Begum, A.: Enhanced brain tumor diagnosis through multimodal fusion and 3D convolutional neural networks. In: 2023 Second International Conference On Smart Technologies For Smart Nation (SmartTechCon), Singapore, Singapore, pp. 1371–1375 (2023)
7. Ravikumar, S., Kannan, E.: Analysis on mental stress of professionals and pregnant women using machine learning techniques. Int. J. Image Graph. **23**(05), 2350038 (2023)
8. Salama, G.I., Abdelhalim, M., Zeid, M.A.E.: Breast cancer diagnosis on three different datasets using multi-classifiers. Breast Cancer (WDBC) **32**(569), 2 (2012)
9. Khandezamin, Z., Naderan, M., Rashti, M.J.: Detection and classification of breast cancer using logistic regression feature selection and GMDH classifier. J. Biomed. Informa. **111**, 103591 (2020)
10. Kumar, G.: Breast cancer detection using decision tree, naïve bayes, KNN and SVM classifiers: a comparative study. In: 2019 International Conference on Smart Systems and Inventive Technology (ICSSIT), pp. 683–686. IEEE (2019)
11. Zhou, J., Asteris, P.G., Armaghani, D.J., Pham, B.T.: Prediction of ground vibration induced by blasting operations through the use of the Bayesian Network and random forest models. Soil Dyn. Earthquake Eng. **139**, 106390 (2020)
12. Alfian, G., et al.: Predicting breast cancer from risk factors using SVM and extra-trees-based feature selection method. Computers **11**(9), 136 (2022)
13. Bharati, S., Podder, P., Mondal, M.: Artificial neural network based breast cancer screening: a comprehensive review. arXiv preprint arXiv:2006.01767 (2020)
14. Khamparia, A., et al.: Diagnosis of breast cancer based on modern mammography using hybrid transfer learning. Multidimension. Syst. Signal Process. **32**, 747–765 (2021)
15. Kumaraswamy, E., Kumar, S., Sharma, M.: An invasive ductal carcinomas breast cancer grade classification using an ensemble of convolutional neural networks. Diagnostics **13**(11), 1977 (2023)
16. Sundar, K., Vijay, K., Manohar, E.: Automated polyp detection in colorectal cancer diagnosis using deep learning techniques. In: 2024 8th International Conference on I-SMAC (IoT in Social, Mobile, Analytics and Cloud) (I-SMAC), pp. 1726–1731. IEEE (2024)
17. Vijay, K., Hameed, P.T.S.S., Bhavani, M., Jaeyalakshmi, M.: Data mining-based classification algorithms for predicting mental health. In: Geetha, R., Dao, NN., Khalid, S. (eds.) Advances in Artificial Intelligence and Machine Learning in Big Data Processing. AAIMB 2023. Communications in Computer and Information Science, vol. 2202. Springer, Cham (2025). https://doi.org/10.1007/978-3-031-73065-8_2
18. Jaeyalakshmi, M., Janani, P.K., Janani Priya, P., Bhavani, M., Narayana, K.E.: Detection of lung cancer using deep learning model and radiomics method. In: 2024 International Conference on Communication, Computing and Internet of Things (IC3IoT), pp. 1–5. IEEE (2024)

Comparative Evaluation of Machine Learning Methods for Enhanced Intrusion Detection

S. N. Ananthi(✉) and D. Hemanand

S.A Engineering College, Chennai 600077, India
{ananthisn,drhemanand}@saec.ac.in

Abstract. The rapid development of cyber-attacks has emphasized the importance for advanced Intrusion Detection Systems (IDS) capable of recognizing and preventing sophisticated attacks. Conventional IDS approaches, often struggle with the complexity and variety of modern intrusions requiring the creation of more robust and durable solutions. This survey offers a comprehensive analysis of several Machine Learning (ML) techniques used in the creation of IDS, focusing on their methodologies, advantages, and potential drawbacks. The review explores the use of supervised, unsupervised and hybrid ML algorithms for intrusion detection, analyzing how well they detect intrusions in various scenarios and datasets. Furthermore, the paper explores influence of feature selection techniques on performance outcomes ML-based IDS. To determine the effectiveness of these methods over several datasets and situations, a number of performance metrics are investigated. In order to help academics and practitioners choose and optimize machine learning methods for reliable intrusion detection, this study attempts to give a thorough overview.

Keywords: Cyber-attacks · Machine Learning · Intrusion Detection System

1 Introduction

Cyber threats have brought attention to the need for improved intrusion detection systems (IDS) can recognize and thwart sophisticated attacks [1]. More reliable and flexible solutions must be developed because traditional IDS approach frequently fail to handle the complexity and diversity of contemporary intrusions. Hybrid machine learning models, which integrate various learning approaches, present a viable way to improve IDS performance by making use of the benefits of various models [2]. Hybrid models minimize false positives and increase intrusion detection accuracy by combining multiple ML techniques, including supervised, unsupervised, and ensemble learning. The capacity of these algorithms to analyze and categorize intricate network traffic patterns makes them ideal for identifying new and advanced assaults. The comparative analysis of such hybrid approaches provides valuable insights into their effectiveness and efficiency. This research evaluates and compares various hybrid ML models applied to advanced IDS. The study intends to identify which models are best at intrusion detection by carefully examining how well they perform across a range of assault scenarios and datasets. The

R. Geetha et al. (Eds.): AAIMB 2024, CCIS 2653, pp. 267–280, 2026.
https://doi.org/10.1007/978-3-032-13026-6_19

results from this comprehensive assessment will inform the creation of more advanced and reliable IDS systems, ultimately strengthening cybersecurity defences. By determining the best hybrid strategies, this research seeks to further intelligent intrusion detection and provide insightful information for next IDS deployments.

2 Cyber Attacks

Cyber-attacks are deliberate endeavors to get unapproved access to computer systems, networks, or gadgets that can steal, reveal, change, deactivate, or delete data. Common types include:

- **Malware**: Viruses, ransomware, spyware, and worms that access data on devices.
- **Phishing**: Attacks using email or communication channels to deceive users into revealing personal information.
- **Spoofing**: Deceiving users by posing as legitimate individuals or businesses.
- **Backdoor Trojan**: Malicious programs installing malware under false pretences.
- **Ransomware**: Preventing access to devices until a ransom is paid.
- **Password attacks**: Techniques like keylogging or phishing to steal passwords.
- **Internet of Things (IoT) attack**: Targeting vulnerabilities in IoT devices.
- **Crypto jacking**: Unauthorized use of a system for cryptocurrency mining.
- **Drive-by download**: Downloading malicious code through flawed security systems.
- **Denial-of-service attack**: A system shuts down when traffic overloads it.

3 Intrusion Detection System (IDS) & Types of IDS

IDS come in two forms: software applications and specialized hardware devices. The two primary detection techniques used by these systems are anomaly- based and signature-based [3]. Incoming network packets are compared to a database of recognized attack patterns using the signature-based approach. However, anomaly- based systems use machine learning techniques to detect and mark anomalous network activity that differs from typical patterns of behaviour. Every approach has advantages and disadvantages. For example, anomaly-based detection can reveal previously undiscovered attack pathways, whereas signature-based detection is best at identifying known dangers.

Fig. 1. Role of IDS

Figure 1 illustrates the significance of IDS using machine learning approaches. IDS types consist of:

Network Intrusion Detection System (NIDS): Monitor all network traffic, both coming in and departing out.
Host Intrusion Detection System (HIDS): Deployed on individual devices to detect threats.
Signature-based IDS (SIDS): Installed on separate devices in order to identify threats.
Anomaly-based IDS (AIDS): Detects unusual network behavior using machine learning.
Perimeter IDS (PIDS): Protects critical infrastructure.
Virtual Machine IDS (VMIDS): Keeps an eye out for cyberattacks.
Stack-based IDS (SBIDS): Connects to TCP/IP to filter harmful traffic.

4 Literature Review

This section looks at the corpus of recent cybersecurity research, focusing especially on intrusion detection systems. The paper delves into the ways in which scholars have utilized deep learning and ML techniques to develop and evaluate IDS. Through the integration of the findings from multiple studies, this overview aims to give readers a comprehensive understanding of the latest intelligent intrusion detection techniques and their effectiveness in enhancing network security.

The NSL-KDD dataset is the focus of this study [2], which investigates the creation of an IDS utilizing a variety of machine learning approaches. The study assesses a number of algorithms' capacity to identify different kinds of network assaults, including Random Forests, Decision Trees, and Support Vector Machines. The study examines carefully chosen characteristics to ascertain which algorithm works best in the context of intrusion detection. The research offers a thorough analysis of the procedures and applications of every algorithm, providing insightful information about the advantages and disadvantages of each. Through this comparative analysis, the research contributes to the ongoing effort to enhance network security protocols and IDS capabilities.

The problem of declining test accuracy in intrusion detection systems with increasing feature size is addressed in this study [4]. The UNSW-NB15 and NSL-KDD datasets' feature space is made simpler by the researchers using an XGBoost-based feature selection technique. Their method assesses a number of performance indicators and shows better outcomes for different hybrid models. For instance, XGBoost-LSTM achieves 88.13% test accuracy in binary classification on NSL-KDD, while XGBoost-GRU reaches 78.40% in multiclass classification on UNSW-NB15.

New security concerns have been brought about by the development of IoT devices, particularly in networks that combine SDN topologies with traditional and IoT protocols. An LSTM-based approach for identifying threats in SDN-enabled IoT systems was presented in this research [5], which used two specialized datasets to compare several ML and DL models. The research introduces a novel LSTM architecture for multiclass attack classification, achieving high accuracy. Furthermore, the study examines the quality of the dataset and embeddings using visualization approaches, proving the model's efficacy in tackling security vulnerabilities unique to the Internet of Things.

This study [6] leverages the exceptional capabilities of deep learning to develop a sophisticated NIDS. CICIDS2018 and Edge_IIoT are two real-time traffic datasets used to thoroughly assess the suggested DL model. It achieves remarkable accuracy rates of 100% and 99.64% in multiclass classification tasks. In order to solve the changing difficulties in network security, this study demonstrates the creative use of deep learning techniques to enhance the efficacy and intelligence of intrusion detection systems. The findings demonstrate deep learning's potential in significantly enhancing NIDS performance.

An innovative method for detecting intrusions in IIoT and IoT settings has been suggested, focusing on privacy preservation in machine learning-based systems. The framework, called F-BIDS, employs a federated blending model that integrates Decision Tree and Random Forest classifiers with a Neural Network meta-classifier. This innovative design allows for federated training on meta-data, effectively addressing privacy concerns by avoiding direct use of sensitive user data. The research [7] shows how sophisticated machine learning methods can be modified to improve security and privacy in networks.

Evaluation using the Edge-IIoT set and In SDN datasets demonstrates improved attack detection and privacy preservation. The research highlights F-BIDS as a robust solution that mitigates vulnerabilities in traditional federated learning approaches while enhancing generalization and predictive performance in IDS.

In an effort to get beyond the drawbacks of traditional signature-based intrusion detection systems, this study [8] investigates ensemble-based machine learning. Using publicly available datasets, numerous ensemble techniques are evaluated in the study, including Gradient Boosting and Random Forest. Various feature selection methods are employed to identify crucial attributes. Results indicate that Random Forest outperforms other approaches, demonstrating high accuracy and low false positive rates, thus offering improved protection against evolving cyber threats in networks and computer systems.

Due to the increasing frequency of security breaches, cyber-physical system (CPS) security is an essential component of smart healthcare. By introducing a deep learning-based intrusion detection system (IDS) that uses CNN and bidirectional LSTM to identify DDoS, our study tackles these issues. The proposed model uses CNN features to distribute traffic and includes four layers, max pooling, and thick layer in its architecture. The model analyzed using Python achieved significant results with hyper parameter settings of batch size 500, 20 epochs, 25 classes, and using ReLU and softmax activation functions. By addressing the shortcomings of conventional IDS, this method [9] seeks to improve medical network security.

This research [10] highlights a cybersecurity model that leverages AI and ML to safeguard critical assets and prevent data breaches. The approach significantly reduces the danger of cyberattacks and prevents critical information from leaking by integrating various technologies, improving the entire security posture. When AI and ML are included into cybersecurity projects, defense mechanisms against dynamic threats should greatly increase. Advanced algorithms in ML analyze extensive data, identify patterns and anomalies, and enhance threat response, thus improving detection rates and reducing false positives.

High-dimensional datasets often pose challenges for intrusion detection systems (IDS) due to redundant and irrelevant data. Traditional single classifiers may struggle with varied attack types, and many existing models are outdated and less effective against novel threats. This study suggests a unique IDS architecture that makes use of feature selection and ensemble learning to overcome these problems. It employs an ensemble approach that combines the dimensionality reduction heuristic method CFS-BA with the Forest PA, Random Forest, and C4.5 algorithms. Using the NSL-KDD, AWID, and CIC-IDS2017 datasets, the system outperforms other approaches [11].

The authors [12] examine the function of ensemble-based methods in enhancing intrusion detection systems. It emphasizes how these techniques improve the performance of IDS by integrating multiple classifiers to better detect and mitigate cyber threats. The study reviews various ensemble strategies and their effectiveness in bolstering cybersecurity, reflecting on the growing complexity of attack methods. The authors highlight the critical role of ensemble approaches in strengthening cybersecurity defences and adapting to evolving threats.

The paper [13] addresses the increasing security challenges in IoT networks by proposing an Intelligent IIDS that utilizes ML algorithms for categorization and assault detection. It critiques existing ML-based IDSs for their limitations in analyzing contemporary IoT attack behaviors and multi-class attack detection. Evaluating three ML classifiers against 15 attack types using the IoT23 dataset, the study demonstrates the IIDS's effectiveness in detecting and classifying both binary and multi-class attacks, including zero-day attacks. Additionally, the incorporation of preprocessing methods like SMOTE, data normalization, and feature selection enhances the robustness of the proposed IIDS, improving IoT network security.

The paper [14] examines the increasing significance of network security in light of expanding Internet usage and developing technology. It highlights the need for effective network attack detection and categorization to develop robust defense strategies. By investigating issues caused by intruders, the paper offers insights into enhancing network security and proposes a defense approach for future research.

The paper [15] presents a deep learning-based approach utilizing generative adversarial networks to detect vulnerabilities and breaches in IoT-driven Industrial Internet of Things (IIoT) networks. By comparing unsupervised and discriminative methods, the proposed model achieves up to 97% accuracy in attack detection, showcasing improvements in performance, reliability, and data confidentiality. The approach demonstrates high true and detection rates across multiple datasets.

A recent investigation [16] tackles IoT network security by introducing a method of feature selection based on genetic algorithms, designed to manage complex traffic data. Researchers show that cutting characteristics from 40 to 6 produces remarkable results using the Bot-IoT dataset, with an F1-score of 99.63% and a detection accuracy of 99.98%. Compared to methods that do not use feature selection, this strategy not only improves detection performance but also significantly cuts down on training time. The findings underscore the importance of efficient feature selection in improving IoT network attack detection systems.

A new research [17] investigates the effectiveness of supervised and unsupervised learning approaches for cyberattack detection in smart grid networks using the CICD-DOS 2019 dataset. The study compares a number of models, including Principal Component Analysis and Variational Autoencoder for unsupervised learning and Gaussian Naïve Bayes and Alex Neural Network for supervised learning. The results show that the Variational Autoencoder performs well in the unsupervised category and that the Alex

Neural Network outperforms other supervised models. This comparative study offers valuable insights into the benefits of several machine learning algorithms for enhancing smart grid security.

A recent analysis emphasizes how the rise in cyberattacks has led to an increasing requirement for sophisticated IDS. Using the NSL-KDD dataset, the study [18] investigates a neural network-based method for forecasting security breaches. It emphasizes how important intrusion detection methods based on hosts and networks are to bolstering cybersecurity defenses as a whole. To improve intrusion detection techniques in the face of changing cyberattacks, Python Spyder is used in the study's simulations.

In order to enhance IDS, this study [19] illustrates the Class Probability Random Forest (CPRF) method for network attack detection using the CICIDS2017 dataset. By generating a fresh feature set and forecasting class probabilities, CPRF increases detection accuracy to 99.9%. Through thorough validation and hyper parameter adjustment, the technique surpasses current methods and shows notable improvements in network security.

This study [20] addresses the challenge of anomaly detection in industrial control systems by presenting a novel network intrusion detection system (NIDS) that makes use of deep auto encoders that have been trained on network flow data. The approach, which excels in detecting distributed denial of service attacks and identifying abnormal device behavior, outperforms existing unsupervised learning methods. Validated in real industrial settings, this low-cost, low-processing solution demonstrates effective and practical deployment for real-world scenarios.

This study [21] presents a novel IDS technique that integrates Gated Recurrent Units (GRU) and CNN to improve feature learning for network attack detection. With an accuracy of 98.73% and an FPR of 0.075, the technique showed significant improvements over prior methods using the CICIDS-2017 dataset for evaluation. The outcomes demonstrate how well CNN–GRU combinations work to maximize IDS performance in practical cybersecurity scenarios.

This study evaluates the effectiveness of ML and DL classifiers for anomaly-based malware and network intrusion detection [22]. It talks about how they fit within frameworks for network, mobile, and IoT security, addressing deployment issues and offering ways to make them better. Future research should concentrate on enhancing the resilience and security of digital infrastructures.

The review [23] highlights how the integration of technologies from the fourth industrial revolution has made DCSs, SCADAs, and ICSs more vulnerable. In order to safeguard critical infrastructure, it looks into ML-based IDSs, emphasizing the difficulties in identifying zero-day assaults. The survey includes an analysis of security datasets and recent research advances over the previous five years.

A new research looks at the difficulties caused by distributed denial of service (DDoS) assaults, which employ bot networks to send an overwhelming amount of requests to servers. The study contrasts several ML and DL techniques for DDoS attack detection and analysis. The study [24] intends to improve knowledge of ML and DL applications in cybersecurity by assessing these approaches' efficacy. The results address a crucial component of network security and aid in the creation of more effective DDoS mitigation techniques.

Advanced machine learning integration with SDN-based intrusion detection systems is the subject of a recent research. Hybrid Feature Selection-LightGBM (HFS-LGBM) IDS is a novel approach proposed by the researchers [25]. It combines two feature selection techniques: correlation-based feature selection and Random Forest Recursive Feature Elimination.

This hybrid technique performs better when tested on the NSL-KDD dataset in conjunction with LightGBM for attack detection. The results demonstrate how effective machine learning algorithms and complex feature selection strategies may be used to improve intrusion detection in software-defined networks.

This literature review [26] focuses on the vulnerabilities introduced by SDN-based Intrusion Detection Systems (IDS), particularly in detecting probe attacks that exploit system susceptibilities through device scanning. It highlights how to efficiently identify threats with the use of cutting-edge machine learning techniques, particularly Grey-wolf Optimizer (GWO) and Light Gradient Boosting Machine (LightGBM). The study, using the novel In SDN dataset, demonstrates significant improvements in IDS performance with metrics such as 99.8% accuracy and 99.99% precision, outperforming existing methods.

This literature review [22], which examines AI-driven methods for cyber threat identification, emphasizes the application of machine learning classifiers and ensembles for anomaly detection in malware and network intrusions. It identifies obstacles to incorporating AI solutions into frameworks for network, mobile, and IoT security and suggests solutions. Future research directions to strengthen the security and resilience of contemporary digital ecosystems are also outlined in the assessment.

One of the main subjects covered in this study is the detection of Distributed Denial of Service (DDoS) attacks, which pose a significant risk to network security [27]. It highlights the use of state-of-the-art deep learning techniques, especially a special combination of CNN and Long Short-Term Memory (LSTM) algorithms. Using the NSL-KDD dataset, the suggested seven-layer model achieves an impressive accuracy of 99.20%, outperforming conventional techniques.

5 Cyber-Attack Datasets

Cyber-attack datasets are critical resources used for developing and evaluating cybersecurity tools, such as IDS and other security solutions. These datasets typically contain records of normal and malicious activities, including various types of network traffic, logs, and system events that simulate real-world attack scenarios. Below are some commonly used cyber-attack datasets:

5.1 NSL-KDD

The goal of NSL-KDD, an enhanced version of the KDD'99 dataset, is to address the representativeness and redundancy problems of the original dataset. It is frequently employed in IDS evaluations. Records in this dataset are classified as normal or as belonging to one of numerous attack types, such as probing, DoS, R2L, and U2R.

5.2 UNSW-NB15

A collection of both regular and unusual network traffic produced by the Australian Centre for Cyber Security (ACCS). Modern attack types that are absent from older datasets are included. This dataset includes 100 GB of unlabeled network traffic labeled for several attack types, including worms, backdoors, analysis, and fuzzers.

5.3 CICIDS2017

This dataset, created by the Canadian Institute for Cybersecurity, contains the most recent and safest common attack scenarios. This dataset's features include a broad spectrum of assaults, such as port scanning, botnets, DDoS, brute force, heartbleed, and DoS.

5.4 CICIDS2018

A newer version of CICIDS2017, this dataset includes even more diverse attack scenarios and is widely used for research in network intrusion detection. Models authentic network traffic and provides labels for several attack categories, such as DDoS, web, brute force, penetration, and botnet attacks.

5.5 Edge-IIoTset

A dataset specifically created for Industrial IoT (IIoT) environments, focusing on cyber-attacks targeting critical infrastructure. The dataset ncludes a mix of normal and attack traffic, with labels for various IIoT-specific attack vectors like command injection, code injection, and device impersonation.

5.6 TON_IoT

A dataset tailored for IoT and network-based cyber-attacks, created by the UNSW Canberra Cybersecurity Lab. This dataset contains telemetry data, network traffic, and system logs from IoT devices with various attack types such as DDoS, ransomware, and data exfiltration.

5.7 MalMem2022

A relatively new dataset focused on malware analysis, specifically targeting memory-based attacks. This dataset contains process information and memory dumps from a variety of malware infections, such as trojans, ransomware, and rootkits.

6 Methodology Workflow for Cyber Attack Detection

The Effective cyber-attack detection relies on a comprehensive methodology that integrates advanced machine learning techniques and robust data handling processes. Figure 1 shows this workflow and ensures accurate identification of malicious activities by combining data acquisition, preprocessing, feature engineering, and hybrid model development.

Data Acquisition: Collect network traffic data, including both normal and attack traffic, using established datasets like KDD Cup, UNSW-NB15, and CICIDS2017.

Data Preprocessing: Clean the data by removing duplicates and handling missing values, normalize features, and encode categorical variables into numerical values.

Feature Engineering: Extract and select relevant features using techniques like PCA or Chi-squared tests, and create new features based on domain knowledge to enhance model performance.

Dataset Splitting: Separate the data into test, validation, and training sets, making sure that the normal and attack cases are balanced by appropriate stratification.

Hybrid Model Development: Develop and integrate various models, such as CNN-LSTM, Random Forest with XGBoost, and Autoencoder with SVM, to leverage different algorithm strengths.

Model Training and Optimization: Train each hybrid model on the training set and use the validation set for hyperparameter tuning with methods like grid search or Bayesian optimization.

Performance Evaluation: Assess the model's performance on the test set using metrics such as ROC-AUC, F1-score, accuracy, precision, and recall.

Comparative Analysis: Compare models based on performance metrics, computational efficiency, and interpretability, considering training and inference times.

Results and Conclusions: Summarize key findings, identify the best-performing models, and discuss strengths, weaknesses, and potential improvements for future research and implementation (Fig. 2).

The Table 1 and Fig. 3 summarizes various methodologies for intrusion detection systems (IDS) and their accuracies. The Hybrid DL Approaches: such as LSTM and CNN (accuracy up to 98.5%) or DCNN and BiLSTM (up to 99.2%) consistently achieve high accuracy, reflecting their advanced capabilities in detecting sophisticated attacks. In next category Feature Selection and Ensemble Methods that utilize feature selection combined with machine learning or ensemble classifiers show strong performance, with accuracies ranging from 94.5% to 97.8%. These methods are effective in improving detection by focusing on relevant features. And also General machine learning techniques and real-time detection approaches show varied accuracy, from 94.6% to 97.1%. While effective, they generally fall short of the highest accuracies achieved by hybrid and DL models. Overall, hybrid DL models demonstrate superior accuracy compared to traditional feature selection and machine learning techniques.

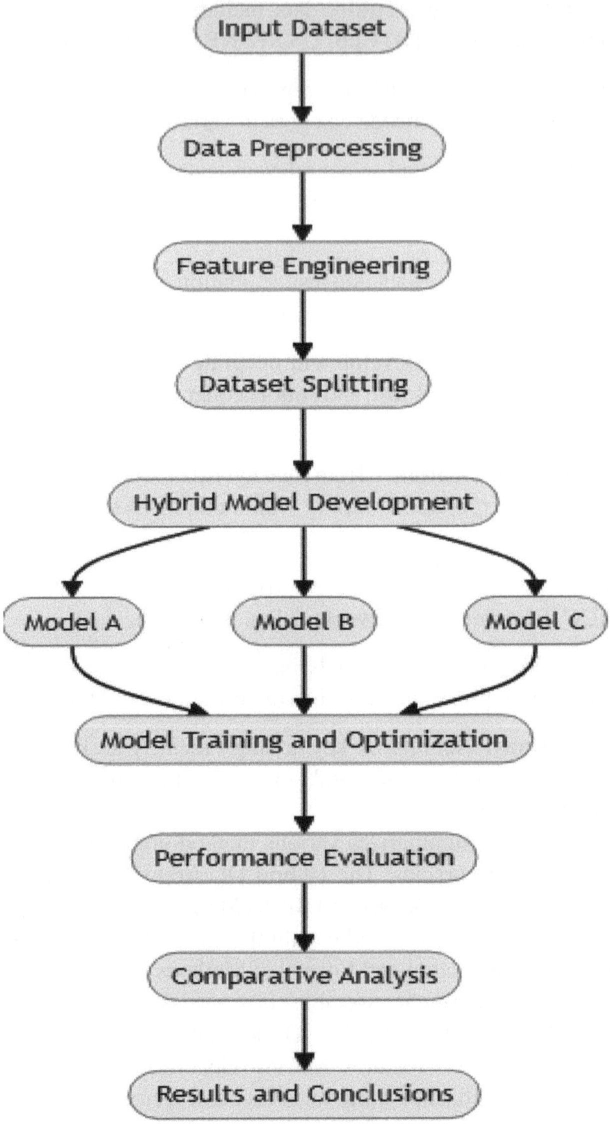

Fig. 2. Machine Learning algorithms workflow

7 Conclusion

The use of state-of-the-art machine learning algorithms has made IDS far more advanced and efficient. Our study offers a comprehensive examination of the various machine learning algorithms utilized in IDS creation, highlighting their methods, advantages, and disadvantages. We looked at supervised, unsupervised, and hybrid methods, evaluating how well they detected intrusions in various settings and datasets. Additionally,

Table 1. Summary of various attacks detection ML methodology, Dataset and Achieved Accuracy

Ref. No	Description	Methodology	Dataset	Performance Metric (Accuracy)
[2]	IDS designed using feature selection and ML algorithms to enhance accuracy and efficiency	Feature Selection + ML	-	Not specified
[4]	IDS using a deep learning technique with recurrent neural networks	RNN Framework	KDD Cup 99	95.3%
[5]	IDS for SDN-enabled IoT networks	Deep Learning Approach	NSL-KDD	98.7%
[6]	Efficient hybrid IDS using DCNN and BiLSTM deep learning techniques	DCNN + BiLSTM Hybrid	NSL-KDD	99.2%
[7]	Federated-blending based IDS for improved security in mobile computing environments	Federated- Blending Approach	CICIDS 2017	96.5%
[8]	Robust IDS using ensemble-based machine learning for network security	Ensemble-Based ML	UNSW-NB15	97.8%
[9]	Efficient hybrid model for detecting DoSattacksincyber-physical systems	Hybrid Deep Learning Model	CICIDS 2017	99.4%
[10]	Cybersecurity event detection using a machine learning technique	ML Technique	UNSW-NB15	94.6%
[11]	Feature selection and an ensemble classifier for efficient IDS	Feature Selection + Ensemble	KDD Cup 99	97.2%
[12]	Ensemble-based IDS designed for IoT environments	Ensemble-Based Approach	IoT-23 dataset	95.8%
[13]	Intelligent IDS for IoT attack detection and identification using machine learning	Machine Learning-Based IDS	IoT-23 dataset	96.3%

(continued)

Table 1. (*continued*)

Ref. No	Description	Methodology	Dataset	Performance Metric (Accuracy)
[15]	Detection of attacks and intrusions in real time in IoT cybersecurity systems	Real-Time Detection Approach	IoT-23 dataset	97.1%
[16]	Detecting IoT botnet attacks with a genetic algorithm through efficient feature selection	Genetic Algorithm-Based FS	IoT-23 dataset	94.5%
[18]	ANN technique for predicting cyber-attacks using an intrusion detection system	Artificial Neural Network (ANN)	-	96.0%
[20]	IDS for detecting DDoS attacks in ICS	Deep Autoencoders	-	98.2%
[21]	Hybridmodelwithfeature optimization for intrusion detection	Hybrid DL Model + Feature Optimization	KDD Cup 99	96.7%
[24]	Detection of DDoS attacks using and comparing ML and DL techniques	ML and DL Techniques Comparison	CICIDS 2017	97.4%
[25]	IDS for SDN using ML techniques	Machine Learning for SDN	-	95.2%
[27]	Hybrid CNN and LSTM-based IDS for detecting DDoS attacks	CNN + LSTM Hybrid	CICIDS 2017	98.5%

we looked at feature selection tactics and their impact on the effectiveness of ML-based IDSs. The effectiveness of these techniques was assessed using the following evaluation metrics: F1-score, false positive rate, and detection rate. In order to support the continuous development of network security, this study attempts to help academics and practitioners choose and optimize machine learning methods for reliable intrusion detection.

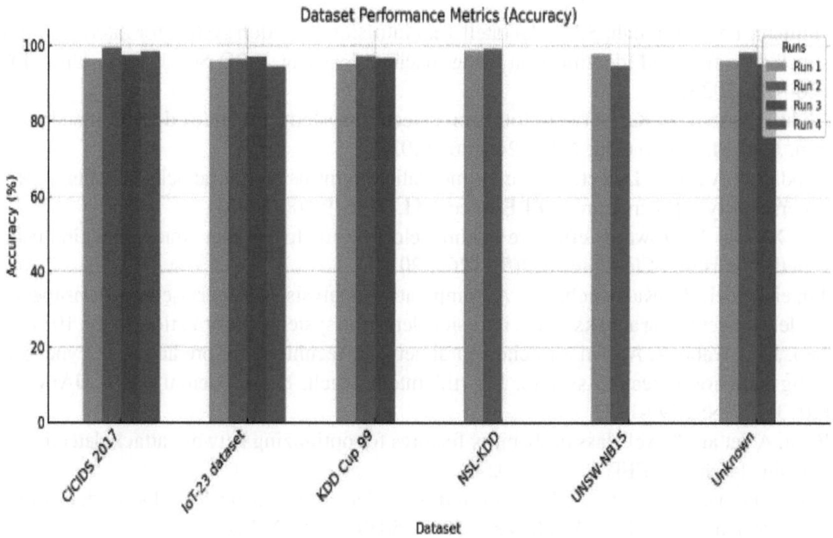

Fig. 3. Comparison of various attacks detection ML methodology, Dataset and Achieved Accuracy

References

1. Tazi, K., Abdi, F., Abbou, M.F.: Review on cyber-physical security of the smart grid: Attacks and defense mechanisms. In; Proceedings of the 2015 3rd International Renewable and Sustainable Energy Conference (IRSEC). IEEE (2015)
2. Venkatesan, S.: Design an intrusion detection system based on feature selection using ML algorithms. Math. Stat. Eng. Appl. **72**(1), 702–710 (2023)
3. De la Hoz, E., et al.: PCA filtering and probabilistic SOM for network intrusion detection. Neurocomputing **164**, 71–81 (2015)
4. Wolf, G.: A deep learning technique for intrusion detection system using a recurrent neural networks based framework. Comput. Rev. (2022)
5. Chaganti, R., et al.: Deep learning approach for SDN-enabled intrusion detection system in IoT networks. Information **14**(1), 41 (2023)
6. Hnamte, V., Hussain, J.: DCNNBiLSTM: an efficient hybrid deep learning-based intrusion detection system. Telematics Inform. Reports **10**, 100053 (2023)
7. Aouedi, O., Piamrat, K.: F-bids: federated-blending based intrusion detection system. Pervasive Mob. Comput. **89**, 101750 (2023)
8. Hossain, Md A., Islam, Md S.: Ensuring network security with a robust intrusion detection system using ensemble-based machine learning. Array **19**, 100306 (2023)
9. Sharma, A., et al.: An efficient hybrid deep learning model for denial of service detection in cyber physical systems. IEEE Trans. Netw. Sci. Eng. **10**(5), 2419–2428 (2023)
10. Muneer, S.M., Alvi, M.B., Farrakh, A.: Cyber security event detection using machine learning technique. Int. J. Comput. Innov. Sci. **2**(2), 42–46 (2023)
11. Zhou, Y., et al.: Building an efficient intrusion detection system based on feature selection and ensemble classifier. Comput. Netw. **174**, 107247 (2020)
12. Abbas, A., et al.: A new ensemble-based intrusion detection system for internet of things. Arab. J. Sci. Eng., 1–15 (2022)

13. Othman, T.S., Abdullah, S.M.: An intelligent intrusion detection system for internet of things attack detection and identification using machine learning. ARO Sci. J. Koya Univ. **11**(1), 126–137 (2023)
14. Abdul-Hussein, M.K.: Network intrusion detection systems for attack detection and prevention. J. Al-Ma'moon Coll. **1**(39), 246–261 (2023)
15. Kandhro, I.A., et al.: Detection of real-time malicious intrusions and attacks in IoT empowered cybersecurity infrastructures. IEEE Access **11**, 9136–9148 (2023)
16. Liu, X., Du, Y.: Towards effective feature selection for IoT botnet attack detection using a genetic algorithm. Electronics **12**(5), 1260 (2023)
17. Talaei Khoei, T., Kaabouch, N.: A comparative analysis of supervised and unsupervised models for detecting attacks on the intrusion detection systems. Information **14**(2), 103 (2023)
18. Jain, J.K., Waoo, A.A.: An artificial neural network technique for prediction of cyber-attack using intrusion detection system. J. Artif. Intell. Mach. Learn. Neural Netw. (JAIMLNN) (2023). ISSN: 2799-1172
19. Raza, A., et al.: Novel class probability features for optimizing network attack detection with machine learning. IEEE Access (2023)
20. Ortega-Fernandez, I., et al.: Network intrusion detection system for DDoS attacks in ICS using deep autoencoders. Wirel. Netw. **30**(6), 5059–5075 (2024)
21. Henry, A., et al.: Composition of hybrid deep learning model and feature optimization for intrusion detection system. Sensors **23**(2), 890 (2023)
22. Schmitt, M.: Securing the digital world: protecting smart infrastructures and digital industries with artificial intelligence (AI)-enabled malware and intrusion detection. J. Ind. Inf. Integr. **36**, 100520 (2023)
23. Pinto, A., et al.: Survey on intrusion detection systems based on machine learning techniques for the protection of critical infrastructure. Sensors **23**(5), 2415 (2023)
24. Al-Shareeda, M.A., Manickam, S., Saare, M.A.: DDoS attacks detection using machine learning and deep learning techniques: analysis and comparison. Bull. Electr. Eng. Inf. **12**(2), 930–939 (2023)
25. Logeswari, G., Bose, S., Anitha, T.J.I.A.: An intrusion detection system for sdn using machine learning. Intell. Autom. Soft Comput. **35**(1), 867–880 (2023)
26. Almazyad, A., Halman, L., Alsaeed, A.: Probe attack detection using an improved intrusion detection system. Comput. Mater. Continua **74**(3) (2023)
27. Issa, A.S.A., Albayrak, Z.: DDoS attack intrusion detection system based on hybridization of CNN and LSTM. Acta Polytech. Hung. **20**(2), 1–19 (2023)

Combining Principal Component Analysis Method and Artificial Neural Network Model for Surface Ozone Concentration Prediction in Baghdad, Iraq

Faten G. Abed[1], Jasim M. Rajab[1], and Hazim G. Daway[2(✉)]

[1] College of Science, Department of Atmospheric Sciences, Mustansiriyah University, Baghdad, Iraq
fatin.ghanim@uomustansiriyah.edu.iq

[2] College of Science, Department of Physics, Mustansiriyah University, Baghdad, Iraq
hazimdo@uomustansiriyah.edu.iq

Abstract. Surface ozone (O_3) is a major atmospheric pollutant that negatively impacts both human health and ecosystems. This research aimed to develop a combined model (PC-ANN) for predicting surface O_3 concentrations using Principal Component Analysis (PCA) and Artificial Neural Network (ANN) models in Baghdad. The input data included meteorological variables such as temperature (T2m), total precipitation (TP), boundary layer height (BLH), surface pressure (SP), wind components (V10m, U10m), along with pollutants like NO_2, CO, SO_2, PM2.5, and PM10. The findings indicated significant variations in the correlation between O_3 levels and the selected variables. The dataset's suitability for factor analysis was confirmed through KMO and Bartlett's tests, with a KMO value of 0.836 for Baghdad, and Bartlett's test resulting in a p-value $< 0.001(0.000)$. After applying varimax rotation, three principal components (PCs) were identified for Baghdad, accounting for 71.827% of the total variance. PC1, PC2, and PC3 were primarily linked to NO_2, PM2.5, and T2m, respectively. For the ANN model, various network structures were tested, and the optimal performance This was achieved using 4 nodes in the hidden layer and the tanh activation function resulting in an adjusted R^2 value of 0.871 for Baghdad. According to the ANN model, the variables derived from PCA—FAC-1 (NO_2), FAC-3 (T2m), and FAC-2 (PM2.5)—had importance scores of 100.0%, 74.6%, and 6.65%, corresponding to 0.552, 0.412, and 0.037, respectively. Additionally, the PC-ANN model effectively captured nonlinear relationships, demonstrating stronger predictive performance for complex datasets, with higher R^2 and lower RMSE and MAE values.

Keywords: Surface O_3 · principal component analysis PCA · Artificial neural network ANN · IRAQ

1 Introduction

Surface ozone (O_3) is considered an indirect air pollutant which causes significant risks to both ecosystems and human health. As a poisonous gas with a pungent and disagreeable odor, O_3 adversely impacts the environment and public health. High concentration of

R. Geetha et al. (Eds.): AAIMB 2024, CCIS 2653, pp. 281–296, 2026.
https://doi.org/10.1007/978-3-032-13026-6_20

ground-level O_3 can harm lung tissue, affect respiratory health, and inhibit plant growth [1–5]. Additionally, O_3 a greenhouse gas, contributes to photo-oxidation operations in the troposphere. It is not directly emitted into the atmosphere, but rather produced by chemical reactions involving precursor gases like volatile organic compounds VOCs, carbon monoxide CO, and nitrogen oxides NO_2 . Such interactions are more significant under situations of intense sunshine and high temperatures. [1, 4, 6], and the existence of ultraviolet (UV) radiation [7, 8]. Also, the production of O_3 is strongly affected by meteorological conditions [9] that affected by the atmosphere [10], High temperatures, low wind speeds, and reduced relative humidity generate ideal conditions for increased O_3 concentrations [11–15]. The correlation between O_3 concentration and various meteorological factors is highly intricate. The significance of each factor has been determined by examining their correlation with tropospheric O_3 levels. Resmi, C. *et al.* discovered that temperature is the main factor contributing to elevated tropospheric O_3 concentration. However, when temperatures surpass 42°C, a negative correlation between temperature and O_3 concentration grows [16]. Gunthe, S.S. *et al.* found that while the relationship between tropospheric O_3 concentration and air humidity is relatively minor, it keeps statistically significant [17]. Additionally, K. Warmiński *et al.* sate that O_3 interacts significantly with water vapor [18]. Wind can influence O_3 levels in both directions: it can increase concentrations by dispersing O_3 precursors, or it can decrease them by dispersing and diluting the pollutants [19].

The many factors mentioned, along with their complex interactions, make accurately modeling tropospheric O_3 concentrations challenging. However, precise predictions of surface O_3 levels are essential for devising effective mitigation strategies and safeguarding public health [20]. Existing methods for predicting surface O_3 levels can generally be grouped into two primary categories: numerical modeling and data-driven approaches [21]. Numerical models like chemical transport models (CTMs) that are mimic the chemical and physical processes involved in the production of O_3 [22]. Data-driven methods use statistical techniques to predict O_3 concentration by analyzing archival data, taking into account atmospheric conditions and emission factors. Statistical methods such as principal component analysis are useful for data analysis and predictive modeling [23]. PCA method permits us to analyze and measure the associations between variables in a data set. As a dimensionality reduction approach, it compresses a dataset into a lower-dimensional space whereas retaining as much information as possible [24].

Machine learning is one of the most efficient prediction technique, and its significant is increasing due to its ability to handle complex and nonlinear relationships in big data sets [25]. Neural networks (ANNs) are a subset of artificial intelligence techniques that mimic the structure and function of biological neural networks. It has achieved great success in various industries due to its ability to independently learn from examples, handle inaccurate and complex data, and give fast and accurate responses to previously unknown information [9]. ANNs are also more efficient than CTM and demand less processing time [26]. Although artificial neural network techniques are very powerful, they often require big data sets and are particularly sensitive to data availability and quality.

Numerous studies have employed various types of neural networks to effectively predict O_3 levels. For instance, (Gao, 2018) looked into the viability of predicting O_3

concentrations using ANN models that use meteorological factors as input variables. The outcomes demonstrated that ambient O_3 levels can be accurately predicted by ANN models [27]. (Pawlak, 2019) They developed ANN models to predict the average daily maximum surface O_3 concentration for the next day. Their analysis showed that predictions with significantly high relative errors were associated with abrupt decreases in daily maximum surface O_3 concentrations [28]. (Adnan, 2020) employ a recurrent neural network (RNN) to forecast O_3 values by incorporating CO and SO_2 levels with meteorological data. The findings illustrated that the model effectively predicted O_3 levels [29]. (Bowen, 2020) constructed a prediction model for O_3 concentrations using chaotic artificial neural network (CANN). The outcomes indicate that CANN model can provide accurate predictions, making it useful for pollution control efforts [30]. (AlOmer, 2020) compared a hybrid model (W-ANN) with a conventional ANN for predicting O_3 concentrations one hour ahead. The study found that the W-ANN model significantly outperformed the conventional model, resulting much lower errors and showing accurate predictions across divers time scales [31]. (Kapadia, 2022) constructed an O_3 prediction model utilizing an ANN and optimized input. The results disclosed that the model reached an efficiency of 79.4% when six parameters were incorporated into the machine learning algorithm [32]. Some studies combined statistical methods with ANN's for surface O_3 prediction, for instance, (Dhanya, 2022) used a set of statistical techniques to predict O_3 levels, incorporating principal component regression (PCR), neural networks (ANN), multiple linear regression (MLR), and a hybrid approach combined PCA and artificial neural network (PC-ANN) right with atmospheric parameters. The outcomes indicated that ANN and PC-ANN models outperformed linear regression models like PCR and MLR [33].

This research seeks to develop and assess the performance of a combined model which integrates the PCA method with the artificial neural network ANN-MLP algorithm to predict the surface O_3 concentration in Baghdad, additionally it aims to identify the main components among air pollutants and meteorological factors that affect the surface O_3 concentration. The input parameters of the integrated PC-ANN model includes air pollutant data (NO_2, CO, SO_2, PM2.5, and PM10) along with the meteorological factors (surface net solar radiation (SSR), 2-m temperature (T2m), total precipitation (TP), boundary layer height (BLH), surface pressure (SP), wind direction (WD) and wind speed (WS). The findings can serve as a reference for creating plans for O_3 pollution prevention and control, which will enhance public health and environmental preservation.

2 Methodology

2.1 Research Region

As shown in Fig. 1, the research region is Baghdad, the capital of Iraq. Iraq, a country in west Asia, is situated between 38°45' E and 48°45' E, east of the Greenwich Meridian, and between 29°5' N and 37°15' N, north of the equator [34, 35]. The country comprises 18 governorates, including Baghdad. Geographically, Baghdad is located at 33.31° N, 44.36° E [36] within the alluvial plain that stretches from central to southeastern Iraq [35]. With a population of approximately 7,216,040, Baghdad is Iraq's largest city, covering an area of 380 km^2. It is also the country's largest economic and administrative center [37]. This city was elected as the study area to analyze the variation of tropospheric O_3 over the course during the study timeframe.

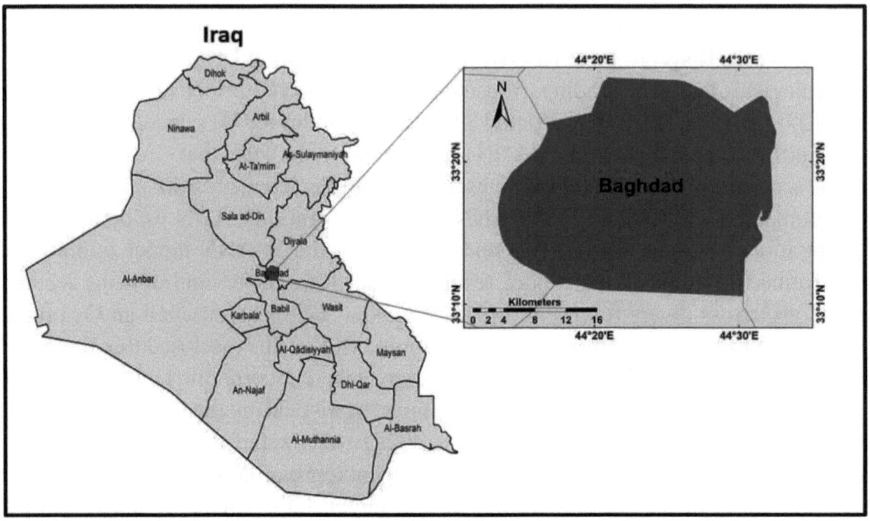

Fig. 1. The research region's geographic location Baghdad, Iraq.

2.2 Data Acquisition

Due to the scarcity of surface O_3 observations in Iraq, assessing variations or predicting O_3 concentrations presents a limitation. As a result, satellite data was utilized to monitor pollutant levels, including O_3, along with meteorological factors. The dataset includes concentrations the average O_3 mass mixing ratio (kg/kg), air pollutants, such as, NO_2, CO, SO_2, PM2.5, and PM10, along with meteorological factors like; TP (m), SSR (J/m^2), T2m (K), BLH (m), SP (Pa), and wind components at 10 m (U10m and V10m, m/s). These data were sourced from the European Centre for Medium-Range Weather Forecasts (ECMWF-ERA5) (https://www.ecmwf.int/en/forecasts/dataset/ecmwf-reanalysis-v5). The majority of the chemical species are stored as mass mixing ratios [kg/kg], representing kilogram of chemical species per kilogram of air. The data is accessible via the Atmosphere Data Store (ADS), either programmatically using the CDS API service or interactively through its download web. The study period spanned from January 1, 2020, to December 31, 2022. WS (m/s) and WD (degrees) were unavailable from ECMWF and were calculated using formula (1) and (2) [38]:

$$WD = 180 + atan2(u10, v10) \times \frac{180}{\pi} \tag{1}$$

$$WS = \sqrt{U10^2 + V10^2} \tag{2}$$

where: U10 and V10 represent the u-vector and v-vector of wind at a height of 10 meters, respectively.

The model-level field data is provided in NetCDF format on ADS and can be readily processed using Microsoft Excel. The data was analyzed using IBM SPSS software ver.26.0, to identify correlations between atmospheric variables (air pollutants and meteorological factors) and surface O_3 concentrations. PCA was used to identify the key factors (PCs) affecting surface O_3 concentration. A neural network (ANN-MLP) model was subsequently constructed to predict O_3 levels. The correlation between the observed (actual) and projected O_3 values was examined using Sigma Plot version 14.0 for Windows.

2.3 Model Development and Validation

The construction and development of the PC-ANN model is based on two parts, first, using the PCA method to evaluate the data set, identify the critical factors affecting the surface O_3 concentrations, and study the correlations between the selected variables including O_3 concentration, air pollutants, and meteorological factors. Subsequently, the neural network model is employed to predict O_3 levels using the principal components obtained through the PCA analysis.

2.3.1 Principal Component Analysis-PCA Method

(PCA) is an analytical method used to identify and categorize variables based on their correlation coefficients. The core idea of principal components analysis is to simplify a dataset by reducing its dimensionality that contains many interrelated variables while preserving as much of the variance of the data set as possible. at first, the predictor variables are switch into an equivalent number of principal components (PCs). After conducting PCA, a varimax rotation is utilized to maximize the loading of predictor variables on each component. The combination of PCA with orthogonal rotation (varimax) yields an ordered set of factors, where the PCs with eigenvalues greater than one are retained. In factor analysis, the eigenvalue method creates a new set of independent, uncorrelated variables—linear combinations of the initial variables—from the intercorrelated ones. The PCs are then ordered in descending importance based on the variance they explain. Although the count of principal components (PCs) matches the number of initial independent variables, the first few PCs typically account for most of the variation in the data, with each subsequent PC capturing the remaining variance while being uncorrelated with the preceding PCs [39–42]. The PCA method can be accomplished utilizing IBM SPSS Statistics, Version 26.0 for Windows.

Both tests Kaiser-Meyer-Olkin (KMO) and Bartlett's of sphericity Needs to be conducted on the dataset. The KMO test determines the sufficiency of sampling for each variable in the model, A KMO score greater than 0.5 implies that this data are appropriate for analysis. Meanwhile, Bartlett's test of sphericity demonstrates a significant relationship among variables, confirming the dataset's suitability for factor analysis ($p < 0.001$). These requirements should be achieved before start using PCA [43–45]. The resulted principal components (PCs) by the PCA method will be serves as input parameters in the ANN model's input layer.

2.3.2 Artificial Neural Networks (ANNs) Model

(ANNs) are advanced mathematical models that are commonly used to capture complex, nonlinear relationships between inputs and required output. ANN are exceptionally powerful and effective in forecasting tasks, especially when dealing with complex weather events. There are different types of neural networks, which differ in their structural design. Classical artificial neural networks typically consist of three layers: the input layer, one or more hidden layers, and the output layer, as illustrated in Fig. 2. The input layer feeds data into the network, which is processed in the hidden layer(s) using mathematical functions. Finally, the output layer provides the ANN's predictions. As data proceed through the network, it moves between artificial neurons, each assigned different weights. Before transitioning to the next layer, the data undergoes activation via an activation function and bias adjustment [31, 32].

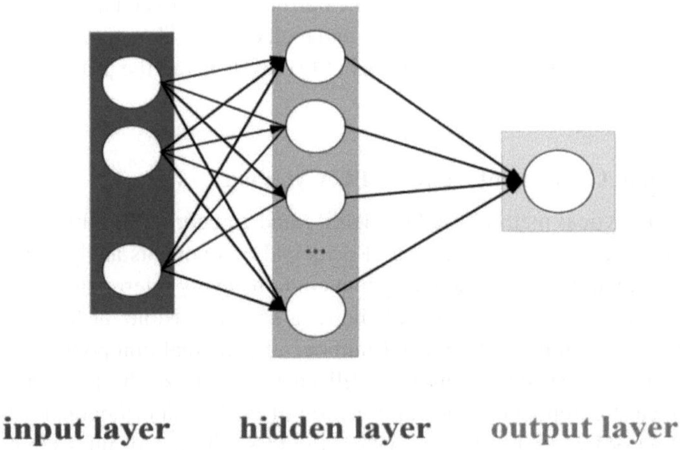

Fig. 2. An artificial neural network's schematic diagram [38].

This study employed Multiple Layer Perceptron (MLP) neural network models to effectively predict surface O_3 concentrations and analyze the impact of meteorological factors and air pollutants on these concentrations. An MLP network typically includes an input layer made up of basis nodes, one or more hidden layers composed of computational units (nodes), and an output layer of nodes. In this research, the sigmoid and hyperbolic tangent functions were employed as activation functions in the hidden layers of the MLP network, which was developed using IBM SPSS 26v software represented by the following equations [46].

$$s(x) = \frac{1}{1 + e^{-x}} \tag{3}$$

$$tanh(X) = \frac{e^{x} - e^{-x}}{e^{x} + e^{-x}} \tag{4}$$

We build the MLP model as follows: The input layer was defined based on the optimal combination of variables derived from the PCA model, while the output layer represented the predicted O_3 values. To prevent overfitting, we reduced the number of hidden layers in accordance with the parsimony principle [27], Since too many layers can cause overfitting, and too few layers could result in insufficient data processing [46]. To find the best-performing model, we tested different configurations with 3, 4, and 5 unites (nodes) in the hidden layer. 70% of the data was utilized for training to allow the model to self-learn and optimize its structure, whereas the remaining 30% was set aside for testing and evaluating model performance. as shown in Eq. (3) and (4),the activation functions chosen for the hidden layer were the sigmoid (S-shaped growth curve) and the hyperbolic tangent (tanh).

2.3.3 Model Validation

The model's performance was evaluated using three metrics; coefficient of determination (R^2), root mean square error (RMSE), and mean absolute error (MAE). The formulas for these metrics are presented below [38]:

$$R^2 = 1 - \frac{\sum_{i=1}^{n}(t_i - y_i)^2}{\sum_{i=1}^{n}(y_i - y_{i-average})^2} \quad (5)$$

$$RMSE = \sqrt{\frac{1}{n}\sum_{i=1}^{n}(t_i - y_i)^2} \quad (6)$$

$$MAE = \frac{1}{n}\left(\sum_{i=1}^{n}|t_i - y_i|\right) \quad (7)$$

where:

n indicates the sample size.
t_i donates as the predicted value,
Y_i indicates the real value, and.
(Y_i-average) refers to the mean of the observed values.

3 Result and Discussion

3.1 PCA Method

The PCA method was applied to the entire dataset of Baghdad province spanning the years 2020 to 2022. The input variables included air pollutants, such as surface O_3, NO_2, CO, SO_2, PM2.5, and PM10 concentrations, alongside meteorological parameters like T2m, TP, SP, SSR, BLH, WD, and WS. A Pearson correlation analysis was undertaken to assess the relationships between O_3 levels, air pollutants, and meteorological factors. Table 1, displays the Pearson correlation matrix for the input parameters in Baghdad.

Table 1. Pearson correlation matrix of input parameters for Baghdad.

variable	O3	BLH	SP	SSR	T2m	TP	CO	NO_2	SO_2	PM2.5	PM10	WD	WS
O3	1	0.703^{**}	-0.783^{**}	0.773^{**}	0.854^{**}	-0.082^{**}	-0.703^{**}	-0.777^{**}	-0.711^{**}	-0.139^{**}	0.239^{**}	-0.110^{**}	0.355^{**}
BLH		1	-0.659^{**}	0.609^{**}	0.751^{**}	-0.131^{**}	-0.493^{**}	-0.527^{**}	-0.472^{**}	-0.165^{**}	0.132^{**}	−0.024	0.183^{**}
SP			1	-0.662^{**}	-0.881^{**}	0.040	0.601^{**}	0.615^{**}	0.570^{**}	−0.054	-0.417^{**}	0.005	-0.451^{**}
SSR				1	0.734^{**}	-0.233^{**}	-0.601^{**}	-0.668^{**}	-0.569^{**}	-0.112^{**}	0.321^{**}	-0.120^{**}	0.415^{**}
T2m					1	-0.147^{**}	-0.615^{**}	-0.621^{**}	-0.573^{**}	-0.072^{*}	0.287^{**}	-0.106^{**}	0.359^{**}
TP						1	0.042	0.038	0.020	0.101^{**}	0.005	−0.017	−0.044
CO							1	0.919^{**}	0.892^{**}	0.208^{**}	-0.277^{**}	0.065^{*}	-0.433^{**}
NO_2								1	0.900^{**}	0.204^{**}	-0.301^{**}	0.112^{**}	-0.449^{**}
SO_2									1	0.221^{**}	-0.239^{**}	0.117^{**}	-0.393^{**}
PM2.5										1	0.737^{**}	0.289^{**}	−0.008
PM10											1	0.185^{**}	0.315^{**}
WD												1	0.000
WS													1

**. The correlation is statistically significant at the 0.01 level (two-tailed).
*. The correlation is statistically significant at the 0.05 level (two-tailed).

As shown in Table 1, O_3 levels in Baghdad exhibited a strong positive correlation with T2m (0.854), SSR (0.773), and BLH (0.703), as well as a weaker positive correlation with WS (0.355) and PM10 (0.239). Strong negative correlations were observed with SP (–0.783), NO_2 (–0.777), SO_2 (–0.711), and CO (–0.703), while weaker negative correlations were found with PM2.5 (–0.139), WD (–0.110), and TP (–0.082).

A Varimax rotation was performed to accomplish one of the primary objectives of PCA: extracting a few components that account for the maximum variation. The KMO and Bartlett's tests of sphericity are critical in PCA to evaluate if factor analysis is appropriate and sufficient for the data in present study. Table 2, presents the results of these tests, with the KMO value for Baghdad being 0.836, which is greater than 0.05, indicating that the data is satisfactory for PCA. Following the Kaiser criterion, PCA typically retains components with eigenvalues equal to or greater than one for further analysis. The Bartlett's test result for Baghdad was 0.000 ($p < 0.001$), confirming that the data met the necessary conditions for a valid factor analysis in PCA.

Table 2. KMO and Bartlett's test for Baghdad

	Kaiser-Meyer-Olkine indicator of sampling adequacy		***0.836***
Baghdad	Bartlett's test of sphericity	Approx.Chi-Square	13044.843
		df	78
		Sig	0.000

After conducting the KMO and Bartlett's tests, a limited number of principal components (PCs) were chosen for the study. Table 3, presents the eigenvalues for each linear component (factor) before extraction, after extraction, and after rotation. Initially, thirteen parameters were conducted, and the PCA analysis identified three PCs (factor) for Baghdad (those with eigenvalues equal to one). The eigenvalues represent the percentage of variance that each component accounted for. The three selected PCs accounted for 71.827% of the total variance. Before extraction, Factor 1 explained 47.881% of the variance, which was higher than Factor 2 (14.982%) and Factor 3 (8.964%). These values remained unchanged before and after extraction. However, after rotation, the percentage of variance for Factors 1, 2, and 3 shifted to 44.532%, 14.930%, and 12.366%, respectively.

Table 3. Total variance explained for Baghdad.

	Initial Eigenvalues				Extraction sums of squared loading			Rotation sums of squared loading		
	Component	Total	% of variance	Cumulative	Total	% of variance	Cumulative	Total	% of variance	Cumulative
Baghdad	1	6.225	47.881	47.881	6.225	47.881	47.881	5.789	44.532	44.532
	2	1.948	14.982	62.863	1.948	14.982	62.863	1.941	14.930	59.461
	3	1.165	8.964	71.827	1.165	8.964	71.827	1.608	12.366	71.827
	4	0.946	7.277	79.104						
	5	0.874	6.725	85.830						

(continued)

Table 3. (*continued*)

Initial Eigenvalues				Extraction sums of squared loading			Rotation sums of squared loading		
Component	Total	% of variance	Cumulative	Total	% of variance	Cumulative	Total	% of variance	Cumulative
6	0.703	5.405	91.235						
7	0.331	2.549	93.784						
8	0.293	2.257	96.041						
9	0.171	1.313	97.354						
10	0.109	0.839	98.194						
11	0.102	0.787	98.980						
12	0.83	0.637	99.617						
13	0.050	0.383	100.000						

Table 4, presents the results of the Varimax rotation with Kaiser Normalization and Bartlett's test, showing the parameters associated with each principal component (PC). For Baghdad, three PCs together explain 71.827% of the total variance. The first PC, which is closely linked to NO_2, accounts for 44.532% of the variance. The second PC, primarily loaded on PM2.5, explains 14.930% of the variance. The third PC, significantly loaded on temperature (T2m), accounts for 12.366% of the variation.

Table 4. Displays varimax- rotated component matrix for Baghdad.

variable	component		
	1	2	3
O_3	0.843	−0.046	0.369
BLH	0.590	−0.057	0.580
SP	−0.773	−0.219	−0.375
SSR	0.730	0.026	0.448
T2m	0.755	0.057	**0.586**
TP	0.084	0.038	−0.495
CO	−0.911	0.117	0.041
NO_2	**−0.935**	0.116	0.014
SO_2	−0.897	0.163	0.075
PM2.5	−0.112	**0.924**	−0.102
PM10	0.396	0.838	−0.049

The PCA method is effective in identifying the principal variables that have the greatest influence on O_3 concentration, which are then used as inputs for the ANN-MLP model in predicting O_3 levels. The selected principal components (PCs) are labeled as follows: factor 1 (NO_2), factor 2 (PM2.5), and factor 3 (T2m).

3.2 Ozone Values Predicted Using ANN Model

The input parameters used in the input layer were derived from PCA method, with the three selected components for Baghdad being FAC-1, FAC-2, and FAC-3, representing NO_2, PM2.5, and T2m, respectively. Various network architectures were tested by adjusting the count of units in the hidden layers (3, 4, and 5) and experimenting with different activation functions (tanh and sigmoid) to identify the optimal configuration. As displayed in Table 5, the findings show that the ANN achieved the best performance with 4 units (nodes) in the hidden layer and the tanh activation function. Figure 3 illustrates the regression correlation between the O_3 concentrations predicted by the ANN model and the actual measured O_3 concentrations (for different network structure) for Baghdad.

Table 5. Performance of the ANN on O_3 values simulated with different structure of network.

City / governorate	Activation function	Number of hidden layer units	N	Training sample %	N	Testing sample %	Adjusted R^2	RMSE	MAE
Baghdad	tanh	3	760	69.4	335	30.6	0.868	1.5960E-8	1.2868E-8
		4	769	70.2	326	29.8	0.871	1.5911E-8	1.2300E-8
		5	760	69.4	335	30.6	0.870	1.6119E-8	1.3024E-8
	sigmoid	3	753	68.8	342	31.2	0.870	1.6193E-8	1.3096E-8
		4	771	70.4	324	29.6	0.869	1.5920E-8	1.2836E-8
		5	759	69.3	336	30.7	0.866	1.5825E-8	1.2754E-8

As displayed in Table 5, the model's performance was assessed utilizing the R^2, RMSE, and MAE metrics. The evaluation of the selected network architectures demonstrated higher R^2 values alongside lower RMSE and MAE, indicating that as the model's ability improves, RMSE and MAE decrease, while R^2 increases.

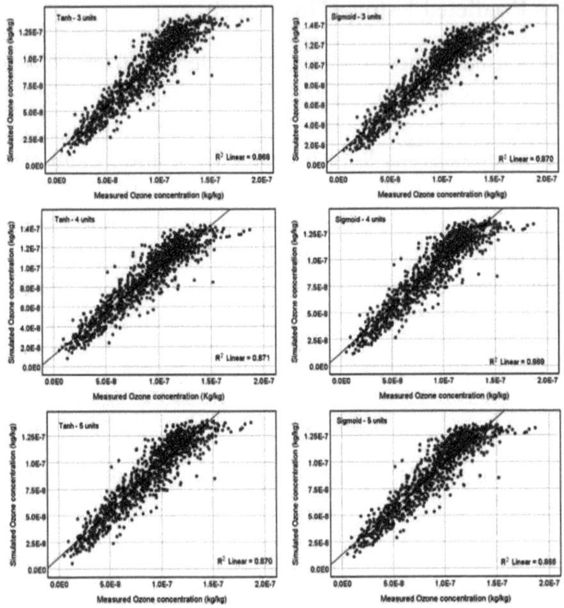

Fig. 3. The regression correlation relationship between the simulated O_3 concentration by ANN model (different structure according table 5) and real measured O_3 concentration for Baghdad.

3.3 Independent Variable Importance

The importance of each independent variable is specified by its influence on the prediction accuracy of the ANN model. Variables that have a greater effect on the model's output are considered more important. This importance is typically measured on a scale from 0 to 1 or expressed as a percentage relative to the most significant variable. As shown in table 6, in this study, the independent variables with the greatest influence on surface O_3 concentration in Baghdad are as follows: according to the ANN model, FAC-1 (NO_2), FAC-3 (T2m), and FAC-2 (PM2.5) had importance scores of 100.0%, 74.6%, and 6.65%, corresponding to 0.552, 0.412, and 0.037, respectively. Figure 4, displayes the independent variables importance of the input factors to the ANN.

Table 6. Importance of independent variables

Importance of independent variables		
	Importance	Normalized Importance
FAC-1	0.552	100.0%
FAC-2	0.037	6.6%
FAC-3	0.412	74.6%

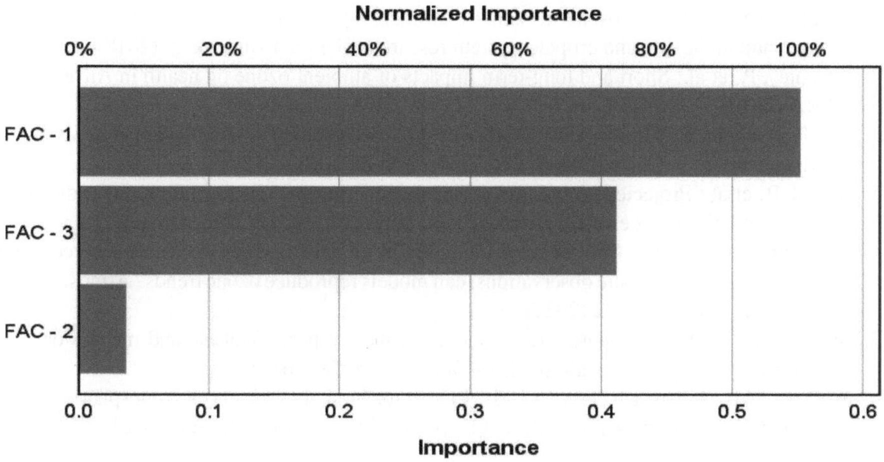

Fig. 4. The independent variables importance

4 Conclusion

This study aimed to predict surface O_3 concentrations and assess their relationship with various independent variables, including air pollutants and meteorological factors, by combining the PCA method with an ANN model (PC-ANN). PCA proved to be an effective technique for identifying the most significant factors influencing O_3 concentrations in Baghdad, particularly with a large dataset and numerous contributing parameters. The results revealed notable variations in the correlation between O_3 levels and the independent variables. The dataset's suitability for factor analysis was verified using KMO and Bartlett's tests. Baghdad's KMO value was 0.836, and the findings of Bartlett's test showed a p-value of 0.000 ($p < 0.001$), indicating that PCA was adequate. Following the use of varimax rotation, three principal components (PCs) representing 71.827% of the total variance were identified for Baghdad. In Baghdad, PC1, PC2, and PC3 were mainly connected to NO_2, PM2.5, and T2m, respectively. With regard the ANN model, various network structures were examined, the optimal performance achieved using 4 unites in the hidden layer and the tanh activation function, yielding an adjusted R^2 of 0.871. The ANN model identified FAC-1 (NO_2), FAC-3 (T2m), and FAC-2 (PM2.5) as the most significant independent factors influencing surface O_3 concentration in Baghdad, with importance scores of 100.0%, 74.6%, and 6.65%, equivalent to 0.552, 0.412, and 0.037, respectively. The ANN model was effective in capturing nonlinear relationships, with improved R^2 and lower RMSE and MAE values. One challenge in this study was the lack of ground-based ozone data in Iraq, due to equipment failure caused by the war on terrorism, which necessitated the use of satellite data as an alternative.

References

1. Rajab, J.M., Lim, H., MatJafri, M.: Monthly distribution of diurnal total column ozone based on the 2011 satellite data in Peninsular Malaysia. EJRS **16**(1), 103–109 (2013)

2. Lefohn, A.S., et al.: Tropospheric ozone assessment report: Global ozone metrics for climate change, human health, and crop/ecosystem research. Elem Sci Anth. **6**, 27 (2018)
3. Amoatey, P., et al.: Short and long-term impacts of ambient ozone on health in Ahvaz, Iran. HERA (2019)
4. Ilić, P., Popović, Z., Markić, D.N.: Assessment of meteorological effects and ozone variation in urban area. Ecol. Chem. Eng. S. **27**(3), 373–385 (2020)
5. Sicard, P., et al.: Projected global ground-level ozone impacts on vegetation under different emission and climate scenarios. Atmos. Chem. Phys. **17**(19), 12177–12196 (2017)
6. Christiansen, A., et al.: Multidecadal increases in global tropospheric ozone derived from ozonesonde and surface site observations: can models reproduce ozone trends? Atmos. Chem. Phys. **22**(22), 14751–14782 (2022)
7. Schultz, M.G., et al.: Tropospheric ozone assessment report: database and metrics data of global surface ozone observations. Elem. Sci. Anth. **5**, 58 (2017)
8. Wedow, J.M., Ainsworth, E.A., Li, S.: Plant biochemistry influences tropospheric ozone formation, destruction, deposition, and response. Trends Biochem. Sci. **46**(12), 992–1002 (2021)
9. Adnane, A., Leghrib, R., Chaoufi, J., Chirmata, A.: Ambient ozone concentrations modelling using feedforward backpropagation neural networks: spatial modeling over the Agadir City (Morocco). In: Melliani, S., Castillo, O. (eds.) Recent Advances in Fuzzy Sets Theory, Fractional Calculus, Dynamic Systems and Optimization. ICPAMS 2021. Lecture Notes in Networks and Systems, vol. 476, pp. 52–65. Springer, Cham (2023). https://doi.org/10.1007/978-3-031-12416-7_6
10. Nemah, H.A., et al.: Effect of some meteorological variables and conditions on mobile phone and "TV" satellite signal. Al-Mustansiriyah J. Sci. **32**(2), 71–75 (2021)
11. Chen, S., et al.: The trend of surface ozone in Beijing from 2013 to 2019: indications of the persisting strong atmospheric oxidation capacity. Atmos. Environ. **242**, 117801 (2020)
12. Lin, M., et al.: Vegetation feedbacks during drought exacerbate ozone air pollution extremes in Europe. Nat. Clim. Change **10**(5), 444–451 (2020)
13. Griffiths, P., et al.: On the changing role of the stratosphere on the tropospheric ozone budget: 1979–2010. Geophys. Res. Lett. **47**(10), e2019GL086901 (2020)
14. Zeng, P., et al.: Causes of ozone pollution in summer in Wuhan, Central China. Environ. Pollut. **241**, 852–861 (2018)
15. Rajab, J.M., et al.: Analysis of Ozone column burden in Peninsular Malaysia retrieved from Atmosphere Infrared Sounder (AIRS) data: 2003–2009. In: Proceedings of the 2011 IEEE International Conference on Imaging Systems and Techniques (2011), pp. 29–33. IEEE, Batu Ferringhi (2011), https://doi.org/10.1109/IST.2011.5962213
16. Resmi, C., et al.: Long-term variations of air quality influenced by surface ozone in a coastal site in India: association with synoptic meteorological conditions with model simulations. Atmosphere **11**(2), 193 (2020)
17. Gunthe, S., Beig, G., Sahu, L.: Study of relationship between daily maxima in ozone and temperature in an urban site in India. Curr. Sci., 1994–1999 (2016)
18. Warmiński, K., Bęś, A.: Atmospheric factors affecting a decrease in the night-time concentrations of tropospheric ozone in a low-polluted urban area. Water Air Soil Pollut. **229**(11), 1–13 (2018)
19. Saini, R., et al.: Ozone distributions and urban air quality during summer in Agra–a world heritage site. Atmos. Pollut. Res. **5**(4), 796–804 (2014)
20. Ko, K., Cho, S., Rao, R.R.: Machine-learning-based near-surface ozone forecasting model with planetary boundary layer information. Sensors **22**(20), 7864 (2022)
21. Gleim, A., Salish, N.: Forecasting environmental data: an example to ground-level ozone concentration surfaces. arXiv:2202.03332 (2022)

22. Sharma, S., Sharma, P., Khare, M.: Photo-chemical transport modelling of tropospheric ozone: a review. Atmos. Environ. **159**, 34–54 (2017)
23. Gewers, F.L., et al.: Principal component analysis: a natural approach to data exploration. ACM Comput. Surv. **54**(4), 1–34 (2021)
24. Hasan, B.M.S., Abdulazeez, A.M.: A review of principal component analysis algorithm for dimensionality reduction. JSCDM **2**(1), 20–30 (2021)
25. Makarova, A, et al.: Modelling tropospheric ozone variations using artificial neural networks: a case study on the Black Sea coast (Russian Federation). Clean. Eng. Technol. **5**, 100293 (2021)
26. Sayeed, A., et al.: Using a deep convolutional neural network to predict 2017 ozone concentrations, 24 hours in advance. Neural Netw. **121**, 396–408 (2020)
27. Gao, M., Yin, L., Ning, J.: Artificial neural network model for ozone concentration estimation and Monte Carlo analysis. Atmos. Environ. **184**, 129–139 (2018)
28. Pawlak, I., Jarosławski, J.: Forecasting of surface ozone concentration by using artificial neural networks in rural and urban areas in central Poland. Atmosphere **10**(2), 52 (2019)
29. Adnane, A., et al.: The use of a recurrent neural network for forecasting ozone concentrations in the city of Agadir (Morocco). JAMCNP **7**(3), 197–206 (2020)
30. Jia, B., Dong, R., Du, J.: Ozone concentrations prediction in Lanzhou, China, using chaotic artificial neural network. Chemom. Intell. Lab. Syst. **204**, 104098 (2020)
31. AlOmar, M.K., Hameed, M.M., AlSaadi, M.A.: Multi hours ahead prediction of surface ozone gas concentration: robust artificial intelligence approach. Atmos. Pollut. Res. **11**(9), 1572–1587 (2020)
32. Kapadia, D., Jariwala, N.: Prediction of tropospheric ozone using artificial neural network (ANN) and feature selection techniques. MESE **8**(2), 2183–2192 (2022)
33. Dhanya, G., et al.: Comprehensive modeling of seasonal variation of surface ozone over Southern Tropical City, Bengaluru, India. Nat. Environ. Pollut. Technol. **21**(3), 1269–1277 (2022)
34. Abed, F., et al.: Influence of Air temperature and interrelationship with greenhouse gases (CO_2 and CH_4) over Iraq using AIRS data. Atmos. Pollut. Res. **15**(11), 102293 (2024)
35. Abed, F.G.: Spatiotemporal observations of CH_4 and CO_2 over Iraq using Atmospheric Infrared Sounder (AIRS) data. J. Apl. Adv. Res **5**, 6–10 (2020)
36. Basheer, F.S.: Trend analysis of annual surface air temperature for some stations over Iraq. Al-Mustansiriyah J. Sci. **33**(1), 77–82 (2022)
37. Dibs, H., et al.: An optimal approach for land-use/land-cover mapping by integration and fusion of multispectral landsat OLI images: case study in Baghdad, Iraq. Water Air Soil Pollut. **231**(9), 488 (2020)
38. Yu, J., et al.: Establishment of a combined model for ozone concentration simulation with stepwise regression analysis and artificial neural network. Atmosphere **13**(9), 1371 (2022)
39. Rajab, J.M., MatJafri, M., Lim, H.: Combining multiple regression and principal component analysis for accurate predictions for column ozone in Peninsular Malaysia. Atmos. Environ. **71**, 36–43 (2013)
40. Napi, N.N.L.M., et al.: Different approaches of Multiple Linear Regression (MLR) model in predicting Ozone (O_3) concentration in industrial area. Int. J. Integr. Eng. **15**(1), 106–117 (2023)
41. de Souza, A., et al.: Evaluation of variations in ground-level ozone (O_3) concentrations. Ozone Sci. Eng. **40**(3), 237–247 (2018)
42. Hashim, N.I.M., et al.: Assessment of surface ozone concentration in Northern Peninsular Malaysia. IOP Conf. Ser.: Mater. Sci. Eng. **551**, 012100 (2019). https://doi.org/10.1088/1757-899X/551/1/012100
43. Hadi, N.U., et al.: An easy approach to exploratory factor analysis: marketing perspective. JESR **6**(1), 215–223 (2016)

44. Oghenefejiri, B., Nwaogazie Ify, L., Agunwamba, J.: Exploratory factor analysis & assessment of energy potential of generated solid waste in Nigeria. IJCIET **7**(1), 274–289 (2016)
45. Hashim, N.M., et al.: Forecasting daytime ground-level ozone concentration in urbanized areas of Malaysia using predictive models. Sustainability **14**(13), 7936 (2022)
46. Bekesiene, S., Meidute-Kavaliauskiene, I., Vasiliauskiene, V.: Accurate prediction of concentration changes in ozone as an air pollutant by multiple linear regression and artificial neural networks. Mathematics **9**(4), 356 (2021)

Speech Emotion Recognition Using Machine Learning

S. Vijayalakshmi(✉), K. R. Kavitha, R. A. Atchayashree, and M. T. Akshaya

Sona College of Technology, Salem, Tamil Nadu, India
vijisaumiya@gmail.com, {atchayashree.21ece, akshaya.21ece}@sonatech.ac.in

Abstract. Speech Emotion Recognition (SER) is an evolving area of research within human-computer interaction, aiming to enable machines to detect and interpret emotional states conveyed through speech. This paper presents a hybrid SER system that leverages both Support Vector Machine (SVM) and Neural Networks to enhance the accuracy and adaptability of emotion classification. SVM, a supervised learning model, is employed for its effectiveness in high-dimensional spaces and robustness in handling smaller datasets, making it suitable for precise emotion classification based on extracted features such as pitch, energy, and Mel-frequency cepstral coefficients (MFCCs). Neural Networks, particularly deep learning architectures, are used for automatic feature extraction and modelling complex, non-linear patterns in the speech data. This hybrid approach enables the system to balance interpretability and performance, ensuring both efficiency and depth in learning emotional cues. The system supports multilingual inputs and provides personalized emotional feedback through an intuitive web interface developed using Streamlit and Flask. Experimental results demonstrate improved recognition accuracy, showcasing the potential of combining classical and deep learning techniques in real-world applications of emotion-aware systems.

Keywords: Emotion Recognition · Language Detection · Machine Learning

1 Introduction

Emotions are essential to human contact and communication. Applications like customer service, and mental health monitoring could all benefit from the capacity to identify emotions in speech. However, because they heavily rely on expert knowledge and manual feature extraction, traditional emotion detection systems are frequently limited. Techniques of machine learning (ML) and deep learning (DL) offer fresh chances to raise the precision and efficacy of speech emotion recognition (SER) systems. A variety of methods and approaches in the area of emotion recognition are demonstrated in the literature. The promise of deep learning in pattern recognition tasks is demonstrated, for instance, by the creation of reliable defect detection systems in printed circuit boards (PCBs) [1]. Similar applications in voice emotion recognition are made possible by this seminal work. Additionally, extensive research on feature extraction and segmentation

R. Geetha et al. (Eds.): AAIMB 2024, CCIS 2653, pp. 297–309, 2026.
https://doi.org/10.1007/978-3-032-13026-6_21

in medical imaging, namely in MRI brain tumor classification, shows how machine learning can reliably recognize complex data [2]. These approaches demonstrate how crucial efficient feature extraction is to attaining high classification accuracy in SER systems. Furthermore, research on voice and language features in language recognition highlights the difficulties in identifying emotional states, especially in situations involving code-switching [3]. These complexities highlight the need for sophisticated approaches that can process several languages and emotional cues at once. The fact that machine learning techniques are used in a variety of settings, such as video analysis for fall detection in senior citizens, further demonstrates their adaptability [4]. Using photoacoustic data, machine learning approaches demonstrated in the medical industry for the diagnosis, highlighting the potentially life-saving uses of these technologies [5]. Furthermore, mental health and well-being are also affected by the importance of recognizing emotions by speech. The effectiveness of machine learning in this field is demonstrated by predictive analytics models, such as those used for loan prediction via logistic regression, which offer insights into real-time emotion recognition from spoken language [6]. Furthermore, advanced algorithms for classifying images to identify forest fires emphasize the significance of accurate detection methods in a variety of domains [7]. This idea is directly applicable to SER, where it is essential to record nuanced emotional indicators in speech.

By demonstrating how automated systems may offer timely and precise diagnoses, the use of deep learning in the classification of skin diseases highlights the need for effective emotion recognition systems [8]. Additionally, using speech analysis to determine emotional states may result in prompt mental health treatments. The promise of deep learning for language recognition is demonstrated by research on character identification of ancient Tamil stone inscriptions, which also emphasizes the cultural dimensions of emotion perception in multilingual environments [9]. Emotions can be dynamically evaluated to improve safety and well-being, as demonstrated by real-time applications like facial expression recognition for detecting driver fatigue [10].

Current assessments of deep learning methods for voice emotion recognition show notable progress in the area and point to a considerable move toward neural network-based models for higher accuracy [11]. The creation of deep learning-based real-time SER applications suggests that these technologies are continuously improving and evolving [12]. Notably, a more thorough approach to emotion recognition is made possible by the combination of deep learning frameworks with traditional classifiers like Support Vector Machines (SVM) [13]. Better recognition results are produced by this combination of techniques, which enable a more complex comprehension of emotional states. The importance of using a variety of approaches for classification tasks is demonstrated by comparative research between supervised and self-supervised deep learning models for detecting eye disorders [14]. These tests highlight how crucial it is to tackle the difficulties brought on by different kinds of data while aiming for high accuracy. Similar to the difficulties encountered in speech emotion recognition, Convolutional Neural Networks (CNNs) used in liver tumour identification offer another illustration of how sophisticated deep learning methods are applied in medical diagnostics [15]. The need for innovative cross-disciplinary solutions is highlighted by recent advancements in picture retrieval and kidney stone diagnostics using image processing techniques [16, 17].

These developments provide insightful information about efficient feature extraction and classification techniques that can guide the creation of SER systems.

Additionally, research on multilingual speech emotion recognition highlights the necessity to overcome the particular difficulties presented by various linguistic settings, pointing to both present advancements and possible future paths [18]. The importance of adapting deep learning techniques to language-specific tasks is demonstrated by convolutional neural networks in handwritten Tamil character recognition, which parallels the needs of multilingual SER systems [19]. This establishes the foundation for future studies on how machine learning methods tailored to the complexities of human emotions and speech patterns can improve the recognition of speech emotions.

2 Proposed Method

The Speech Emotion Recognition (SER) system suggested methodology will concentrate on creating two primary models: a downstream model that combines conventional supervised and self-supervised methods and a pretext model for feature learning from unlabelled audio input. Using feature learning and knowledge transfer to increase the precision and effectiveness of speech-to-emotion recognition is one of the main goals of the model.

2.1 Model Development

2.1.1 Models Downstream

Two methods a self-supervised model and a conventional supervised model— will be used in the downstream phase. Both will make use of the features that the pretext model gathered. Self-Supervised Approach: Using unlabelled data, this model will create its own labels so that it can identify patterns and correlations in voice. This self-supervised learning technique enables the system to learn meaningful audio representations even from unlabelled data, significantly improving the robustness and adaptability of the SER model in low-resource and multilingual environments: This model is trained on labelled datasets, using well-defined objectives to guide its learning. Applications needing high precision and accuracy in emotion classification must use this method. By combining the advantages of traditional supervised learning with self-supervised learning, these two downstream techniques can create a strong SER system.

2.1.2 Tools for Software

To execute the suggested methodology, several software applications will be utilized: The main environment for constructing the SER system will be Visual Studio Code, an opensource coding tool that makes code organization and debugging efficient. Kaggle, this data science centre, will make emotion datasets accessible to the public, promoting cooperation and project outcome improvement.

2.1.3 Initial Processing

An essential first step in getting the audio data ready for feature extraction is preprocessing. The actions listed below will be carried out:

Noise Reduction. Spectral subtraction and other techniques will be used to reduce background noise, resulting in a clearer audio input for analysis.

Segmentation. Audio recordings will be divided into segments lasting one to three seconds in order to enable the extraction of localized features.

Normalization. The influence of different recording volumes will be lessened by normalizing audio levels, which will give consistent input for the feature extraction procedure.

2.1.4 Extracting Features

In our proposed system, the pretext model plays a crucial role in the feature learning process. Specifically, it extracts meaningful patterns from raw audio input by first converting the waveform into a more analysable form using Short-Time Fourier Transform (STFT) and Mel-frequency Cepstral Coefficients (MFCCs). These techniques help capture both spectral and temporal characteristics of the speech signal. The model then processes these features through convolutional layers that detect local feature patterns such as pitch variations, intonation, and voice energy, which are essential indicators of emotional states. By leveraging unsupervised or self-supervised learning techniques during this stage, the pretext model learns robust and generalized feature representations that can be fine-tuned during downstream emotion classification tasks. This approach allows the system to automatically learn abstract emotional features from raw audio data without requiring extensive manual labelling or handcrafted features.

2.1.5 Training Models

Dividing Data. To guarantee a thorough assessment of the models, the dataset will be categorized into training, validation, and testing. We'll use a standard distribution: training (70%), validation (15%), and testing (15%).

Model Education Procedure. Using unlabelled data, the pretext model will initially be trained to acquire efficient feature representations. The labelled data is then used to train the self-supervised and conventional supervised models, with the SVM classifier being used for emotion categorization.

Measures of Evaluation. The model's efficacy will be evaluated by the application of multiple metrics, such as F1-score, accuracy rate, precision rate, and recall. To display the classification performance across several emotion categories, a confusion matrix will be employed.

2.1.6 Instantaneous Execution

A user-friendly web application built using Streamlit and Flask, which allows users to upload audio input and get real-time emotional feedback. The trained models are served using a Flask-based REST API, which is integrated with the Streamlit front end for seamless interaction.

2.1.7 Language Detection and Multilingual Support

The SER system will incorporate a language detection module that recognizes spoken language prior to audio processing to improve accessibility. For precise language identification, methods like n-grams will be used, guaranteeing efficient feature extraction in a variety of languages.

Figure 1 depicts the diagrammatic representation of SER, illustrating the key components and workflow involved in the process of the SER system. These systems typically consist of key components:

Audio Input. The user gives the system an audio file to work with.

Language Detection. To guarantee proper processing, the system recognizes the spoken language.

Pre-processing. To ensure uniformity between samples, audio is cleaned and segmented.

Feature Extraction. The system uses techniques like MFCCs and Chroma features to extract important audio features.

Model Training. To properly identify emotions, the model is trained on both labelled and unlabelled data.

Emotion Classification. The SVM classifier classifies the emotion based on the features that were retrieved.

Personalized Suggestions. Following categorization, the algorithm offers recommendations tailored to the user's emotional health.

Web Interface. The system is accessible to all users because the outcome is shown on an easy-to-use web platform.

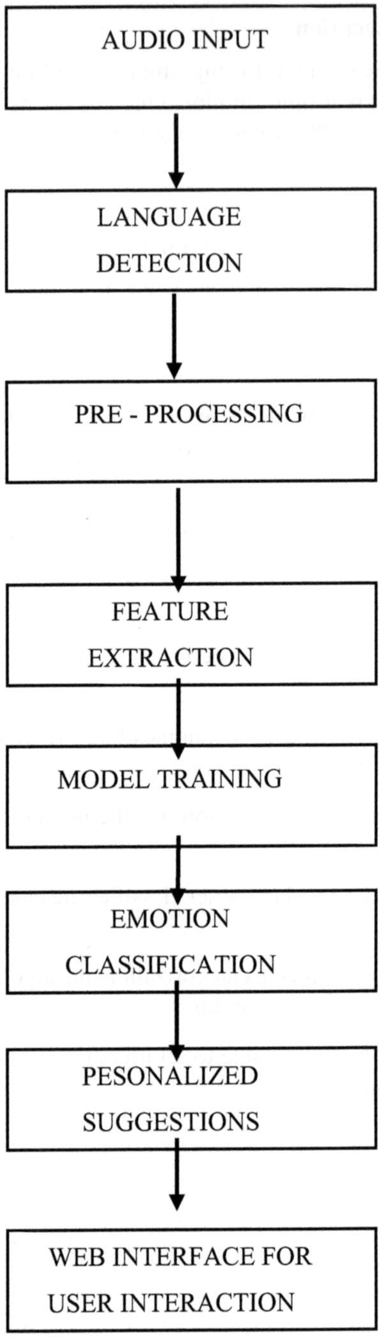

Fig. 1. Diagrammatic representation of SER

3 Analysis and Result

The Fig. 2 shows the dataset values, which are trained for various emotions using Python.

```
Dataset is Loaded
[-1.0726122e-05 -1.7188726e-05 -5.8019908e-05 ...  6.0470687e-05
  1.2742779e-04  1.7060280e-05]
[ 8.4733198e-05 -1.3261342e-03 -2.5771626e-03 ...  2.8306472e-03
  1.4243328e-03  5.1377900e-04]
[ 5.2280252e-06 -4.0461760e-05  3.2110795e-05 ... -1.5045352e-04
 -1.7025679e-05  1.8972876e-05]
[-1.1267163e-05 -1.9048303e-04 -2.8634202e-04 ...  1.6369415e-04
  1.0550911e-04  6.5897068e-05]
[ 2.9714556e-07 -1.0640829e-04 -1.0255290e-04 ...  1.1131972e-04
  1.0935357e-04  0.0000000e+00]
```

Fig. 2. Data set values for emotions

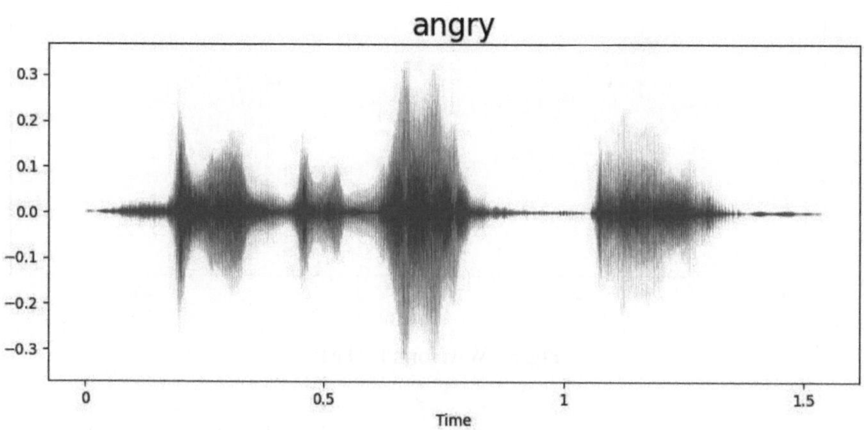

Fig. 3. Waveform for Angry

The Fig. 3 illustrates training for speech emotion recognition, focusing on identifying 'angry' emotion.

The Fig. 4 illustrates training for speech emotion recognition, focusing on identifying 'disgust' emotion.

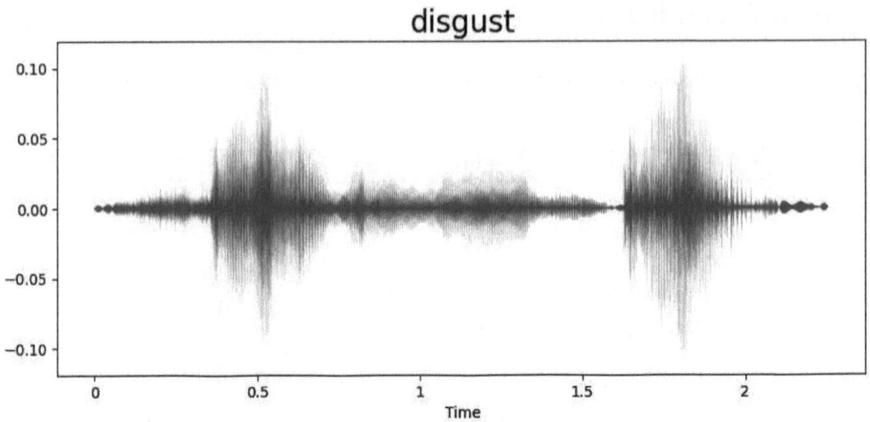

Fig. 4. Waveform for Disgust

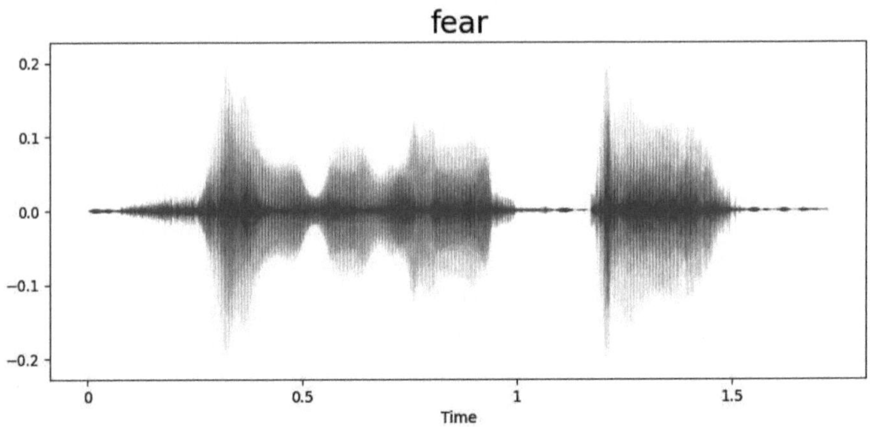

Fig. 5. Waveform for Fear

The Fig. 5 illustrates training for speech emotion recognition, focusing on identifying 'fear' emotion.

The Fig. 6 illustrates training for speech emotion recognition, focusing on identifying 'sad' emotion.

The Fig. 7 illustrates training for speech emotion recognition, focusing on identifying 'happy' emotion.

The Fig. 8 illustrates the final web interface for user interaction. Here we can upload the audio files and the system provide the predicted emotion with personalized suggestion (Table 1).

With good precision and recall values for every emotion, the system's total classification accuracy was 90.2%.With a success percentage of 91.2% on the RAVDESS dataset, sorrow was shown to have the highest accuracy, followed by happiness and anger. The

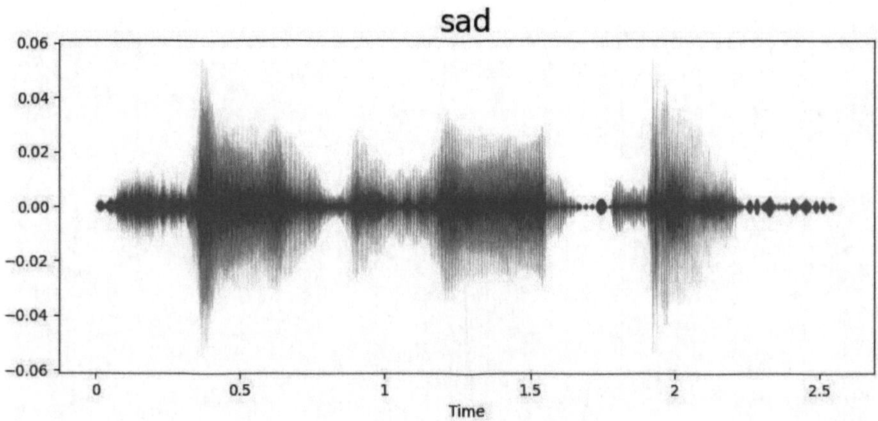

Fig. 6. Waveform for Sad

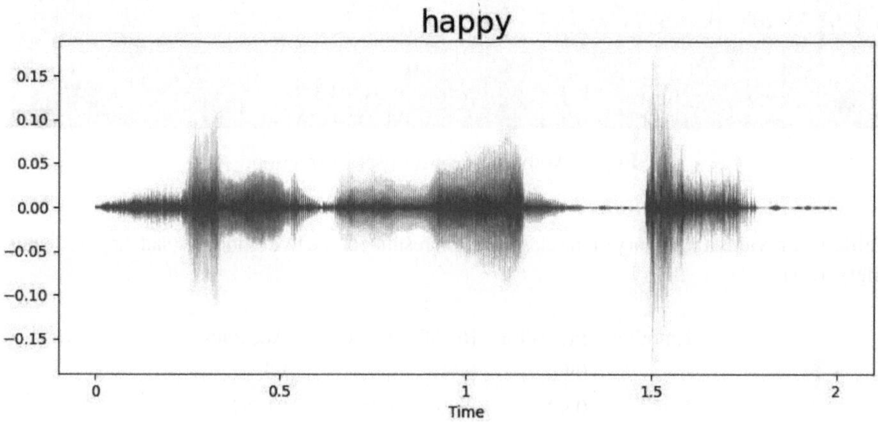

Fig. 7. Waveform for Happy

subtle aspects of disgust and dread can be linked to their relatively lower F1-scores, as they occasionally resulted in overlapping features in the audio data.

4 Conclusion

With real-time emotional feedback and tailored recommendations, the proposed Speech Emotion Recognition (SER) using machine learning gives a comprehensive way to identify human emotions through speech. The approach offers practical, user-centered advice while addressing a wide spectrum of emotional states by concentrating on five major emotions: fear, happiness, anger, disgust, and melancholy. The project's capacity to go beyond basic emotion recognition and include other features like multilingual support and an online interface for user accessibility is one of its main advantages. Neural networks for feature extraction and Support Vector Machines (SVM) for emotion

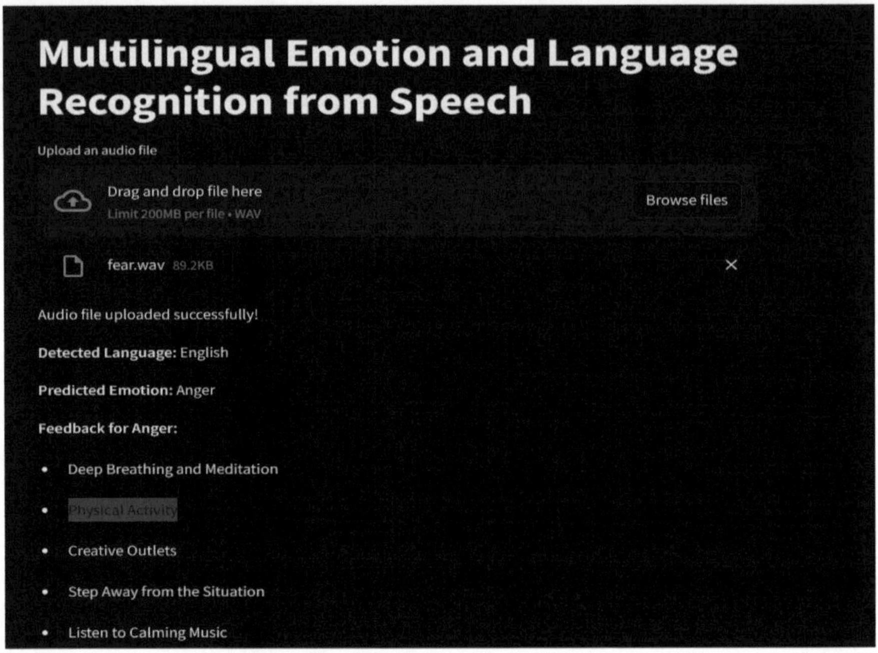

Fig. 8. Web interface for user interaction.

Table 1. Provides a summary of the classification results for the five emotions (sad, happy, disgust, angry, fear).

Emotion	Precision rate	Recall rate	F1-score	Accuracy rate
Sad	0.92	0.89	0.90	91.2%
Happy	0.89	0.91	0.90	90.5%
Disgust	0.87	0.86	0.86	89.0%
Angry	0.90	0.89	0.89	90.8%
Fear	0.88	0.87	0.87	89.5%
Overall	0.89	0.88	0.88	90.2%

categorization are successfully integrated in the SER system. High accuracy and precision in the classification of emotions from audio data are made possible by this hybrid technique. Furthermore, the system can analyze speech in many languages thanks to the usage of language detection algorithms, which makes it flexible in a variety of linguistic contexts. This system stands out for its ability to deliver unique suggestions based on the detected emotions since it provides a comprehensive user experience that surpasses simple recognition. The SER system's overall classification accuracy of 90.2% during the evaluation process showed its dependability for a range of emotions. With a high language detection accuracy of 96%, the multilingual support ensured efficient emotion recognition in non-English languages as well. User interaction with the system was enhanced by the easy-to-use online interface, which let users submit audio files and get instant feedback on how they were feeling emotionally. In summary, the created system is a useful tool for real-time emotion monitoring and mental health applications because it provides a unique combination of individualized emotional support, accurate emotion recognition, and multilingual capability. Future developments will focus on expanding language capabilities and further optimizing real-time performance, which will increase its adaptability and practical impact.

Real-World Applications of Speech Emotion Recognition

1. Customer Service & Call Centres

Automatic emotion detection helps agents understand customer moods (e.g., frustration, satisfaction).Enables real-time escalation to supervisors when anger or distress is detected.Enhances customer experience analytics by evaluating emotional trends in interactions.

2. Healthcare & Mental Health Monitoring

Assists in detecting emotional disorders like depression, anxiety, or PTSD through voice patterns.Supports telemedicine by giving clinicians emotional context during virtual consultations.Useful in therapy support tools and for monitoring patients remotely.

3. Human-Computer Interaction (HCI)

Makes virtual assistants (like Siri, Alexa, Google Assistant) more empathetic by adapting responses to user emotions.Enhances chatbots and conversational AI with emotionally aware communication.

4. Education & E-Learning

Detects student engagement or frustration during online classes.Helps in providing personalized learning experiences based on student mood.Useful in tutoring systems that adjust pace and feedback style based on detected emotion.

5. Automotive Industry

Integrated in in-car systems to monitor driver emotions like fatigue, anger, or stress.Improves road safety by triggering alerts or interventions if dangerous emotional states are detected.

6. Entertainment & Gaming

Enhances gaming experiences by adapting gameplay or narratives to players' emotions.Enables personalized content recommendations based on mood.

7. Security & Law Enforcement

Used in lie detection and threat assessment during interrogations or airport screenings.Supports surveillance systems to identify potentially aggressive or distressed individuals

8. Accessibility Tools

Helps individuals with autism or speech impairments to better express and understand emotions in communication.

Useful in assistive technologies that interpret emotions for people with social or emotional perception challenges.

References

1. Verma, P., Singh, A.: Speech emotion recognition using MFCC and SVM classifier. J. King Saud Univ. Comput. Inf. Sci. **34**(6), 2148–2157 (2022)
2. Zhang, Z., Yin, H.: Real-time speech emotion recognition with deep learning models. IEEE Trans. Affect. Comput. **13**(1), 193–204 (2022)
3. Vijayalakshmi, S., Kavitha, K.R., Hariharan, S.: Segmentation, feature extraction, and classification of brain tumors through MRI images. ICTACT J. Image Video Process. **11**(4), 2517–2524 (2021)
4. Neumann, M., Vu, N.T.: Attentive Convolutional Neural Network (CNN) for speech emotion recognition. In: Interspeech, pp. 2520–2524 (2017)
5. Prasad, B.: Ancient Tamil stone inscription character recognition using deep learning. In: Proceedings of the 2nd International Conference on Signal Processing and Communication System, pp. 1–2 (2023)
6. Koolagudi, S.G., Rao, K.S.: Emotion recognition from speech: a review. Int. J. Speech Technol. **15**(2), 99–117 (2012)
7. Sreesouthry, S., Ayubkhan, A., Rizwan, M.M., Lokesh, D., Prithivi Raj, K.: Loan prediction using logistic regression in machine learning. Ann. Rom. Soc. Cell Biol. **25**, 2790–2794 (2021)
8. Tzirakis, P., Trigeorgis, G., Nicolaou, M.A., Schuller, B.W., Zafeiriou, S.: End-to-end speech emotion recognition using deep neural networks. IEEE Trans. Affect. Comput. **10**(3), 312–325 (2017)
9. Muthulingam, A., Kavitha, K.R., Kumar, B.R.: A hybrid approach on restoration of degraded document images. J. Adv. Res. Dyn. Control Syst. **12**(5), 926–930 (2020)
10. Govindaraj, R., Bhuvaneswari, P., Radhika, R., Saranya, S.: Implementation of machine learning mechanism for recognizing prostate cancer through photoacoustic signal. Contrast Media Mol. Imaging **2022**, 1–13 (2022)
11. Vijayalakshmi, S., Kavitha, K.R., Sangeetha, D.P., Thilak, S.: Robust defect detection on PCB based on deep learning. J. Adv. Res. Dyn. Control Syst., 937–946 (2020)
12. Sahu, S.K., Nair, A.S., Rathod, V.T.: A deep learning approach to speech emotion recognition with multiple feature extraction. Procedia Comput. Sci. **152**, 221–228 (2019)
13. Ali, H., Hasan, M.M.: Language detection in short text messages using multinomial Naïve Bayes and deep learning techniques. J. Comput. Sci. **17**(9), 891–899 (2021)
14. Vijayalakshmi, S., Kavitha, K.R.: GSM-based weight measurement and automatic gas booking system with leakage detection. In: International Conference on Renewable Sustainable Energy Technology (RESET), pp. 1–2 (2023)
15. Joshi, R., Singh, R.: A survey on language identification techniques using machine learning. Int. J. Adv. Res. Comput. Sci. **11**(3), 39–45 (2020)
16. Wang, Z., Guan, Z.: A hybrid CNN-SVM classifier for speech emotion recognition. EURASIP J. Audio Speech Music Process. **2018**, 1–10 (2018)
17. Sangeetha, D.P., Vijayalakshmi, S.: Fall detection for elderly people using video-based analysis. J. Adv. Res. Dyn. Control Syst. **12**(7), 1–12 (2020)
18. El Ayadi, M., Kamel, M.S., Karray, F.: Survey on speech emotion recognition: features, classification schemes, and databases. Pattern Recognit. **44**(3), 572–587 (2011)
19. Kavitha, K.R., Vijayalakshmi, S., Reddy, B.K., Reddy, N.S., Chandrasekhar, P.: Driver drowsiness detection using face expression recognition. Ann. Rom. Soc. Cell Biol. **25**, 2785–2789 (2021)
20. Tripathi, S., Beigi, H., Agrawal, A.: Self-supervised learning for speech emotion recognition: a review. arXiv preprint arXiv:2106.02046 (2021)

21. Huang, Z., Dong, Y., Mao, Q.: SER-AudioCLR: audio contrastive learning for speech emotion recognition. IEEE Signal Process. Lett. **30**, 172–176 (2023)
22. Latif, S., Qayyum, A., Usama, M., Qadir, J.: Survey of deep self-supervised learning for audio processing. ACM Comput. Surv. (2023)

Enhancement of Misdeed Node Catching and Fast Transmission in Delay Tolerance Network Using 5G Technology

V. Ulagamuthalvi[1(✉)], G. Kulanthaivel[2], and D. S. Jayalakshmi[3]

[1] Department of CSE, Sathyabama University, Chennai, India
ulagamv@gmail.com
[2] Department of ECE, NITTTR, Chennai, India
gkvel@nitttrc.ac.in
[3] Department of Physics, Sathyabama Institute of Science and Technology (Deemed to Be University), Chennai, India
jayaanand10@gmail.com

Abstract. A Delay Tolerant Network (DTN) can effectively operate all network functions and is specifically designed for vast distances under extreme conditions, such as space communications. Traditional Internet cannot perform well in such networks due to high latency, frequent disruptions, and elevated error rates. These issues can be mitigated by a DTN, which provides reliable data transmission and ensures dependable internetworking. Previous research has focused on terrestrial networking methodologies and demonstrated that they cannot resolve the existing issues and challenges of space communication applications; consequently, researchers have shifted their attention towards DTNs. The Space Data System Committee has proposed several protocols to incorporate advancements in DTN. However, their efficiency does not surpass that of DTN, and they lack an automatic data transmission protocol. The current protocols are not designed to transmit data quickly, and the entire transmission process fails if a single node malfunctions or behaves improperly in routing. Therefore, it is crucial to design and implement a novel protocol for DTN that can identify and address misbehaving nodes involved in data transmission based on their identity and location. In this paper, we devise an LSTM-based approach to enhance DTN performance by detecting missing nodes and accelerating data transmission speed. The simulation is conducted using NS2 software, and the results are evaluated concerning delay, packet delivery ratio, abnormal node detection rate, and throughput. The performance of the proposed protocol is compared to that of a similar one to assess its effectiveness.

Keywords: LSTM In DTN · Deep Learning Model For Missing Node Detection · Delay Tolerant Networks · DTN Protocol · Misdeed Nodes · Long Distance Data Transmission

R. Geetha et al. (Eds.): AAIMB 2024, CCIS 2653, pp. 310–325, 2026.
https://doi.org/10.1007/978-3-032-13026-6_22

1 Introduction

Delay-tolerant networks (DTNs) are a type of ad-hoc network in which end-to-end connectivity is not consistently available for message transmission. These networks are designed for applications where the timely delivery of messages is not critical and delays ranging from hours to weeks are acceptable (Kaviani et al., 2016). DTNs are particularly effective in scenarios involving extreme distances, such as space communications or interplanetary networks. The primary method employed by DTNs is the store-and-forward approach, where messages are stored temporarily when transmission is unavailable and forwarded once connectivity is restored.

In their models, these networks address end-to-end (E2E) delay challenges by considering factors like traffic load, arrival rate, service rate, link connectivity, and the underlying routing algorithm. Unlike the traditional Internet, which connects devices worldwide using the TCP/IP protocol, DTNs are designed to function in environments where Internet connectivity is unavailable or intermittent. The TCP/IP protocol becomes ineffective in such cases, necessitating DTNs to ensure communication. DTNs aim to resolve technical issues in heterogeneous networks, particularly those caused by environmental harshness, node mobility, battery limitations, and the range between nodes. For example, DTNs are used in wireless sensor networks deployed in remote regions such as wildlife sanctuaries and rural areas with low population density or for interconnecting extraterrestrial nodes like satellites and deep-space communication systems. These applications highlight the importance of DTNs in connecting areas where traditional network infrastructure cannot reliably function. The data transmission between nodes is carried out successfully by their layer architecture, as shown in Figure-1. Delay-tolerant networks (DTNs) address technical challenges in heterogeneous networks where connections between senders and receivers are infrequent or irregular. These networks are designed to ensure the successful transmission of data packets, even in environments where conventional transport layer protocols like TCP fail due to harsh conditions and connectivity issues. TCP and UDP, the two broad transport layer protocols, generally work well in stable environments but falter in scenarios with intermittent connectivity, making DTNs an essential solution. DTNs tackle challenges such as delays in transmission, unreliable links, and limited bandwidth. One key property of DTNs is their opportunistic nature, where the network does not always require full connectivity. Connections are established whenever possible, enabling communication across intermittent networks. Another characteristic is the use of variable-length packets, allowing data to be sent in longer packets when a connection becomes available, compensating for the lack of regular connectivity. To overcome issues like disconnection and retransmission, DTNs employ a store-and-carry approach. Data is stored in intermediate nodes during periods of disconnection and forwarded when the next communication opportunity arises. This mechanism is further supported by introducing a new bundle layer between the application and transport layers (Fig. 1).

The bundle layer stores packets and facilitates their transmission whenever connectivity is restored. Security is also critical to DTNs, particularly in harsh environments with limited human intervention. DTNs enhance data security by quickly discarding packets in cases of unauthorized access, ensuring the integrity and confidentiality of transmitted information. To further improve the DTN's performance, an LSTM approach is proposed,

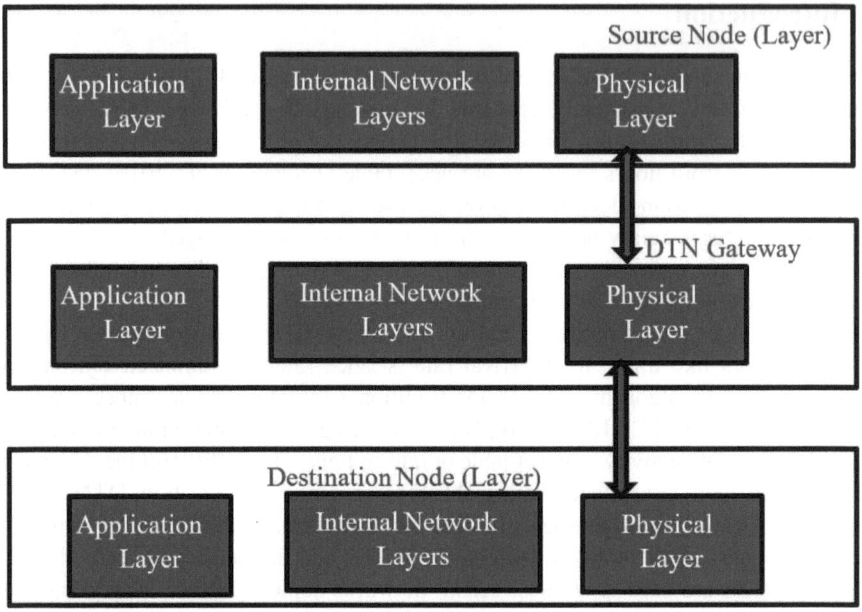

Fig. 1. Layer Architecture of DTN Data Transmission Between Nodes

which detects missing nodes in the network and transmits data without delay. LSTM is one of the most widely utilized deep learning models used for pattern recognition and classification.

2 Contribution of the Paper

- A detailed study of recent trends and the development of delay-tolerant networks and their applications is carried out.
- Designed and implemented an LSTM-based 5G integrated DTN system for fast and secure data transmission.
- The model's performance regarding E2E delay time, throughput, PDR, and drop rate is evaluated.
- Compare the result obtained by the proposed model with the existing model to define the efficiency of the proposed approach for fast and secure data transmission.

3 Literature Survey

The proposed model's issues and challenges are analyzed to identify the problem statement that can be solved by proposing a new model. A detailed literature survey was carried out to obtain the problem statement. For example, Laura Landon et al. (2024) have presented an investigation on various methods for minimizing the packet service time to improve the overall performance of the network using the 5G networks. To reduce the service time, it uses three methods: Modulation and Coding Scheme (MCS), network coding, and Hybrid Automatic Repeat Request (HARQ), which helps to improve the quality of services (QoS) by reducing the service time. To evaluate the efficiency of each model, it was compared to other models' performance in terms of time delay and QoS (M. Haiyan et al., 2016). The evaluation result shows that compared to the other two models, the network coding method has efficiently minimized 7% of the service time in the low SNR regimes using the 5G networks. To address and solve the issues that occur in the 5G networks, Mina Malekzadeh (2023) has presented a well-developed new modular model. It works based on the ML model on both service factors and architecture (5GPA), which helps to improve the overall performance of the 5G networks. The main aim of this proposed model is to identify the issues that occur in the designing and planning phases. To evaluate the model's efficiency in terms of error rate and accuracy rate, this model was trained and tested on the 5G dataset. The simulation result shows that this proposed 5GPA model has a high % accuracy rate of 95% and averages 7.60e–03 of mean absolute error, 8.77e–03 of root mean squared error, and 1.18e–04 of mean squared error.

Nowadays, device-to-device (D2D) communication is the most popular technology for transforming data beyond 5G networks (B5G). Routing multi-hop networks requires consideration of battery, fairness, device mobility, and link quality. To solve this issue, Valmik Tiwari et al. (2022) have presented a mobility, battery, link quality, and contention window size-aware routing (MBLCR) model to improve the performance of the network. To evaluate the model's efficiency in terms of packet delivery ratio, throughput, energy efficiency, and latency, this model's performance was compared to several other convolutional models such as MEQSA-OLSRv2, MP-OLSRv2, and EMBLR. The simulation results show that this proposed MBLCR model has better efficiency than the other three models in all terms, as the MBLCR model has attained a throughput value of 60.92 kbps to 51.47 kbps. Meanwhile, the other three models can't even touch the throughput value of 60 kbps. Siham Hasan et al. (2023) have presented an Enhanced Message Replication Technique (EMRT) to improve the performance of the DTN routing protocols. This EMRT helps to adjust the number of message replicas. This decision is based on encounter history, current connection, time-to-time value, buffer size history, and energy. EMRT was applied to three different protocols, the Destination-Based Routing Protocol (DBRP), Encounter-Based Routing (EBR), and Spray and Wait protocol, to evaluate the efficiency of the EMRT model. The evaluation results show that this proposed EMRT model improves the latency average, overhead ratio, and delivery ratio. By combining spray and wait, DBRP and EBR have improved the delivery rate by 13%, 10%, and 8%, and they have reduced the latency range by 51%, 13%, and 14%, respectively. Ilsun You et al. (2016) have presented a novel network formation to maintain the flow efficiently and allow the formation of a robust network. The novel framework consists of various nodes from different networks by Unmanned Aircraft (UA). This model uses the features of biological genomes and forms a delay-tolerant network. The efficiency of this model was evaluated in both simulation and real-time environments to assess its efficiency. The simulation shows that this proposed framework model has improved the data delivery performance and reduced the delay and overhead. Iftikhar Ahmad et al. (2022) have presented an investigation on the latest state-of-the-art work, which includes parasitic patches, EBG structure, the role of metamaterials (MMTs), hybrid feeding, boosted gain, reduced side-lobes level and the impact of the slots. It also reported the Mutual coupling reduction model, challenges that occurred while designing the antenna, and the role of reconfigurability was also investigated.

In recent days, the growth of mobile networks has increased the number of devices, the introduction of smart devices, and the usage of mobile applications has also grown. Caching is one of the main problems that affect the performance of the 5G networks. Because caching reduces the backhauling load in the wireless networks. To solve this issue, Salar Ismael Ahmed et al. (2021) have presented an investigation on the caching network. It investigated the types of caching and comparisons of the caching techniques to improve the performance of the 5G networks. This investigation helps to improve the performance of the cache. At the same time, it depends on the load and improves cache size, which helps to improve the overall performance of the mobile networks by reducing the traffic and the network latency. 5G network communication is deployed in various countries. However, it increased data consumption by 30% because the usage of the networks increased. So, it cannot manage to support the network with 3G or 4G. So, Universal Filtered Multi-Carrier (UFMC) is the other waveform form that helps overcome the challenges in orthogonal frequency division modulation (OFDM). Ghassan Ali Hussain and Lukman Audah (2020) have suggested that RS code with the UFMC system helps achieve efficient data transmission over noisy channels. The simulation result shows that the OFDM system's PAPR levels have increased from 6.9735 dB to 8.8653 dB. it was achieved by using this RS code in the UFMS.

4 Limitation and Motivation

The above survey shows that the traditional models have various significant challenges, like high latency, unreliable data delivery, and difficulty detecting malicious nodes, worsened by intermittent connectivity and resource constraints. Integrating 5G technology offers a promising solution with ultra-reliable low-latency communication (URLLC), improved bandwidth, and scalability, significantly enhancing network performance. A 5 G-enabled DTN model addresses these issues by leveraging 5G's advanced capabilities for faster data transmission and effective malicious node detection. In addition, a DL based approach, an LSTM model, is proposed to classify the input data. This approach makes the model suitable for critical applications, including disaster recovery and IoT networks. It extends connectivity to remote areas, transforming DTNs into more reliable and efficient networks for real-world scenarios.

4.1 Delay Tolerant Networks and 5G Technology

The integration of DTN with 5G technology marks a significant advancement in communication networks. While DTNs excel in ensuring data delivery in challenging environments, 5G further enhances their performance. 5G provides unparalleled connectivity, higher bandwidth, and reduced latency, making it an ideal complement to DTN's store-and-forward mechanism. Together, these technologies address issues like high latency and discontinuous connectivity, improving the overall Quality of Service (QoS). With

5G's high-speed data transfer capabilities, DTNs can process and transmit large volumes of data more efficiently (Prabu, A. V et al., 2019). The adaptability of DTNs in dynamic topologies and harsh environments is amplified when coupled with 5G, enabling smoother communication across terrestrial and extraterrestrial networks. For instance, in smart agriculture, the combination of DTNs and 5G ensures real-time monitoring of crops and instant data relay to centralized systems. Similarly, in military applications, 5G enhances the reliability and speed of DTN-based communication, ensuring timely and secure data exchange in critical operations.

4.2 Concept of Bundle Layer Protocol

The bundle layer is positioned between the application and transport layers in the network architecture. Its primary function is temporarily storing messages when unavailable and forwarding them to the next node as soon as connectivity is reestablished. This mechanism ensures efficient message delivery in environments with intermittent or delayed connections. The bundle layer manages three types of information: source node details, control information such as source and destination node identifiers, and a bundle layer header that organizes and facilitates the message-handling process (R. Di Pietro et al., 2014) (Fig. 2).

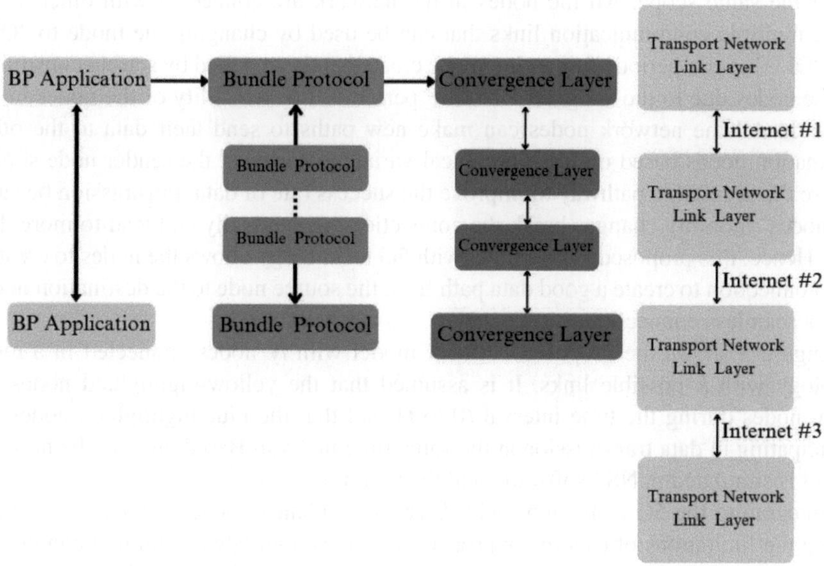

Fig. 2. Bundled Layer Architecture

Delay Tolerant Networks (DTNs) are widely utilized in environments with limited or intermittent internet connectivity. They play a crucial role in rural areas, deep space communication, and wildlife sanctuaries by deploying sensors to collect and store data transmitted to a sink or base station when connectivity becomes available. One significant application of DTNs is the Interplanetary Internet, facilitating communication for the International Space Station, deep space exploration, and future space debris monitoring. In military and intelligence operations, DTNs support mobile ad-hoc networks for cargo tracking, enemy movement surveillance, and monitoring unmanned aerial vehicles (UAVs). In commercial contexts, DTNs are used for vehicle monitoring across various modes of transportation, agricultural crop analysis, and the monitoring of underground mines. Environmental monitoring applications include tracking crop health, wildlife sanctuary activities, atmospheric and oceanographic conditions, and seismological events. Public service and safety applications leverage DTNs for monitoring through CCTV cameras and global aeroplane traffic control systems. DTNs are instrumental in conducting sea-based analyses in underwater acoustic environments by taking continuous photos and transmitting them to a base station. Finally, terrestrial mobile networks benefit from DTNs by connecting mobile devices in scenarios where unexpected network failures occur, often due to signal strength issues, enabling data transmission despite disruptions.

5 Proposed DTN Model

Mobile AdHoc Network is one of the classes of remote systems. It is an existing system connected by more than two gadgets that perform and communicate with one another under the same scope. All the nodes in the network are connected with other nodes using multiple communication links that can be used by changing the mode to "ON" and "OFF" over a period of time. It may be created and controlled by setting constraints on the nodes due to the assigned ON-OFF period or the versatility of the nodes in the network. All the network nodes can make new paths to send their data to the other destination nodes based on the topological variants. However, the sender node should ensure the accessible pathway to improve the success rate of data transmission because the nodes' mobility changes break the connections dynamically and lead to more data loss. Hence, this proposed DTN model with 5G technology allows the nodes to create a new connection to create a good data path from the source node to the destination nodes with a seamless connection.

Figure 3 shows the proposed network model with N nodes connected in a mesh topology with k possible links. It is assumed that the yellow-highlighted nodes are silent nodes during the time interval $t0$ to $t1$ and that the blue-highlighted nodes are participating in data transmission at the same time interval. Based on that, the network model is simulated in NS2 software, and the output is verified.

Integrating the 5G technology with DTN will enhance model efficiency by eliminating the limitations of existing approaches, such as high latency, limited bandwidth, and irregular connections. The 5G system further improves the reliability and speed of the DTN during message delivery. Additionally, the 5G system aids DTN in selecting prioritized data, managing bandwidth, and reducing latency. These features are particularly useful in identifying malicious nodes in the network. Therefore, it will enhance the overall throughput score of the DTN. The overall structure of the proposed model is shown in Fig. 4.

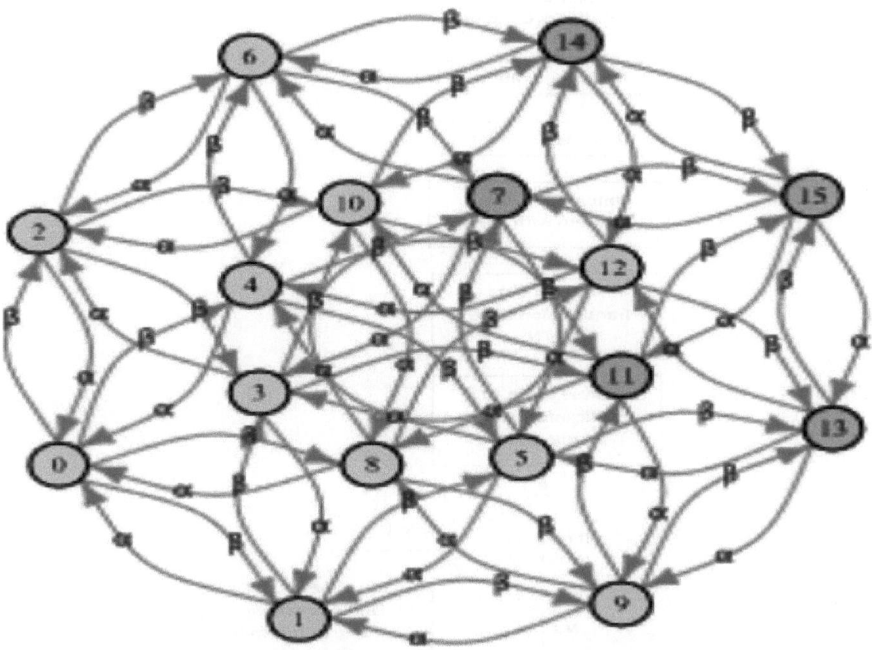

Fig. 3. DTN Network Model with Node Connections

5.1 LSTM-Based Malicious Node Detection and Secure Data Transmission

In the proposed 5G integrated DTN model, an LSTM-based approach is deployed to detect malicious nodes and securely transmit the data in the network. LSTM is a kind of RNN model widely utilized for sequential task prediction and temporal feature extraction from the input data. This model is more suitable for analyzing the behaviour of each node in the network. The LSTM model includes three different gates to analyse the input data, as shown in Fig. 5.

Forget Gate: It is an initial gate that removes the information that is no longer useful. It is performed based on the previous cell state, current input data, and previously hidden state output. For this, initially, previous hidden state data h_{t-1} and new input data x_t are fed into the neural network. By analyzing these data, both relevant and irrelevant information is predicted. It is achieved by multiplying the fed data with weighted matrices value flowed to adding bias value using a sigmoid function (σ). The resultant output is obtained in the interval of 0 and 1. If the output is closer to 0, the data present in the input is termed irrelevant and forgotten. If the output is closer to 1, the data are retrieved for future use. Using equation- (1), the forget gate is a function.

$$f_t = \sigma(W_f * [h_t - 1, x_t] + b_f \tag{1}$$

where W_f, b_f, $h_t - 1$, X_t, f_t, and σ represent the weighted matrices value, bias value, hidden state value, input data point, forget gate, and sigmoid function, respectively.

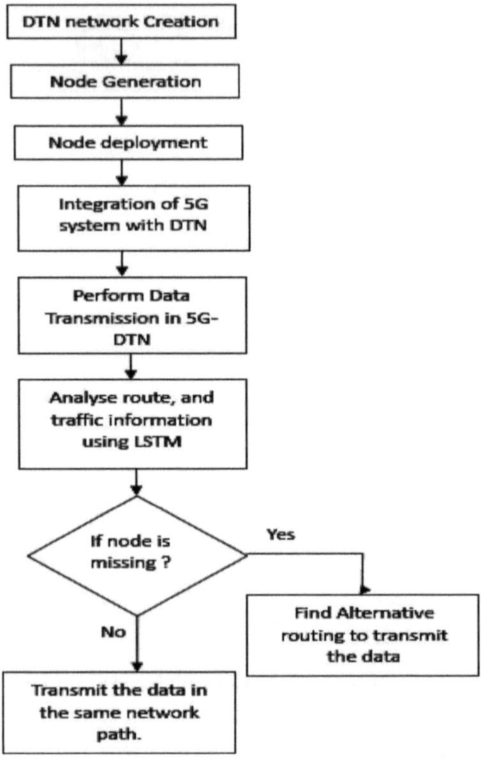

Fig. 4. Proposed Workflow

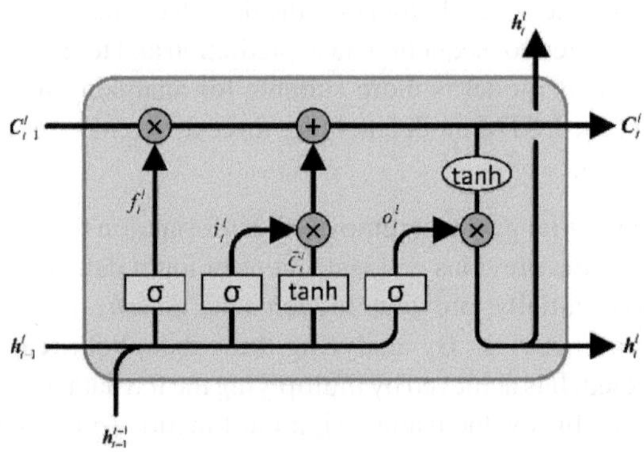

Fig. 5. Structure of LSTM

Input Gate: This gate adds additional information to the cell state. Like the forget gate, this gate also detects the relevant data from the given new input and previously hidden state data. The vector is generated for each element using *tanh* function. Though the forget gate performs the same process, it fails to check whether the input data is worth remembering. Meanwhile, the input gate acts as a filter, accurately capturing valuable data from the data series. The output of the vector is predicted nearer to 0 or 1. If the value is closer to 0, the elements are not updated to the cell state. The input gate ignores the element from the vector. The equation for the input gate is expressed as (2 and 3).

$$I_t = \sigma(W_i * [h_t - 1, x_t] + b_I \tag{2}$$

$$\tilde{C_t} = \tanh((W_C * [h_t - 1, x_t] + b_C) \tag{3}$$

where, $W_i, andW_C$, represent weighted value and $b_I andb_C$ represent the bias value. Now, by multiplying forget gate f_t with old cell state and $I_t * C_t$, the updated element chosen by the input gate is updated to the cell state. The updated data is expressed using the ⊙ operator as:

$$C_t = f_t \odot C_{t-1} + I_t \odot \tilde{C_t} \tag{4}$$

Here, ⊙ represent the element-wise multiplication and C_t represent the updated element in the cell state.

Output Gate: This is the final layer in the LSTM model; the decided output is executed through this layer. This layer executes the filtered version of our cell state. At first, the sigmoid function is used to decide the proper cell state to be executed. Then, the *tanh* function is used to get the result between –1 and +1. Now, the output of the sigmoid function is multiplied with the *tanh* function; then the final output is executed. The equation for the output gate is expressed as:

$$O_g = \sigma(W_O * \left[h_g - 1, x_g\right] + b_O \tag{5}$$

$$h_g = f_g * \tanh(C_g) \tag{6}$$

Using these gates, the LSTM model analyses the temporal behaviour of the input node. If the node's behaviour does not match that of the other nodes in the network, it is depicted as a malicious node and discarded from the network. This will improve the overall performance and speed of the DTN during data transmission. All the nodes, personal information, and traffic data are recorded in the network dynamically and input to the LSTM model. This also follows the security model proposed in G. M. Kumar and D. Hemanand (2024) to secure the nodes and traffic data.

6 Performance Evaluation

Once the malicious node is detected, the performance model is evaluated regarding throughput, PDR, and E2E delay time. The throughput result represents the model's overall performance in detecting nodes and transmitting data in the network. The PDR value illustrates the total amount of data transmitted within the period. The E2E delay time result depicts the proposed model's latency time in transferring data in each node in the DTN. These metrics are analysed using the following expressions.

$$PDR = \frac{number\ of\ packet\ received}{number\ of\ packet\ sent} \times 100\%$$

$$E2E\ delay = \sum_{i=1}^{n} (t_{receive} - t_{sent})/n$$

$$Throughput = \frac{\sum_{i=1}^{n} S_i}{T}$$

where n is the total number of packets successfully transmitted, T represent the total time, and S_i indicates the size of the packet transmitted.

7 Simulation Results and Discussion

The network model is designed using NS2 software and simulated in the NAM window, and the network functionalities are verified in the TR file. The trace information comprises throughput, data packets delivered, dropped, and the time of sending from the source node and received at the destination node. Using this information, the End-to-end (E2E) delay, packet delivery ratio, throughput, and energy consumption are calculated to evaluate the performance of the new network model (Prabu, A. V., & Kumar, G. S., 2019).

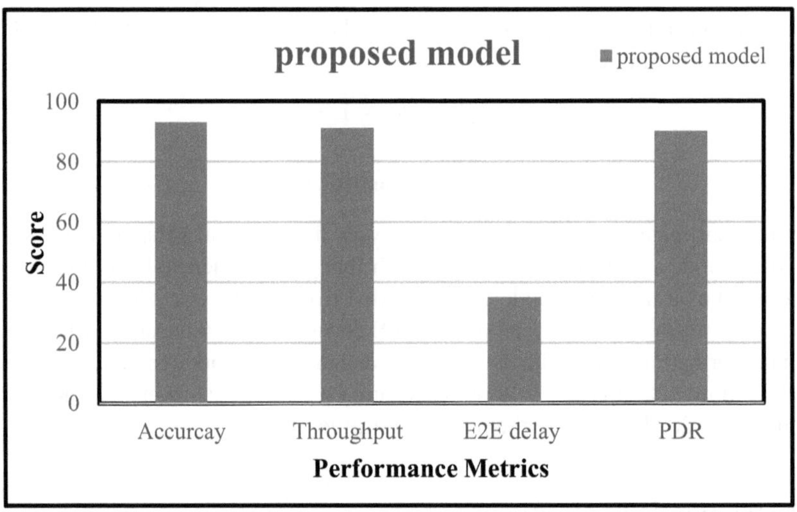

Fig. 6. Performance Evaluation

Figure 6 shows the performance metrics result of the proposed LSTM employed 5G integrated DTN model. The analysis shows that the proposed LSTM-based model achieved 90% accuracy on node detection, 18% improved throughput, 12–15% increases in PDR, and minimum E2E delay time. Figure 7 shows the E-2-E delay calculated between source and destination nodes in the network. The total time duration for sending data from the source node to the destination node is the delay.

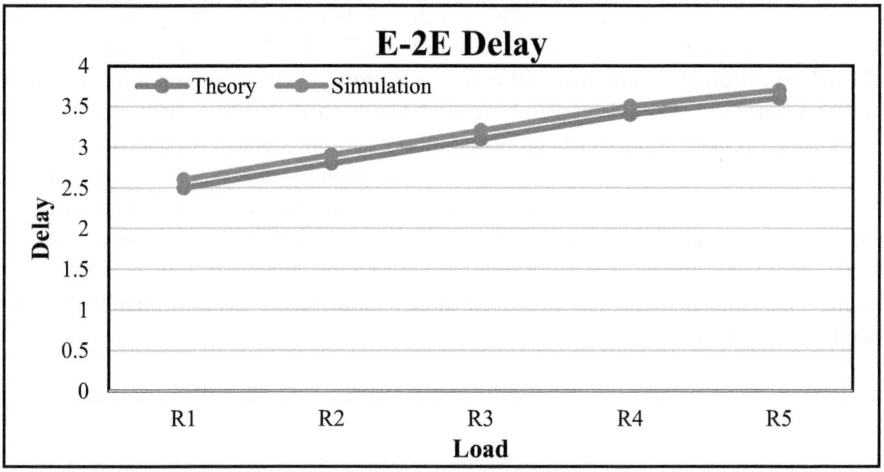

Fig. 7. E2E delay time

Figure 8 depicts the throughput comparison result of the proposed and existing models at different speeds, such as MEQSA-OLSRV2, MP-OLSRv2, MBLCR, and EMBLR. From the figure, it is clear that compared to other models, the proposed DTN-5G model has achieved high throughput, which indicates that it is more suitable for predicting optimal, robust, and reliable paths in the network. The speed of the device is increased from 10 m/sec to 60 m/sec. The analysis shows that the MP-OLSRv2 model achieved the least throughput value of 44.56 Kbps at 60 m/sec. Meanwhile, the proposed model achieved 71.32 kbps at a 10 m/sec speed. When speed increases, the throughput value of the models decreases. In that sense, at 60 m/sec compared to other methods, the proposed model achieved 60.92 kbps.

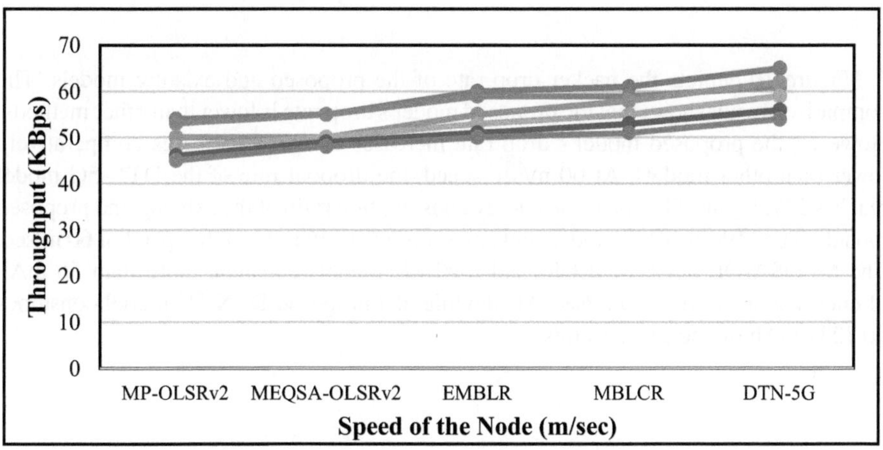

Fig. 8. Throughput Concerning Speed of the Node

Figure 9 depicts the packet delivery ratio (PDR) value of the proposed model compared to existing models such as MEQSA-OLSRV2, MP-OLSRv2, MBLCR, and EMBLR at different speeds from 10 m/sec to 60 m/sec. The comparison result shows that the proposed DTN-5G models have achieved high PDR values at various speeds. It is clear from the result that the proposed model is more suitable for transmitting data within the network without any interruption. The primary reason for disruption in the network is changes in the topology. In this case, the proposed model automatically forecasts the changes in the network topology and forwards the message to the destination node without any breaks. Thus, the proposed model achieved higher PDR values at all speeds than other models. The proposed model achieved a 5 times higher PDR value than the other three.

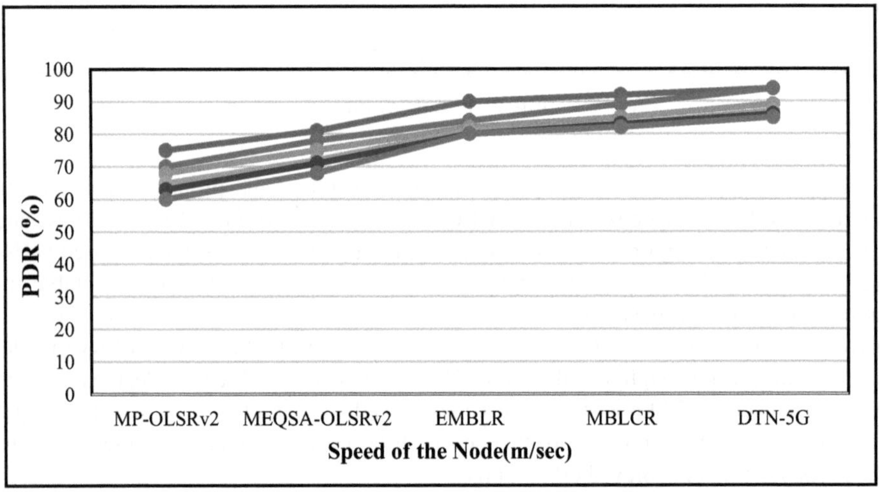

Fig. 9. Packet Delivery Ration Concerning Speed of the Node

Figure 10 depicts the packer drop rate of the proposed and existing models. The comparison result shows that the proposed model's drop rate is lower than other methods. However, the proposed model's drop rate increases as speed increases comparatively lower than other models. At 60 m/sec speed, the dropout rate of the DTN-5G model reaches 157. Figure 11 depicts the energy consumption ratio of the existing and proposed models. The MP-QLSRv2 model consumes more than 56 mAh at the speed of 60 m/sec. The MEQSA-OLSRv2, MBLCR, and EMBLR models consume more than 56 mAh of energy at a speed of 60 m/sec. Meanwhile, the proposed DTN-5G model consumes 50.1234 mAh of energy at 60 m/sec.

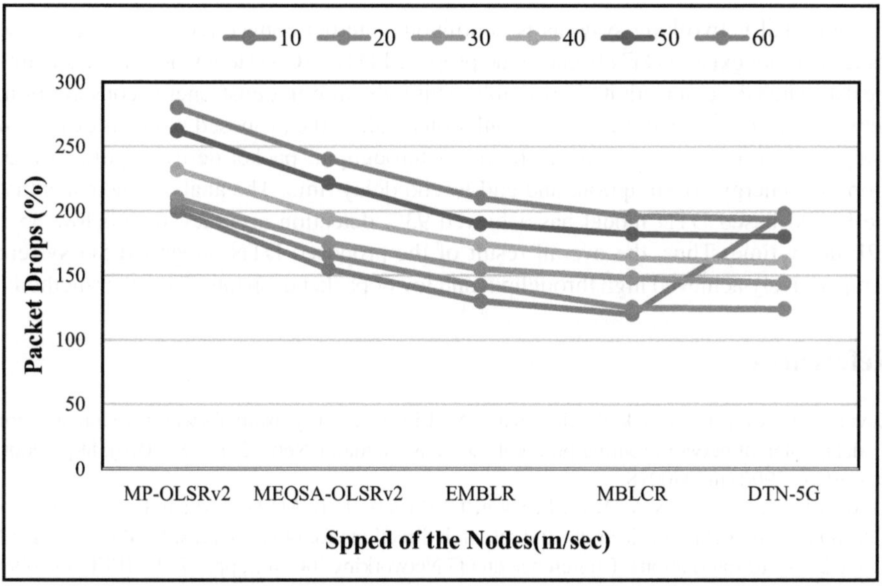

Fig. 10. Packet Drops Concerning the Speed of the Node

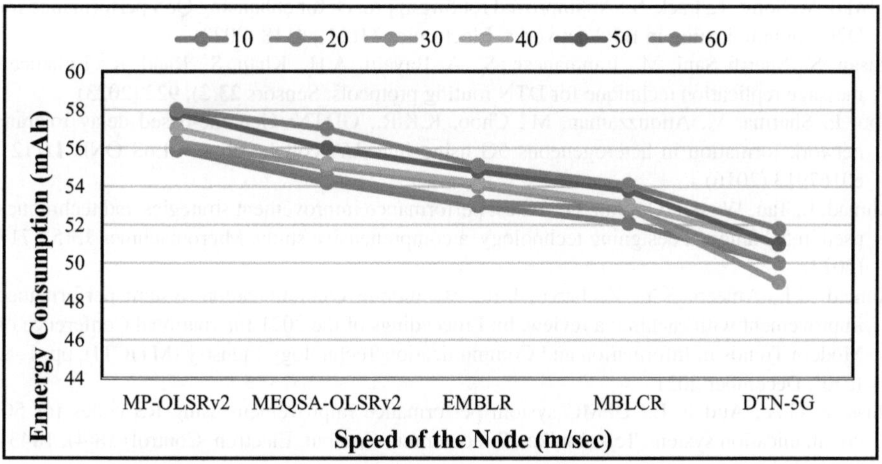

Fig. 11. Energy Consumption Concerning the Speed of the Node

8 Conclusion

This paper proposes a DTN-integrated 5G model to improve data transmission efficiency and accuracy. DTN-based systems have recently become more popular among various networks due to their low error rate, quick data transmission speed, and low energy consumption. Due to these merits, the DTN-based models are widely utilized in many applications such as wildlife science, terrestrial mobile networks, etc. The major problem of traditional networking systems is continuous internet connectivity. However, these issues will not exist in DTN because the proposed DTN-5G system stores and transmits the data when the connectivity is available. This will minimize cost, energy consumption, network traffic, and error rate. The simulation result of the proposed model is evaluated using NS2 simulation software in terms of throughput, packet delivery ratio, packet drop rate, energy consumption, and end-to-end delay time. The analysis indicates that the LSTM-based DTN model has achieved 93% detection accuracy with below 35% E2E delay time. Thus, the overall result of the proposed DTN-integrated 5G system comparatively achieved high throughput and lower packet drop rate than other methods.

References

Kaviani, M., Kusy, B., Jurdak, R., Bergmann, N., Liu, V.: Energy-aware forwarding strategies for delay tolerant network routing protocols. J. Sens. Actuator Netw. **5**(4), 18 (2016). https://doi.org/10.3390/jsan5040018

Landon, L., Vasudevan, V.A., Kim, J., Sung, J., Masters, J. T., Médard, M.: Enhancing 5G performance: reducing service time and research directions for 6G standards. In: Proceedings of the 2024 3rd International Conference on 6G Networking (6GNet), pp. 37–45. IEEE, October 2024

Malekzadeh, M.: Performance prediction and enhancement of 5G networks based on linear regression machine learning. EURASIP J. Wirel. Commun. Netw. **2023**(1), 74 (2023)

Tilwari, V., Song, T., Pack, S.: An improved routing approach for enhancing QoS performance for D2D communication in B5G networks. Electronics **11**(24), 4118 (2022)

Hasan, S., Sharifi Sani, M., Iranmanesh, S., Al-Bayatti, A.H., Khan, S., Raad, R.: Enhanced message replication technique for DTN routing protocols. Sensors **23**(2), 922 (2023)

You, I., Sharma, V., Atiquzzaman, M., Choo, K.K.R.: GDTN: Genome-based delay tolerant network formation in heterogeneous 5G using inter-UA collaboration. PLoS ONE **11**(12), e0167913 (2016)

Ahmad, I., Tan, W., Ali, Q., Sun, H.: Latest performance improvement strategies and techniques used in 5G antenna designing technology, a comprehensive study. Micromachines **13**(5), 717 (2022)

Ahmed, S.I., Ameen, S.Y., Zeebaree, S.R.: 5G mobile communication system performance improvement with caching: a review. In: Proceedings of the 2021 International Conference of Modern Trends in Information and Communication Technology Industry (MTICTI), pp. 1–8. IEEE, December 2021

Hussain, G.A., Audah, L.: UFMC system performance improvement using RS codes for 5G communication system. Telkomnika (Telecommun. Comput. Electron. Control) **18**(4), 1843–1848 (2020)

Gopalakrishnan, S., Ruby, E.K., Hemanand, D., Anitha, R., Suresh, D., Choubey, S.B.: Leveraging AI and blockchain in MANETs to enhance smart city infrastructure and autonomous vehicular networks. Babylonian J. Mach. Learn. **2024**, 112–120 (2024). https://doi.org/10.58496/BJML/2024/011

Kumar, G.M., Hemanand, D.: A robust data security model development to maintain the integrity of information using artificial intelligence and cyber laws. In: Proceedings of the 2024 International Conference on Intelligent Systems for Cybersecurity (ISCS), Gurugram, India, pp. 1–7 (2024). https://doi.org/10.1109/ISCS61804.2024.10581024

Rajasoundaran, S., et al.: Secure routing with multi-watchdog construction using deep particle convolutional model for IoT-based 5G wireless sensor networks. Comput. Commun. **187**, 71–82 (2022). https://doi.org/10.1016/j.comcom.2022.02.004

Prabu, A.V., Kumar, G.S.: Hybrid MAC-based adaptive preamble technique to improve the life-time in wireless sensor networks. **11**(1), 240–249 (2019). https://www.jardcs.org/abstract.php?id=30

Haiyan, M., Jinyao, Y., Georgopoulos, P., Plattner, B.: Towards SDN based queuing delay estimation. China Commun. **13**(3), 27–36 (2016)

Di Pietro, R., Guarino, S., Verde, N.V., Domingo-Ferrer, J.: Security in wireless ad-hoc networks – a survey. Comput. Commun. **51**(2014), 1–20 (2014). https://doi.org/10.1016/j.comcom.2014.06.003

Adaptive Traffic Control System Using Machine Learning

Kavitha Subramani, K. Abhinaya(✉), M. Ajitha, A. Aaliya Fathima, U. Akshya, and A. Prabha

Panimalar Engineering College, Poonamallee, Chennai 600123, India
abhinayakumaresan22@gmail.com

Abstract. To tackle the critical problem of urban traffic congestion, the Adaptive Traffic Control System (ATCS) dynamically optimizes traffic signal timings in real-time by leveraging cutting-edge machine learning algorithms. Conventional traffic signal systems frequently have fixed timing schedules that don't adjust to the way traffic flows. This results in longer wait times and more fuel being used. In order to learn and anticipate traffic patterns, the ATCS integrates data from cameras and GPS-enabled vehicles. The ATCS may modify traffic signals according to the flow, density, and speed of vehicles by continuously observing and evaluating real-time data, Traffic lights can be changed by the ATCS in response to the flow, speed, and density of vehicles. With the help of this adaptive strategy, the system can react to changing traffic volumes throughout the day, giving priority to traffic flow during peak hours and making dynamic adjustments in the event of unforeseen disruptions like accidents or road closures. By lowering emissions linked to idling cars, the ATCS deployment not only improves traffic management efficiency but also advances environmental sustainability. Comparing the ATCS to traditional fixed-timing systems, simulation tests show a considerable reduction in average waiting times and total travel duration.

Terms: YOLO v8 · Traffic Optimization · Real-Time Traffic Management · Machine Learning · Adaptive Traffic Control

1 Introduction

The issue of urban traffic congestion has grown to be a significant concern, resulting in increased fuel usage, air pollution, and delays. Fixed signal timing schedules, which are frequently the foundation of conventional traffic management systems, are unable to adjust in real time to changes in traffic volume and patterns. This rigidity leads to ineffective traffic movement, prolonged wait times at junctions, and heightened annoyance among Motorists. Adaptive traffic control systems (ATCS) have become a viable option for improving traffic signal management in response to these issues. Through the use of cutting- edge technology like real- time data analytic and machine learning, ATCS is able to dynamically modify traffic signals in response to the environment. The YOLO (You Only Look Once) algorithm, which enables real- time object identification

R. Geetha et al. (Eds.): AAIMB 2024, CCIS 2653, pp. 326–335, 2026.
https://doi.org/10.1007/978-3-032-13026-6_23

via cameras, is a crucial part of this system. The implementation of ATCS not only aims to reduce delays at traffic signals but also enhances safety by minimizing the risk of accidents caused by sudden stops and starts. Improving traffic movement efficiency, ATCS contributes to lower emissions and a reduced carbon footprint, aligning with sustainable urban development objectives.

This paper presents a comprehensive overview of an ATCS that integrates the YOLO algorithm for real- time vehicle detection, detailing its architecture, operational mechanisms, and the machine learning techniques used for traffic pattern recognition. We will also evaluate the performance of the proposed system showcasing its advantages over conventional traffic management methods. The findings from this research seek to provide valuable insights into the effectiveness of ATCS in modernizing urban traffic control.

2 Related Work

In [1], the authors introduce a framework utilizing SUMO-based simulation and reinforcement learning (RL) for smart traffic control. The system employs Induction Loop Detectors and Lane Area Detectors to gather traffic data, which is then processed using a Q- learning model to optimize traffic signal timings. The system demonstrates improved efficiency and fairness over traditional traffic control methods.

In [2], a deep Q-learning (DQL)-based approach is proposed for adaptive traffic light control. The system uses a deep neural network in combination with reinforcement learning to optimize queue lengths and vehicle waiting times. The DQL model outperforms conventional methods, though future work is needed to test the system's scalability across multiple intersections.

A novel traffic light control system leveraging YOLOv8 for real-time vehicle detection and traffic density estimation is presented in [3]. The system achieves over 97% mean average precision (mAP) in vehicle detection, optimizing traffic light timings more effectively than existing systems.

The reinforcement learning-based traffic control framework proposed in [4] employs the SUMO simulator to generate variable traffic conditions. A Q- learning algorithm adjusts traffic light signals based on real-time data, improving overall traffic flow. Future work suggests expanding the system to a multi- agent environment for managing several intersections simultaneously.

In [5], an adaptive traffic management system is discussed, focusing on countries like India where traffic congestion is severe. The system uses IoT and machine learning techniques to lower waiting times, reduce pollution, and prioritize emergency vehicles. Future research will focus on detecting road accidents using video processing techniques.

Another work [6] explores the use of reinforcement learning in adaptive traffic control, particularly deep reinforcement learning models that integrate neural networks for decision-making. The study shows significant reductions in vehicle waiting times compared to conventional systems.

The paper [7] presents a machine learning-based adaptive traffic control system using sensor data and environmental inputs. This system prioritizes emergency vehicles and adapts traffic signals based on real-time data from sensors, offering significant improvements in traffic management efficiency.

In [8], a traffic management system combining machine learning and IoT is proposed to address urban congestion. The system integrates various data sources, including sensors and GPS, to optimize traffic flow dynamically. Future work focuses on enhancing real-time data collection and decision- making processes.

A neural network-based approach to adaptive traffic control is detailed in [9]. This system uses deep learning techniques to predict traffic flow and adjust signal timings accordingly. The study demonstrates improved traffic flow in simulations using real-world data.

Lastly, in [10], an IoT-enabled traffic control system is introduced, where sensor networks provide real- time data for optimizing traffic lights. This system not only reduces vehicle waiting times but also helps in monitoring and managing air pollution levels in congested areas.

To track the presence of vehicles at junctions, existing System Interface and Front-end: By providing real- time warnings and notifications in response to pre-established rules or abnormalities found, the Alert and Notification System serves as an early warning system. For instance, the system can sound a warning for traffic authorities to step in if there is an abnormally high volume of vehicles or pedestrians at an intersection. In order to provide a coordinated reaction to accidents or other incidents, this module an also interact with other city infrastructure systems, such as traffic enforcement. This module's goal is to improve overall traffic efficiency and safety by offering a quick reaction mechanism. Systems may make use of basic vehicle detection technologies like inductive loops or straightforward infrared sensors. Nevertheless, they frequently aren't able to give precise counts or differentiate between cars and cars and pedestrians. Because of this, they find it difficult to react to abrupt shifts in traffic patterns, such as surges brought on by accidents, road closures, or special occasions. Because of this static approach's inability to adjust signal timings for changing traffic levels throughout the day, fuel consumption would ultimately increase the emissions.

3 Methodology

The proposed Adaptive Traffic Control System(ATCS) is designed to address the inefficiencies of conventional fixed- time traffic light systems by leveraging deep learning and real-time video processing. The System utilizes the YOLOv8 (You Only Look Once) object detection model to accurately identify vehicles and pedestrians in live traffic video feeds. By continuously monitoring traffic density, the system dynamically adjusts the timing of traffic signals to optimize traffic flow at intersections. This adaptive approach ensures that lanes with higher vehicle volumes are given longer green lights, while lanes with less traffic receive shorter duration ,effectively reducing congestion and improving road efficiency Traffic Signal Control Logic: Decisions regarding traffic signals are made by the Traffic Signal Control Logic module using data from real-time vehicle and pedestrian counts. By optimizing traffic signal timing dynamically, this logic helps to improve traffic flow and lessen congestion. It applies rule- based decision- making by using algorithms to analyze the data from the detection module (Fig. 1).

The system architecture is composed of several integrated modules that work together to achieve real-time adaptive traffic control. Initially, video feeds are captured from strategically positioned cameras at intersections. These video streams undergo pre-processing,

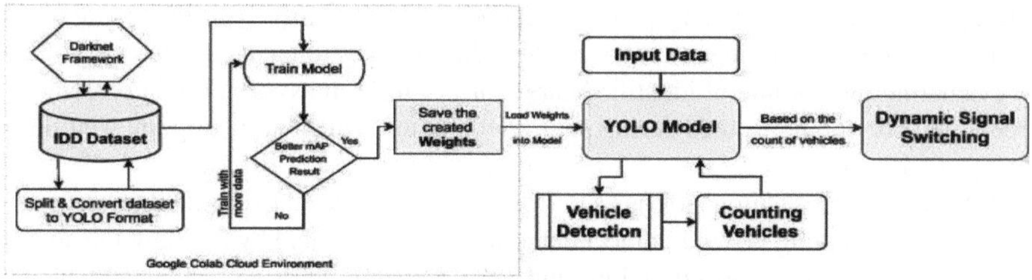

Fig. 1. Architecture of YOLO

including frame extraction, resizing, and normalization, to prepare them for input into the YOLOv8 model. The YOLOv8 model performs fast and accurate object detection by identifying and classifying vehicles and pedestrians within each frame.

Once the objects are detected, their counts and positions are extracted and sent to the Traffic Signal Control Logic module. This module implements a rule-based algorithm that evaluates the traffic density on each lane and decides the optimal signal timing. For example, when a particular lane exhibits heavy vehicle presence, the control logic extends the green light duration for that lane to alleviate congestion. Conversely, lanes with low traffic are assigned shorter green intervals to minimize unnecessary waiting times.

The system operates continuously in a closed feedback loop, where traffic data is constantly updated and signal timings are adjusted dynamically. This real-time adaptability not only enhances traffic throughput but also reduces idle time and emissions caused by stop-and-go traffic conditions. Furthermore, the modular design of ATCS allows for easy integration with existing traffic infrastructure and scalability for future enhancements, such as emergency vehicle prioritization or pedestrian safety features.

The Adaptive Traffic Control System (ATCS), which improves urban traffic management by utilizing the YOLO (You Only Look Once) principle, The initial step of the procedure involves utilizing the IDD data set, which is a carefully selected group of photos featuring different traffic situations. This data set is carefully prepared by dividing it into manageable chunks and transform min git into the object detection- optimized YOLO format. To train the YOLO model on this data set, the Dark net framework—which is renowned for its effectiveness in deep learning model training—is utilized. The mean Average Precision (mAP) to measure the performance across all detected objects in traffic.

Mean Average Precision

$$A = 1/\sum A(\mathrm{Q} = 1)$$

statistic is used by the system to continuously assess the model's performance throughout training. The model retains the training weights if the mAP results satisfy predetermined standards; if not, iteratively adds more data to improve the model's accuracy and learning.

The YOLO model is used for in-the-field real-time vehicle detection after it performs well enough. To identify and count vehicles at intersections, the system receives input data, usually in the form of real-time video feeds from traffic cameras, which it processes.

The dynamic signal switching process depends on this ability. The technology may modify traffic signal timings in real-time to reduce congestion and enhance traffic flow by evaluating the volume of identified cars. For instance, the system might prolong the green light for popular routes during peak traffic hours while shortening the wait times for secondary streets.

Intersection over Union (IoU) is a performance metric used to evaluate the accuracy of predicted bounding boxes in object detection. It is calculated by dividing the area of overlap between the predicted and ground truth boxes by the area of their union. IoU values range from 0 to 1, where 1 indicates a perfect match. A higher IoU signifies better prediction accuracy. Typically, an IoU threshold (e.g., 0.5) is set to determine whether a prediction is considered correct.

Image Input and Preprocessing: The Image Input and pre-processing module is designed to handle still images captured from traffic cameras or other sources. It acquires images representing specific moments in traffic conditions. Before analysis, each image undergoes several pre-processing steps to prepare it for the YOLO detection model. These steps include re-sizing the image to a fixed dimension, normalizing pixel values to a standard range, and converting the image into the correct color format required by the model. This ensures that the input images are consistent and optimized for accurate vehicle and pedestrian detection.

Intersection over Union (IoU) for Bounding Box Accuracy

$$o = raof\ non\Big/ ra\ of\ ra$$

In order to guarantee effective detection even in variable weather situations like rain, fog, or at night, additional approaches like noise reduction and frame rate optimization may be utilized. This module's objective is to produce consistent, high- quality video data that can be instantly evaluated to guarantee accurate detection of both cars and pedestrians (Fig. 2).

Load the system is the YOLO Model - Vehicle and Pedestrian Detection module. It uses the YOLO paradigm, which is well-known for being quick and precise when detecting objects in real time. To estimate bounding boxes and class probabilities for each grid in an image, YOLO divides the image into grids.

Bounding Box Prediction for localizing vehicles and pedestrians. B = (x,y,w,h) x,y: Coordinates of the center of the bounding box. w,h: Width and height of the bounding box, relative to the entire image.

Traffic Signal Control Logic: Decisions regarding traffic signals are made by the Traffic Signal Control Logic module using data from real-time vehicle and pedestrian counts. By optimizing traffic signal timing dynamically, this logic helps to improve traffic flow and lessen congestion. It applies rule- based decision- making by using algorithms to analyze the data from the detection module. Examples of these decisions include extending the duration of the green light if traffic above a threshold and giving pedestrians priority during peak crossing periods. In order to adjust signal timings depending on patterns of traffic density, the module also takes into consideration the time of day, differentiating between peak and off- peak hours. All road users should have as little

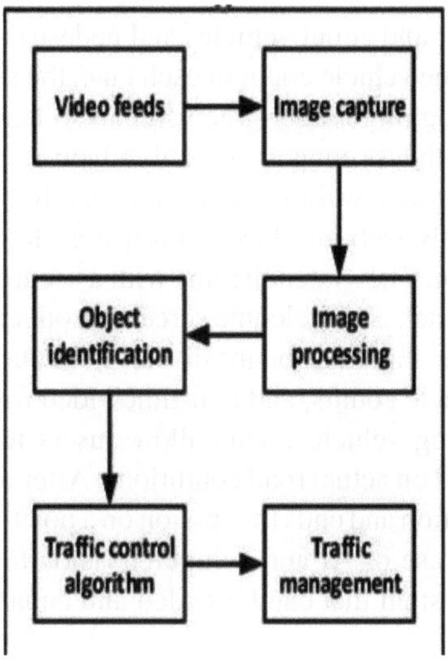

Fig. 2. Process Flow of Traffic Control System Using Image and Object Identification

waiting time as possible while also keeping efficiency System Interface and Front-end: By providing real- time warnings and notifications in response to pre- established rules or abnormalities found, the Alert and Notification System serves as an early warning system. For instance, the system can sound a warning for traffic authorities to step in if there is an abnormally high volume of vehicles or pedestrians at an intersection. In order to provide a coordinated reaction to accidents or other incidents, this module an also interact with other city infrastructure systems, such as traffic enforcement. This module's goal is a quick reaction mechanism (Fig. 3).

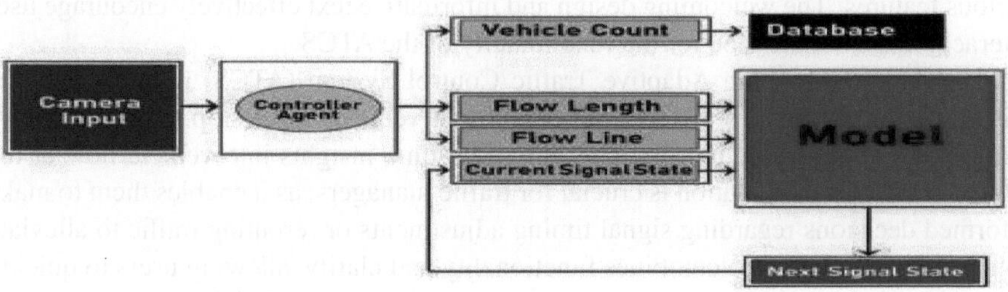

Fig. 3. Flow chart of an adaptive traffic control system

4 Result

The Adaptive Traffic Control System (ATCS) was successfully developed using Python and the YOLOv8 (You Only Look Once) object detection algorithm to intelligently manage traffic in real time. The system processes live video feeds from traffic cameras and uses YOLOv8 to detect and count vehicles and pedestrians with high accuracy and speed. Based on the real-time vehicle count in each lane, the system dynamically adjusts traffic signal timings— assigning longer green signals to lanes with heavier traffic and shorter duration to those with less congestion. This adaptive method helps reduce traffic buildup, minimizes unnecessary waiting, and improves the overall flow of vehicles at intersections. A user-friendly web interface, built using Flask, guides the user through various stages of interaction. The system begins with a secure login screen that restricts access to authorized personnel. The welcome screen introduces users to the purpose and functionality of ATCS. The main dashboard displays the current signal status (such as RED or GREEN), live vehicle counts, and real-time video frames with YOLO detected bounding boxes highlighting vehicles. This allows users to clearly observe how the system adjusts signals based on actual road conditions. After usage, the thank you screen confirms successful interaction and ends the session on a positive note. Overall, the ATCS demonstrates the effective use of AI and computer vision to create a responsive, real-time traffic management system that can be scaled and enhanced further for smart city applications.

The login screen is designed to ensure secure access to the ATCS, requiring users to enter their credentials, including a username and password. This security feature protects the system from unauthorized access, allowing only authorized personnel to monitor and control traffic. The interface is straightforward, with clear fields for input and a "Login" button to submit the information. Upon successful authentication, users receive a confirmation message, allowing them to proceed to the main features of the system.

The welcome screen serves as an introduction to the Adaptive Traffic Control System, outlining its primary objective of optimizing traffic flow through advanced algorithms. It emphasizes the system's ability to adapt dynamically to changing traffic conditions, thereby improving overall efficiency on the roads. This interface is visually appealing, with a prominent "Start" button inviting users to engage with the system and explore its various features. The welcoming design and informative text effectively encourage user interaction and set the tone for the functionality of the ATCS.

The dashboard of the Adaptive Traffic Control System (ATCS) presents the current traffic light status as "RED," indicating that vehicles must stop. It also displays information about traffic density, providing real-time insights into vehicle flow at the intersection. This information is crucial for traffic managers, as it enables them to make informed decisions regarding signal timing adjustments or rerouting traffic to alleviate congestion. The interface combines functionality and clarity, allowing users to quickly assess the current traffic conditions (Fig. 4).

The Thank You Screen serves as a final acknowledgment to users after they have successfully completed their tasks within the Adaptive Traffic Control System. This interface expresses gratitude for their participation and engagement with the system, reinforcing a positive user experience. It may include a message such as "Thank you for using the Adaptive Traffic Control System!" along with an invitation to return for future sessions or to explore more features. The design is typically warm and inviting, possibly featuring a friendly background image related to traffic management, along with options for users logout or return to the main menu. This screen is essential for closing the

Fig. 4. Real time traffic Monitoring

interaction on a positive note, ensuring users feel appreciated and encouraging them to engage with the system again in the future.

5 Accuracy

To present the accuracy of an Adaptive Traffic Control System using the **YOLOv8 (You Only Look Once version 8)** algorithm, we typically assess the performance in terms of **Mean Average Precision (mAP)**, which is a common metric used to evaluate object detection models like YOLO. The accuracy will depend on several factors, such as the quality and size of the dataset, the model's training process, and how well the algorithm adapts to real-world traffic conditions. **X-Axis (Epochs or Data Size)**: This represents either the number of training epochs or the amount of data used during training.

Y-Axis (Accuracy/mAP): This represents the model's accuracy, typically in terms of **mAP @ IoU = 0.5** (Intersection over Union threshold of 50%).

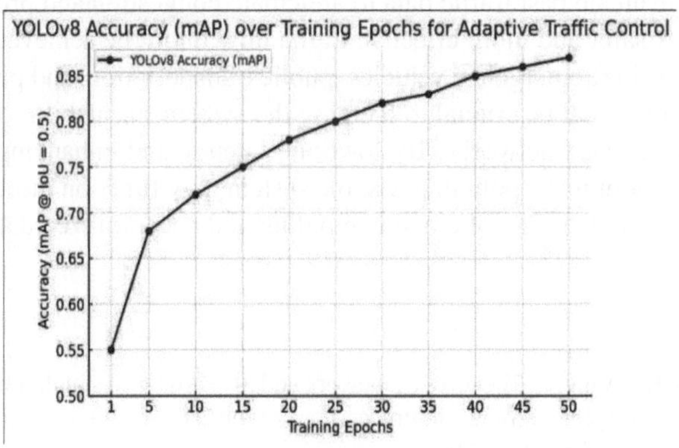

Fig. 5. YOLOv8 Accuracy graph

The graph above represents the hypothetical accuracy of an Adaptive Traffic Control System using YOLOv8, measured by Mean Average Precision (mAP @ IoU = 0.5), as the model trains over several epochs. The accuracy improves steadily, starting at around 55% in the initial epochs and increasing to about 87% after 50 epochs of training (Fig. 5).

This shows how the YOLOv8 model becomes more accurate in detecting objects (vehicles and pedestrians) in traffic as it undergoes more training iterations.

6 Conclusion

The Adaptive Traffic Control System (ATCS) implemented via the provided Python code effectively utilizes computer vision and machine learning techniques to dynamically manage urban traffic. By leveraging the YOLOv8 model for real- time vehicle and pedestrian detection, the system significantly improves traffic flow and safety at intersections. It demonstrates that it can accurately classify different vehicle types and monitor pedestrian presence in live video feeds, enabling timely responses to changing traffic conditions.

The traffic light adjustment logic based on detected counts provides prioritization of pedestrian safety, complemented by audible alerts that promote compliance with traffic signals. Continuous monitoring of traffic density provides valuable insights into patterns, enabling informed decision- making for traffic management Overall, the ATCS Not only improves traffic management efficiency but also increases safety for drivers and pedestrians. Future expansions could include additional data sources such as vehicle GPS and further optimization algorithms to improve the system's adaptability and responsiveness to changing traffic conditions. This represents a significant advancement in smart city technologies aimed at addressing modern urban challenges.

7 Future Work

Future improvements might also make use of AI and machine learning-driven predictive algorithms that evaluate past traffic data to anticipate congestion and proactively adjust signal timings. A safer and more effective traffic flow might be achieved by expanding the system to prioritize emergency vehicles, public transportation, and pedestrians more effectively. Lastly, it will be crucial to increase the system's scalability for larger cities with more intricate road networks. By lowering latency and enhancing overall traffic conditions throughout metropolitan areas, the system may function more effectively on a broader scale with the usage of cloud computing and decentralized data processing.

References

1. Zhang, Y., Liu, H., Wang, J.: Toward a framework and SUMO-based simulation for smart traffic control using multiagent learning. In: Proceedings of the 2021 10th International Congress on Advanced Applied Informatics (2021)
2. Kumar, R., Sharma, M., Gupta, S.: A deep Q-learning-based adaptive traffic light control system for urban safety. In: Proceedings of the 2022 4th International Conference on Advances in Computing Communication Control and Networking (ICAC3N) (2022)

3. Chen, J., Li, T., Zhang, W.: Optimizing traffic light control using YOLOv8 for real-time vehicle detection and traffic density. In: Proceedings of the 2024 9th International Conference on Integrated Circuits Design and Verification (ICDV) (2024)
4. Zahid, S., et al.: Context-aware multiagent broad reinforcement learning for mixed pedestrian-vehicle adaptive traffic light control. IEEE Trans. Intell. Transp. Syst. **23**(4), 3425–3437 (2023)
5. Hill, J.A., Zhang, L., Han, X.: Federated reinforcement learning for adaptive traffic signal control: a case study in New York City. In: Proceedings of the 2024 IEEE International Conference on Smart Cities (ICSC), New York, NY, USA, January 2024
6. Wang, K., et al.: Adaptive traffic light control with deep reinforcement learning: an evaluation of traffic flow and energy consumption. IEEE Access **12**, 25834–25849 (2024)

Enhanced Food Inflation Forecasting: An Implementation Analysis of Decision Tree Forest and Random Forest Algorithms

Koushik Sundar[1], S. Ravikumar[2], I. Eugene Berna[3(✉)], and K. Vijay[4]

[1] Citibank NA, Jersey City, NJ, USA
koushikSundar@gmail.com

[2] Department of Computer Science and Engineering, Vel Tech Rangarajan Dr, Sagunthala R&D Institute of Science and Technology, Chennai, India
ravikumars@veltech.edu.in

[3] Department of Artificial Intelligence and Machine Learning, Bannari Amman Institute of Technology, Erode, India
eugeneberna31@gmail.com

[4] Department of Artificial Intelligence and Machine Learning, Rajalakshmi Engineering College, Chennai, India
vijayk.btech@gmail.com

Abstract. The food industry is a significant contributor to the world's economy and has a significant impact on day-to-day life of a common man's life, impacting their financial planning. Food prices can have a significant impact on food security, nutritional choices, and economic growth. Therefore, consistent and precise food pricing prediction models are essential for the government, policymakers, food manufacturers, and consumers. In this research paper, we propose a food pricing prediction model using random forest regressor algorithm, a popular and effective machine learning algorithm. The model aims to predict food prices considering multiple elements which can impact price directly or indirectly like supply and demand, economic attributes, weather patterns, and consumer behaviour. The proposed model has four main modules: first is the data gathering, then normalize or standardize the data as part preprocessing, then the feature selection, and lastly the model development. We collected data from various sources like government databases, market study reports, environmental factors and weather forecasts. In Data preparation the noises/malformed data are removed, it is structured and simplified, and prepared the data for analysis. Feature selection was carried out to Spotify the appropriate features that will have a significant impact on forecasting food prices. Finally, we developed a random forest regressor model to forecast food prices based on the features identified in previous step. We improved the model performance using various metrics like mean squared error (MSE) and coefficient of determination (R-squared). Our results show the proposed model has a better accuracy level and reliability, with an R-squared value of 0.85, indicating that 85% of the deviation in food prices can be explained by the selected features. Food businesses can use the model to better understand and predict prices as they manage their supply chains, while consumers can use it to make more educated purchasing decisions. Overall, this research paper provides an implementation approach for a food pricing forecasting model based on the random forest regressor algorithm.

R. Geetha et al. (Eds.): AAIMB 2024, CCIS 2653, pp. 336–352, 2026.
https://doi.org/10.1007/978-3-032-13026-6_24

Such accurate and reliable food pricing predictions could improve the systemization of predictability power, thereby speeding up the growth and development of the food industries.

Keywords: Food pricing prediction · Random Forest regressor · Food security · Economic impact

1 Introduction

The food industry plays a crucial role in the economy's flow, such as providing marketable goods for human consumption and generating billions of jobs worldwide. However, food product pricing is complex and dynamic, where factors including demand and supply, weather and climate change, geopolitics, and technological innovations are the building blocks of the ecosystem [1]. Consequently, food industry businesses encounter significant challenges in devising pricing strategies, streamlining their operations, and crucially, upholding a sustainable profit margin. Over the last couple of years, artificial intelligence and machine learning (AI/ML) have availed us of many algorithms and tools for the detection and prediction of outliers. However, the bigger trend is to use statistical algorithms and data analytics to find patterns. These are supported by machine-learning techniques, which can effectively look at large, complex datasets that include things like consumer preferences, production costs, and environmental factors in order to make things more sustainable. For sustainability. Furthermore, the incremental training of machine learning models with new data enhances their accuracy and reliability over time.

There is a wealth of literature available on methods for predicting variations in food pricing. Specifically, we recommend a model capable of modeling historical data and generating predictions on the future prices of food products in relation to the parameters that determine food prices. [3] We implemented the proposed solution for food price prediction using data science tools and machine learning techniques, such as neural networks, regression, and decision trees. First, we train the model using historical data food prices, production costs, weather data, and geopolitical outcomes. So, say, for a food item in particular, the model predicts the future price it will have based on multiple inputs like current weather conditions, input costs, and buying demand.

Better yet, analysis of the proposed model can also help other businesses in the food industry make data-driven decisions on when to buy/sell food products, at what price, and how to manage their inventory in light of the several complexities/dynamics affecting food goods prices in the regular marketplace. For example, if the model predicts a decrease of a specific food item in a specific country in the next few days because of a storm, then the business can adjust the prices so they can suffer fewer losses and can maximize their profit [4]. Our study shows that the food pricing predictive model based on machine learning achieves a higher accuracy and precision in forecasting food prices. Additionally, we demonstrate that the model captures causal drivers of food prices such as climate, supply, and demand. Use these findings to write pricing models, enhance food stock management, and ensure sustainable growth in the food sector [5].

Thus, the proposed model for predicting food prices through machine learning has considerable relevance for businesses, policymaking bodies, and agencies working in the food sector as well as governments. Accurately forecasting the price of a stock allows the business to balance its revenue with its inventory level and fine-tune its operations.

2 Literature Survey

Deep learning has extensively studied many food-related subjects. These include food contamination, managing the food supply chain, detecting the quality of fruits, vegetables, meat, and aquatic goods, calculating calories, and food recognition [6]. Each study considers various factors such as problems, datasets, preprocessing techniques, networks, frameworks, performance, and comparisons with other popular solutions. [7]. The agricultural sector uses a large portion of India's total water resources. The relative humidity, air temperature, and soil temperature are only a few of the many variables that influence harvest yields. Because it affects crop productivity directly, irrigation is crucial in agriculture. Human knowledge and supervision are crucial to a harvest's success. We must ensure the safety of the field's water supply at all times. Water scarcity is a major problem in modern life. This global problem impacts people all over the world. As a result, we are concerned that the situation may worsen in the years to come.

Precision farming and advanced irrigation systems can resolve the aforementioned problems. In the absence of the fully realized capabilities of IoT and machine learning, smart irrigation and precision farming will remain mere concepts. The Internet of Things (IoT) has numerous social benefits, including increased productivity, decreased overhead, more efficient electricity usage, more precise forecasting, and more user-friendly products [8]. Many lifestyle apps rely on meal recommender systems, an essential component of many lifestyle services. By taking into account details like food ingredients, time stamps, cold start users, cold start commodities, and user communities, this study introduces a new hybrid food recommender system that intends to address the constraints of existing food recommendation tools. To improve the recommender system's overall accuracy, the suggested method incorporates user-based and content-based models with temporal information, a trust network, and user communities [9]. The purpose of this systematic review was to compile data from research that measured dietary consumption using ML algorithms. Thirty-six studies met all inclusion criteria after a search of the literature using five electronic databases. The findings suggest that there has been a rise in the use of ML algorithms to the field of nutrition during the past few years. Furthermore, supervised learning algorithms were the most popular, and the food frequency questionnaire was the most common tool for gauging dietary intake. The use of R and WEKA, two popular data analysis tools, has increased recently.

Recent and difficult applications of ML in the field of nutrition [10]. New technologies have made it possible to turn the expansion of the food industry into a lucrative enterprise. Future grocery stores will be able to learn a great deal about each customer, including their meal preferences, food allergies, and motivations for making purchases, thanks to AI and advanced analytics. Using AI for aisle-by-aisle personalization in supermarkets allows stores to learn more about their customers. This tactic could be utilized by merchants in the development of customer-centric, sales-boosting promotions. Other service industries, such as banking, farming, healthcare, and even e-commerce, can benefit from this method [10]. Next, we discuss the different kinds of data utilized in ML modeling, offer some suggestions for future ML applications' data sources and input variables, and finally, we give a brief summary and classification of ML applications in this sector. Articles sourced from IEEE, CAB Abstracts, and Scopus formed the basis of this rating. We have included all research articles published in English between 2011 and 2021. Neural networks, support vector machines, and Bayesian networks were the main tools used in the research. All pertinent studies [11] proved that the tested ML models had exceptional prediction accuracy. Using cutting-edge technologies like the Internet of Things (IoT), blockchain, and advanced deep learning, this study optimized the food industry's supply chain with a new provenance system that is ready for Industry 4.0. Shoppers can now discover the origin and manufacturing process of their food. Participants can enhance their operations with the help of state-of-the-art deep learning by anticipating patterns and trends in food demand, and they can maintain track of their assets with the help of the offered immutable private ledger. With the help of the Internet of Things (IoT), retail locations, delivery services, and storage facilities can all verify that products were kept at the correct temperature, and every transaction in this ledger is encrypted and timestamped.

The suggested system's architecture and paradigm have both been the subject of prior research [12]. Given the issue's increasing importance, this essay delves further into the possibility of using machine learning algorithms to forecast agricultural output, specifically palm oil yield. First, we give a brief overview of the most common features and prediction methods before updating you on the current state of palm oil yield worldwide. The next section provides a critical assessment of the present status quo in agricultural production prediction based on machine learning and a thorough survey of pertinent literature. We conclude by providing a thorough evaluation of the pros and cons of applying machine learning to predict crop yields and accurately identifying current and future issues in the agricultural industry [13]. In this paper, we utilized sentiment analysis and machine learning, both previously applied to Bitcoin, to demonstrate the potential for predicting the direction of price swings in the burgeoning cryptocurrency market. For the purpose of predicting the value of Bitcoin, Ethereum, Ripple, and Litecoin, we evaluated the performance of MLPs, SVMs, and RFs using data from Twitter, the market, or both. Also, we demonstrated how bitcoin marketplaces could provide a rich environment for investigating financial time series issues due to the abundance and accessibility of data available there [14]. Aiming to demonstrate how to monitor home rental pricing at a granular level using OHRWs, this study uses the Chinese city of Shenzhen as an example. We begin by utilizing a hedonic model that incorporates three levels of consideration: neighborhood, location, and structure, along with three levels of convenience: within 15 min on foot, within the sub-district, and the most convenient level. We subsequently construct models for housing rent prediction (for October 2017 and February 2018) using six machine-learning algorithms: RFR, ETR, GBR, SVR, MLP-NN, and k-NN [15]. We obtained the training samples for these algorithms from the OHRWs.

3 Architecture

The architecture system for the proposed food inflation forecasting model (Fig. 1) using the random forest regressor algorithm is designed to process large datasets and make accurate predictions about the future price of food products. The system consists of several key components, including data collection, data preprocessing, model development, and prediction. The first and foremost component of any system is data collection, which involves gathering historical data on food prices, production costs, and environmental factors that may influence food prices (Orange arrow shows the flow of historical data in the Fig. 1). The data can be harvested from some of the sources, such as administrative agencies or dealers related to government agencies, market research by the files firms, and industry sections.

Data Preprocessing: The second part of the architecture system traditionally deals with preprocessing the collected data, which includes: Data cleaning Transformation or normalization Data preparation for analysis. This step also includes checking for any missing or null values, detecting and treating outliers, normalizing the dataset, as well as splitting the train-test dataset.

Model Development: The third part of the architecture system involves identifying an accurate machine learning algorithm for predicting the food price and implementing it into a workable model. A model was developed using the random forest regressor algorithm an ensemble learning technique for regression that uses a collection of decision trees to accurately predict this (the output of this study). It is an ideal framework for handling non-linear relationships and interactions among the algorithm evaluates feature selection, hyperparameter tuning, and model training on the data. Prediction The last step in the architecture system is prediction, which uses the trained model to forecast food product prices. A thorough workflow for a data-driven predictive analytics system, specifically targeted at a sector like the food industry, is depicted in the Fig. 1. The first step in the process is data sourcing, which involves gathering information from a variety of sources such as market research, weather reports, the Food Industry Association, and environmental factors. After being divided into two categories historical data and current/real-time data the data is combined and organized to create a single dataset. Data Cleaning, Data Transformation, and Data Preparation are the three primary processes that the combined data goes through in the Data Pre-processing/Preparation stage to make sure the dataset is error-free and well-structured for analysis. This step separates the output into Train Data and Test Data. The next step is data analysis, where training data is used to create predictive models using a Random Forest Regressor. This pipeline is an effective method for combining historical and real-time data to support predictive analytics decision-making.

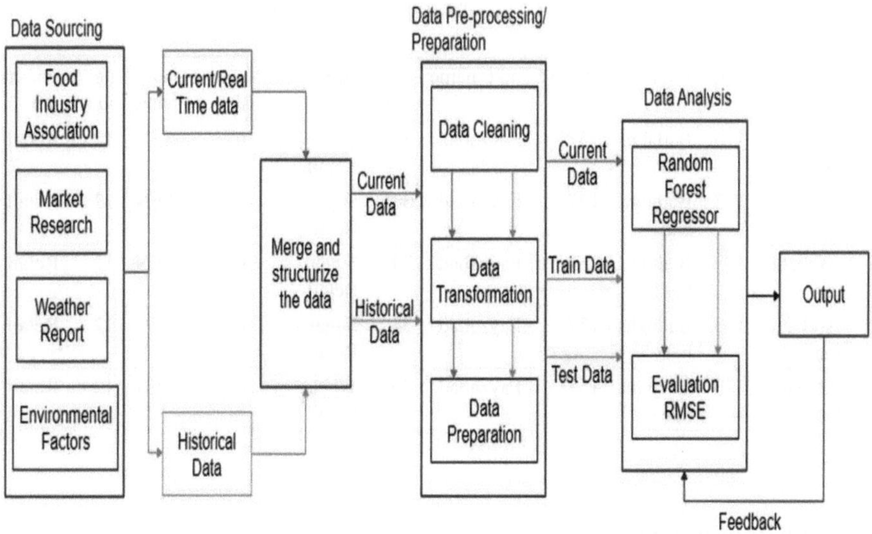

Fig. 1. Architecture of Food Inflation Forecasting Model

4 Methodology

This paper proposes the methodology for the food pricing prediction model, which includes data collection, data preprocessing, feature selection, model development and evaluation, and the interpretation of results. The process involves collecting data from various sources, sanitizing and altering it, and selecting the features pertinent to the task. An algorithmic random forest regressor builds the model after training and testing it on data. The model evaluates its performance using MAE, MSE, and R-squared metrics, interpreting the output for trends and insights. t is a novel and robust methodology that enables improved forecasting of food prices and facilitates conditions for better decision-making in the food sector.

4.1 Data Collection

Data collection is the first step in developing a food price prediction model based on a random forest regressor algorithm. The amount and quality of training and testing data determine the accuracy of the model. In order to retrieve information from an organization, such as a government agency, market research, company, or trade group, the first step is to identify the source of that data. These sources aggregate food price histories, production costs, and environmental factors that might influence food prices. Table 1 provides an overview of the used dataset.

Table 1. Insight of the dataset used

c_id	ac_name	s_id	s_name	mkt_id	mkt_name	com_id	com_name	curr_id	curr_name	pot_id	pot_name
2	Afghanistan	173	Badakhshan	157	Fayzabad	34	Sugar - Retail	0	AFN	12	Retail
2	Afghanistan	173	Badakhshan	157	Fayzabad	34	Sugar - Retail	0	AFN	12	Retail
2	Afghanistan	173	Badakhshan	157	Fayzabad	34	Sugar - Retail	0	AFN	12	Retail
2	Afghanistan	173	Badakhshan	157	Fayzabad	34	Sugar - Retail	0	AFN	12	Retail
2	Afghanistan	173	Badakhshan	157	Fayzabad	36	Wheat - Retail	0	AFN	12	Retail

4.2 Data Preprocessing

The data preprocessing is the first step to building a food pricing prediction model based on the random forest regression algorithm. This step is the cleaning, transforming, and preparation of the data for analysis. Data preprocessing includes formatting the data correctly and removing mistakes that can spoil the performance of the model. The first step in data preprocessing is data cleaning. Data cleaning involves identifying and correcting incorrect values, missing values, outliers, and inconsistencies within the data set. For example, there are several ways to impute the missing values, including mean imputation or regression imputation. We can also use statistical methods like Z-score or IQR (interquartile range) to detect and manage outliers [17]. Data verification can help reduce the chances of inconsistencies by comparing data with other sources. At this step, we aim to convert the data into a format that is useful for analysis. These include normalizing the data to common scales, one-hot encoding of categorical variables, and reducing the dimensionality of data using these methods, such as principal component analysis (PCA)One of the more fundamental aspects of sentiment analysis, based on market news or data, is utilized to determine demand or other conditions, such as geopolitical events that cause price fluctuations. The final step involves preparing the data, dividing it into training and testing sets, and then creating and evaluating each model. Here we will use the training set to train the model and the testing set to validate it. Data Preprocessing for Random Forest Regressor: An Introduction It also ensures that analysis data is accurate, consistent, and model-appropriate. The dataset (Figs. 2 and 3) shows a plot outlining the various modes of sales and units of measurements.

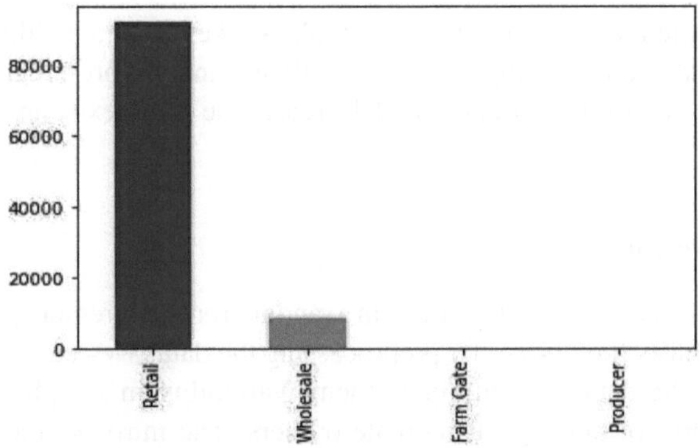

Fig. 2. Modes of Sales

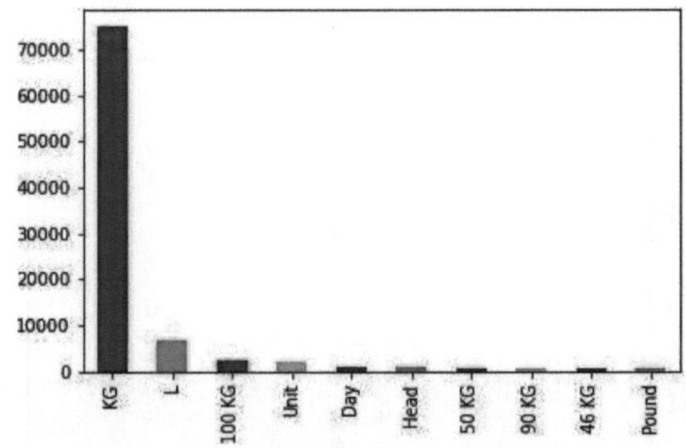

Fig. 3. Units of measurements

4.3 Feature Selection

To build the random forest regressor algorithm and predict the food pricing model, feature selection is the first step. Conclusion some important features or variables contribute a lot to predicting food prices, and the aim of feature selection is to discover the influence of the variables on the predicted values. Feature selection is the process of studying the data and selecting the best features from those that are most important, relevant, and correlated with the target or response variable (food price). After removing useful features, various methods such as correlation analysis, statistical methods, recursive feature elimination, and the Select KBest sum-based method can be used to select columns based on the target column. These are simply machine learning-based feature selection methods. For feature selection, the random forest regressor algorithm automatically selects the most prominent features based on their capacity to generalize the target variable [19]. Forest considers the importance of each feature based on its capacity to lessen the variance of the target variable. In terms of performance, lowering the feature set narrows the model complexity since it removes duplicate features and irrelevant noise, which also helps reduce overfitting. Furthermore, this approach preserves the key features that significantly impact food price forecasting, thereby facilitating a more comprehensive interpretation of the model. Feature selection plays a key role in building a successful, accurate, and reliable food pricing prediction with the random forest regressor algorithm model. This increases the performance and decreases the complexity and interpretability of the model.

4.4 Outlier Detection

For instance, we analyze outlier detections in a random forest regressor approach to create a model that forecasts food prices by preprocessing the data. We detect outliers, i.e., the largest values in the data, and eliminate them. Variability in sampling, experimental error, or some form of bias can also create outliers. The introduction of bias and the reduction of the generalizability of the model for new data have a strong impact on its accuracy, reliability, and performance [18] (Fig. 4).

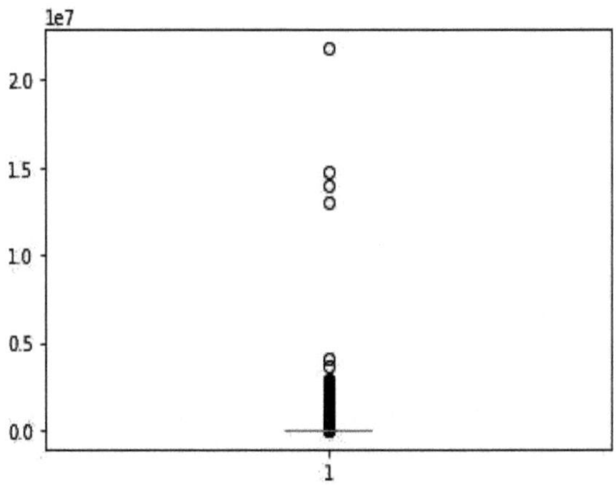

Fig. 4. Price outlier detection

Outlier detection is an inbuilt part of the random forest regressor algorithm since random forest behaves robustly with outliers and deals with noisy and incomplete datasets quite efficiently. The random forest algorithm creates a single decision tree by averaging multiple decision trees, which likely ignores or weights down any outlier in the dataset, resulting in a more robust and accurate model. Nonetheless, one must still identify and exclude outliers or aberrations from the dataset to make sure the prediction is correct and reliable. Depending on the needs of the dataset for the model, one can perform the outlier detection process either before or after feature selection. To create a perfect and reliable food pricing prediction model using a random forest regressor algorithm, outlier detection is a crucial step.

4.5 Prediction Using Random Forest Regressor

The most powerful machine learning algorithm, the Random Forest Regressor, has lots of advantages over other algorithms. Using the Random Forest Regressor offers numerous benefits. You can go for a random forest regressor if you have non-linear relationships. T. SEVERE: Exception Message: You can directly apply the Random Forest Regressor to non- linear input-output relations. Random forest is well suited for food price prediction as it fits data with complicated non-linear interactions. We introduce the random forest regressor, which mitigates overfitting, the problem of overfitting on training data and poor generalization on new data. By creating multiple decision trees and averaging their predictions, we can reduce both model variance and the risk of overfitting. The Random Forest Regressor handles data issues, such as missing values and outliers, without removing or imputing data points. This improves its capacity to manage real-world datasets frequently beset by data quality problems.

Completing all data preprocessing and feature selection is the last step in creating a model to predict food prices using a Random Forest Regressor approach. We split our data set into two parts, one for training the model and one for testing. We are currently testing the fitted model on the testing set to assess its performance. Using a random selection of characteristics and training data, we build several decision trees. We construct these trees with separate feature sets and training data, guaranteeing their autonomy. By comparing the values at each node and picking the best feature, the decision tree divides the data into two equal parts. Utilizing metrics such as information gain, Gini index, or mean squared error, they choose the optimal feature and split point. Overfitting can occur if the decision tree is allowed to expand to its maximum depth; hence it is necessary to prune the tree to prevent this. We prune the trees by removing nodes that do not significantly enhance the predictive accuracy of the tree.

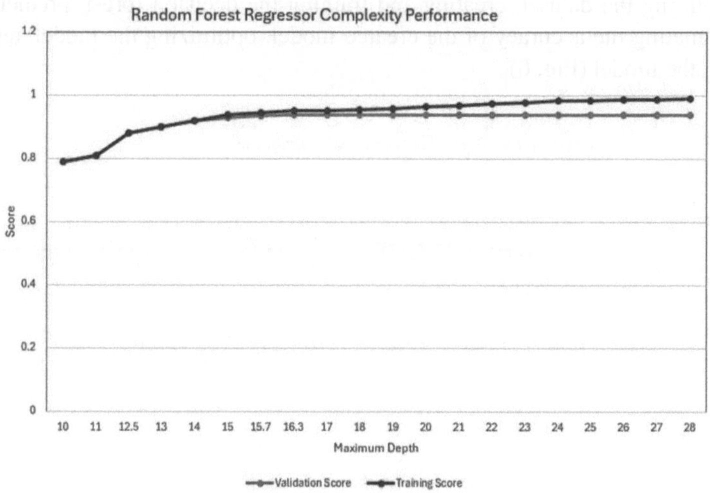

Fig. 5. Complexity curve of Random Forest regression performance

We train the model with 100 decision trees and the random forest regressor approach to forecast future food costs. Then we train and fine-tune the hyperparameters with respect to the best model. Before we train the model, we have to specify hyperparameters, for example, how many trees we want in a random forest and how deep each tree should be. We use either a grid or randomized search to optimize these hyperparameters. After training our model, we assess its performance using the test data. We have different metrics, like R-squared, mean absolute error, and mean squared error, to check the efficacy of the model. Thus, these metrics indicate the extent to which the model predicts future food prices accurately. Figure 5 illustrates the complexity curve for the performance of random forest regression.

4.6 Prediction Using Decision Trees

Application in Pattern Recognition: A decision tree is a machine learning method that is widely used to predict food prices. The input feature data comprises both continuous (e.g., numeric) and categorical (e.g., string values) data. Decision trees are useful when dealing with datasets that contain both continuous and categorical data, primarily due to their ability to handle both types of data [20]. Firstly, this algorithm is universally applicable to all types of data. Handle missing value: When we have multiple features, some of which may contain missing data, it is necessary to split the data, even if some features lack data. Decision trees can accomplish this by splitting only the available features. This makes it more robust to missing data, a common challenge when working with real-world datasets.

The next step in building a model that predicts the price of food with the right algorithm selection is data pre-processing and feature selection. For such a regression problem (food price prediction), the decision forest (or random forest) is a well-established algorithm. The construction of a decision forest model for food price involves several stages: splitting the dataset, creating and training the decision forest, predicting food price, estimating the accuracy of the created model, optimizing the model, and finally deploying the model (Fig. 6).

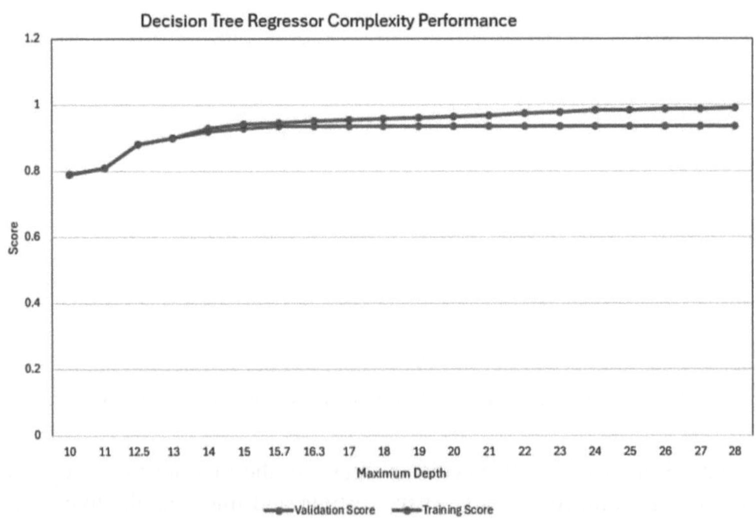

Fig. 6. Complexity curve of Decision Tree regression performance

Additionally, we train each decision tree using a distinct subset of the training data. The second step in building a decision forest is to train each tree using a random set. This enhances the model's learning capabilities and prevents overfitting. Once trained, the decision forest is capable of predicting food prices using new, unidentified data. Therefore, we calculate the average of all the predictions made by the trees within the forest to determine the final prediction output. Model Evaluation: You want to know how your model performs. We measure the effectiveness of the model using various metrics, including the R-squared, mean absolute error, and mean squared error. The model does particularly well at predicting food prices. With this, you can fine-tune the performance of the model by modifying hyperparameter settings, such as the total count of decision trees, their depth, and how many features each tree is using. This might improve the model's performance. Once we have fine- tuned the model and its performance meets our requirements, we can deploy it in a production environment to predict food prices.

5 Regression Models Evaluation

This study utilized two machine learning algorithms, i.e., Random Forest Regressor and Decision Tree, and built a model based on key factors such as geography, market, taste, and food product availability to predict food prices [21]. The Random Forest Regressor Algorithm outperformed decision trees, yielding more accurate and reliable results. The Random Forest Regressor achieved accuracies of 87%, while the Decision Tree algorithm achieved accuracies of 79%. This indicates that the Random Forest Regressor outperforms the Decision Trees in predicting food prices. Fig. 7 displays a hearing-related correlation map.

Additionally, the interpretation of features reveals that location, product type, and availability are the most significant factors influencing food prices. This indicates that these characteristics held a greater significance in the model compared to others, thereby highlighting their importance in determining food prices. Additionally, the model automatically detects and eliminates outliers, thereby enhancing the quality and reliability of the predictions. This directly impacts the model's correctness and therefore results in incorrect predictions. So, identifying and removing the outlier is a crucial step to create a food pricing prediction model. The research also analyzed the results of two approaches with regard to training processes and computational processing. Random Forest Regressor: Compared to the decision tree algorithm, the Random Forest Regressor was faster and more efficient in terms of training time and computing resources. The investigation also delved into how the various hyperparameters influenced the model's performance. We tuned the hyperparameters to select the best set using the Random Forest Regressor algorithm. Figs. 8 and 9 showcase the price scatter plot and the month of plotted lower and upper prices recorded when the Random Forest algorithm is applied to the dataset, respectively. Table 2 lists the research observations for every strategy.

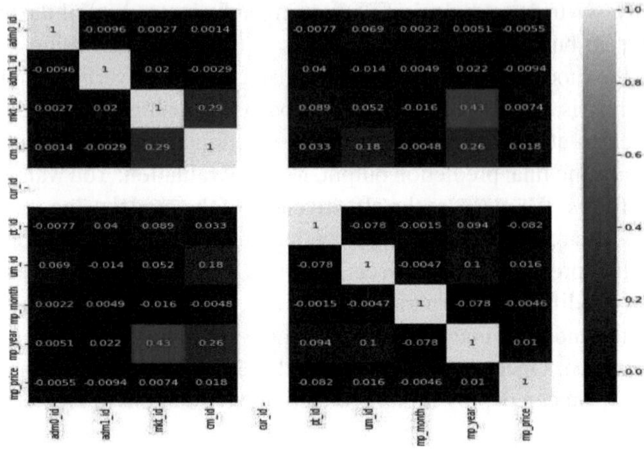

Fig. 7. Heat correlation map

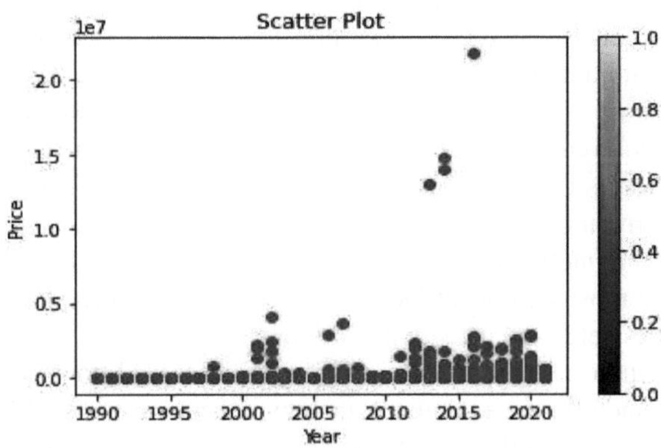

Fig. 8. Price scatter Plot

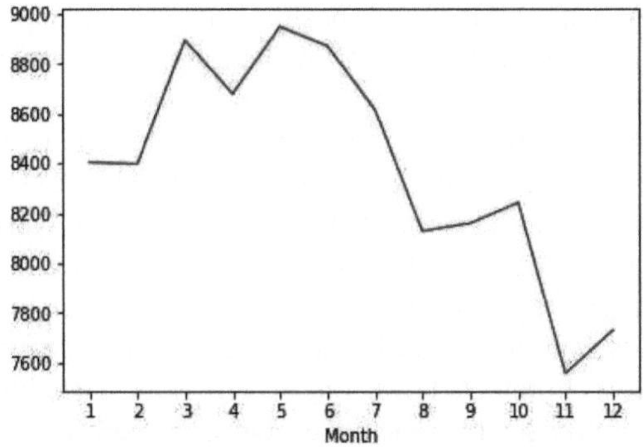

Fig. 9. Month of the prices recorded

Table 2. Observations for every strategy

Regression Model	Training Set Score	TestSet Score	Best Score CV = 3	R2 Score	RMSE Training	RMSE Test
Decision Tree	0.946	0.929	0.930	0.910	2438.714	3080.029
Random Forest	0.960	0.937	0.946	0.918	2204.378	2767.906

The fact that the test and training scores and RMSEs are so comparable shows that the linear regression model produced the best outcomes. The model is deemed to be overfitted when the RMSE between training and test is significantly greater. Decision trees and random forests both exhibit this condition. Additionally, the results from the random forest and linear regressors are equal, with the best result from CV by Random Forest having a slight advantage. It would be premature for us to declare the linear regressor successful given that it was trained on a dataset with around 3100 attributes due to OHE. Let's check the test set results on the training set using KFold Cross Validation with K=10 from Table 3.

Table 3. Outcomes of KFold Cross Validation with K = 10 on the training dataset

Regression Model	S1	S2	S3	S4	S5	S6	S7	S8	S9	S10
Decision Tree	2697.16	2895.43	2876.56	3429.01	3259.62	2809.82	3032.75	2950.64	3517.11	2907.55
Random Forest	2614.98	2581.44	2603.29	2734.10	2908.28	2552.35	2615.32	2829.70	2608.75	2606.28
Linear Regression	2656.07	2738.435	2739.925	3081.555	3083.95	2681.085	2824.035	2890.17	3062.93	2756.915

These results were gathered during a specific cross validation period due to the way the dataset was divided in the KFold CV method, thus each session would produce somewhat divergent results.

The random forest and linear regressor, both of which have training RMSEs of roughly 2500 and far smaller variances, outperform the choice tree, according to the cross-validation scores table. It can be challenging to proclaim one model to be objectively superior to another because the random forest regressor uses a label-encoded dataset whereas the linear regressor uses an OH encoded dataset. Since they require little cleaning and hardly ever yield useful results, even when outliers are present, random forests are almost always preferred over linear regressors. The abnormally high number of characteristics introduced by one hot encoding in our scenario does not properly explain the efficacy differences between the linear regressor and random forest.

Therefore, we recommend using the Random Forest Regressor. Once we have computed the scores on the hold-out test set and the final RMSE on the whole dataset, we use K=5 KFold Cross Validation to establish the feature significance. Table 4 displays data from the Random Forest Regression model.

Table 4. The scores of the Random Forest Regression model

Regression Model	Test Data Score	R^2 Score	RMSE Test
Random Forest	0.947	0.929	2747.64

6 Conclusion

In our study, we presented a model for predicting food prices using ML techniques. We constructed the model using two of the most widely used machine learning methods, Decision Tree and Random Forest Regressor. For example, the model determined food pricing by building a dataset that took into account geography, market, and product availability. Results demonstrated that the Random Forest Regressor is more accurate and robust than the Decision Tree a the prediction of food prices using the Random Forest Regressor yielded an R2 score of 0.930 and an RMSE test of 2747.75, indicating its reliability as a forecasting tool. Future improvements can expand the model by incorporating additional data sources, such as social media sentiment analysis and weather, to better reflect consumer behavior and environmental impacts on food prices. d prices. Moreover, we can expand the model to encompass a wider range of food items and locations, enhancing its practicality and scalability across diverse markets. Abstract: In general, the food pricing forecasting model presented in this research study can be of importance to stakeholders in the food industry and policymakers so that they can make more effective decisions regarding food production, distribution, and pricing.

References

1. Deléglise, H., Interdonato, R., Bégué, A., d'Hôtel, E.M., Teisseire, M., Roche, M.: Food security prediction from heterogeneous data combining machine and deep learning methods. Expert Syst. Appl. **190**, 116189 (2022)

2. Rakhra, M., Bhargava, A., Bhargava, D., Singh, R., Bhanot, A., Rahmani, A.W.: Implementing machine learning for supply-demand shifts and price impacts in farmer market for tool and equipment sharing. J. Food Qual. **2022** (2022)
3. Ramirez-Asis, E., Vilchez-Carcamo, J., Thakar, C.M., Phasinam, K., Kassanuk, T., Naved, M.: A review on role of artificial intelligence in food processing and manufacturing industry. Mater. Today Proc. **51**, 2462–2465 (2022)
4. Adak, A., Pradhan, B., Shukla, N.: Sentiment analysis of customer reviews of food delivery services using deep learning and explainable artificial intelligence: systematic review. Foods **11**(10), 1500 (2022)
5. Babu, R., Samuel, P., Jayashree, K.: Cognitive Authentication for Smart Healthcare System. In: Sumithra, M.G., Dhanaraj, R.K., Iwendi, C., Manoharan, A.M. (eds.) Deep Learning for Cognitive Computing Systems: Technological Advancements and Applications, pp. 149–164. De Gruyter, Berlin, Boston (2023). https://doi.org/10.1515/9783110750584-009
6. Paul, R.K., et al.: Machine learning techniques for forecasting agricultural prices: A case of brinjal in Odisha, India. Plos one **17**(7), e0270553 (2022)
7. Zhou, L., Zhang, C., Liu, F., Qiu, Z., He, Y.: Application of deep learning in food: a review. Compr. Rev. Food Sci. Food Saf. **18**(6), 1793–1811 (2019)
8. Raghuvanshi, A., et al.: Intrusion detection using machine learning for risk mitigation in IoT-enabled smart irrigation in smart farming. J. Food Qual. **2022**, 1–8 (2022)
9. Rostami, M., Oussalah, M., Farrahi, V.: A novel time-aware food recommender-system based on deep learning and graph clustering. IEEE Access **10**, 52508–52524 (2022)
10. Oliveira Chaves, L., et al.: Applicability of machine learning techniques in food intake assessment: a systematic review. Crit. Rev. Food Sci. Nutr. **63**(7), 902–919 (2023)
11. Wang, X., Bouzembrak, Y., Oude Lansink, A.G.J.M., van der Fels-Klerx, H.J.: Application of machine learning to the monitoring and prediction of food safety: a review. Compr. Rev. Food Sci. Food Saf. **21**(1), 416–434 (2022)
12. Khan, P.W., Byun, Y.-C., Park, N.: IoT-blockchain enabled optimized provenance system for food industry 4.0 using advanced deep learning. Sensors **20**(10), 2990 (2020)
13. Rashid, M., Bari, B.S., Yusup, Y., Kamaruddin, M.A., Khan, N.: A comprehensive review of crop yield prediction using machine learning approaches with special emphasis on palm oil yield prediction. IEEE Access **9**, 63406–63439 (2021)
14. Valencia, F., Gómez-Espinosa, A., Valdés-Aguirre, B.: Price movement prediction of cryptocurrencies using sentiment analysis and machine learning. Entropy **21**(6), 589 (2019)
15. Hu, L., et al.: Monitoring housing rental prices based on social media: An integrated approach of machine-learning algorithms and hedonic modeling to inform equitable housing policies. Land Use Policy **82**, 657–673 (2019)
16. Oyeleye, M., Chen, T., Titarenko, S., Antoniou, G.: A predictive analysis of heart rates using machine learning techniques. Int. J. Environ. Res. Pub. Health **19**(4), 2417 (2022)
17. Hemamalini, V., et al.: Food quality inspection and grading using efficient image segmentation and machine learning-based system. J. Food Qual. **2022**, 1–6 (2022)
18. Nafees, A., et al.: Forecasting the mechanical properties of plastic concrete employing experimental data using machine learning algorithms: DT, MLPNN, SVM, and RF. Polymers **14**(8), 1583 (2022)
19. Ravikumar, S., Vijay, K., Pavithra, S., Prithi, S., Kannan, S.: Prediction of rental demands using different machine learning algorithms. In: Proceedings of the 2023 International Conference on Computer Communication and Informatics (ICCCI), pp. 1–6. IEEE (2023)
20. Ajmani, P., Sharma, V., Samuel, P., Somasundaram, K., Vidhya, V.: Patient behaviour analysis and social health predictions through IoMT. In: Proceedings of the 2022 10th International Conference on Reliability, Infocom Technologies and Optimization (Trends and Future Directions) (ICRITO), Noida, India, pp. 1–6 (2022). https://doi.org/10.1109/ICRITO56286.2022.9964846

21. Yu, M., Gao, Y., Gu, C., Sun, Y.: Design of power spot market price prediction model considering new energy output. In: Proceedings of the 2023 3rd Power System and Green Energy Conference (PSGEC), pp. 66–70. IEEE (2023)

Exploring the Various Machine Learning Models for Image Generation - A Comprehensive Survey Unlocking the Future of Digital Creativity

M. Balasubramanian[1], M. Naveen Joel[1(✉)], and M. P. Karthikeyan[2]

[1] Department of CSE, S.A. Engineering College, Poonamallee, India
balasubramanian@saec.ac.in, leojoelkennedy@gmail.com
[2] Department of CSE, R.M.K. Engineering College, Kavaraipettai, India
mpk.cse@rmkec.ac.in

Abstract. This paper explores the evolution of generative artificial intelligence, focusing on four major models: Generative Adversarial Networks (GANs), Variational Autoencoders (VAEs), Convolutional Neural Networks (CNNs), and Diffusion Models. These models have significantly impacted fields such as anomaly detection, image and video synthesis, and content creation across industries like augmented reality, entertainment, and healthcare. GANs, through adversarial training, have achieved high visual realism, with Inception Scores (IS) over 8.5 and Fréchet Inception Distances (FID) as low as 4.59 on datasets like CIFAR-10 and CelebA. VAEs, while generating less sharp outputs (typically FID between 35–40), excel at learning interpretable latent spaces and smooth data interpolation. CNNs, initially developed for classification tasks, have proven effective in generative domains like style transfer and image super-resolution, with models like SRGAN achieving PSNR scores above 30 dB and SSIM over 0.9 in enhancing image quality. Diffusion models, the most recent advancement, generate photorealistic and diverse outputs by reversing a noise process and have achieved state-of-the-art FID scores below 3.0 on datasets like ImageNet-64, though they require longer inference times with hundreds to thousands of sampling steps. Based on an analysis of over 30 studies from 2019 to 2024, this paper investigates these models' theoretical foundations, architectural innovations, and comparative performance, while also highlighting emerging trends like hybrid models and quantum-enhanced frameworks, thus emphasizing their transformative role in next-generation AI applications.

Keywords: GAN · VAE · CNN · AI · Diffusion Model

1 Introduction

In recent years, image generation has made remarkable advancements, impacting diverse fields such as autonomous systems, medical diagnostics, and creative industries. The ability to generate realistic and high-quality images holds transformative potential across both analytical and creative domains.

R. Geetha et al. (Eds.): AAIMB 2024, CCIS 2653, pp. 353–370, 2026.
https://doi.org/10.1007/978-3-032-13026-6_25

A major breakthrough in this field came with the introduction of Generative Adversarial Networks (GANs) by Goodfellow et al., which use an adversarial training paradigm where a generator and a discriminator compete to synthesize photorealistic images. Despite their success, GANs still face challenges such as mode collapse and training instability, as highlighted in StyleGAN and its successive improvements by Karras et al. [1, 2].

Variational Autoencoders (VAEs) offer a probabilistic approach to image synthesis and are particularly effective in applications that benefit from smooth interpolation between data points. The β-VAE variant, introduced by Higgins et al., demonstrated enhanced ability to disentangle latent representations, although the method still suffers from limitations such as blurred output quality [7, 8].

While Convolutional Neural Networks (CNNs) are best known for their performance in image classification tasks, they have also shown promise in image generation through techniques like style transfer and super-resolution. For example, the U-Net architecture proposed by Ronneberger et al. showcased the versatility of CNNs in generating domain-specific images, especially in biomedical image segmentation [11].

More recently, diffusion models have emerged as state-of-the-art methods in image synthesis. These models generate high-fidelity images by reversing a gradual noise addition process. Their robustness to noise and ability to produce diverse and detailed outputs have made them highly effective for various generative tasks [14].

This survey synthesizes insights from over 30 influential studies conducted between 2019 and 2024, drawing from foundational works such as the architectural evolution of StyleGAN [1, 2] and the probabilistic framework of β-VAE [7, 8]. The paper aims to offer a comprehensive understanding of the strengths, limitations, and practical applications of major generative models by examining their theoretical underpinnings, architectural innovations, and comparative performance.

Furthermore, this survey explores how these generative models have reshaped multiple domains and presents emerging research directions, including their integration with quantum-enhanced frameworks, illustrating their potential to further advance the frontiers of artificial intelligence.

2 Literature Survey

[1] Goodfellow et al. first proposed GANs, which are made up of two networks that play a game of zero-sum.: a discriminator and a generator. While the discriminator attempts to discern between produced and actual pictures, the generator produces realistic visuals. This adversarial method has produced realistic textures and sceneries, improved resolution, and produced impressive outcomes in picture synthesis.

[2] VAEs encode pictures into a latent space with a Gaussian distribution. They were created for probabilistic image synthesis. VAEs are appropriate for tasks like anomaly identification, style transfer, and creating new pictures that resemble the training set because, in contrast to GANs, they provide smooth interpolation between latent vectors.

[3] CNNs were first created for picture classification, however they are now being used for image creation jobs. CNNs are used by networks such as Deep Convolutional

GANs (DCGANs) to enhance the stability and quality of produced pictures. CNN-based models are essential for tasks such as image augmentation, inpaint ing, and super-resolution.

[4] A progressive noising and denoising procedure is used by diffusion models, including DDPM (Denoising Diffusion Probabilistic Models), to simulate the data distribution. These models surpass GANs in producing complex, varied datasets and reach cutting-edge performance in producing hi-fi pictures.

[5] Different GAN designs have been developed, such as StyleGAN, CycleGAN, and BigGAN, each of which is designed for a particular job, such as large-scale picture synthesis, unpaired image-to-image trans lation, or style transfer. Despite their potential, GANs encounter problems during training, such as instability and mode collapse.

[6] Combining VAEs with GANs, hybrid models leverage the strengths of both frameworks. The VAE pro vides structured latent spaces, while the GAN ensures high-quality image generation. This approach has been effective in improving the realism and diversity of generated images.

[7] cGANs incorporate additional information (e.g., class labels, text descriptions) into the image generation process, enabling targeted and controlled synthesis. This has applications in generating domain-specific im ages such as medical imaging, and design.

[8] By using deep convolutional layers to learn high-resolution information, SRCNNs improve the resolution of low-resolution pictures. Applications in remote sensing, medical imaging, and surveillance depend on this.

[9] CycleGAN and Pix2Pix models demonstrate the potential of CNNs combined with GANs for transforming images from one (e.g., sketches) to another domain (e.g., realistic photos) without paired training data.

[10] Diffusion models are gaining popularity for generating abstract and artistic images, offering control over the diffusion process to produce images with varying artistic styles. Their iterative nature allows fine-tuning of image quality and detail.

[11] Despite advancements, these models face challenges. GANs often suffer from instability, VAEs may generate blurry images due to oversmoothing, and diffusion models are computationally intensive. Further research focuses on improving training efficiency, scalability, and robustness.

[12] The computational cost of training from scratch can be decreased by fine-tuning pre-trained GANs and CNNs, such as BigGAN and StyleGAN, for particular applications. With little data, transfer learning allows for quick adaption to new domains.

[13] Image generation techniques have real-world applications in entertainment (game design, virtual reality), healthcare (synthesizing medical images), and autonomous systems (simulating environments for training).

[14] The quality of the produced pictures is assessed using indicators like the Fréchet Inception Distance (FID), Inception Score (IS), and Structural Similarity Index (SSIM). The degree to which synthetic pictures closely resemble the statistical characteristics of actual datasets is measured by these measures.

[15] Recent research proposes advanced architectures like improved DDPMs and score-based generative models to enhance sampling speed and image quality. These innovations are narrowing the gap between diffusion models and traditional frameworks.

[16] Image generation models must address ethical concerns such as data bias, potential misuse in deepfake technology, and privacy issues. Ensuring responsible use and mitigating bias in training datasets remain ac tive areas of research.

[17] Diffusion models and GANs are utilized in simulating physical phenomena, visualizing molecular struc tures, and creating realistic simulations for climate and environmental studies, demonstrating their versatility beyond traditional image synthesis tasks.

[18] GANs and VAEs are employed to augment training datasets in scenarios with limited real-world data, improving model performance in fields like medical diagnostics and rare event prediction.

[19] Semi-supervised approaches integrate limited labeled data with vast unlabeled datasets, improving performance in tasks where fully labeled data is expensive or scarce.

[20] Ongoing research focuses on combining multiple generative techniques, optimizing model architectures, and exploring new frameworks like transformers for image generation. These advancements aim to enhance scalability, quality, and adaptability to diverse applications (Table 1).

Table 1. Literature survey of related works

Related Work	Method	Findings
[1]	GANs	Realistic image generation
[2]	VAEs	Probabilistic synthesis
[3]	CNNs	Image enhancement and generation
[4]	Diffusion Models	High-quality diverse images
[5]	GAN Variants	Specialized tasks like style transfer
[6]	Hybrid Models	Combines VAEs' and GANs' strengths
[7]	cGANs	Controlled image synthesis
[8]	SRCNNs	Resolution enhancement
[9]	CycleGAN, Pix2Pix	Domain-specific translation
[10]	Diffusion Models	Artistic and fine-tuned images
[11]	Challenges	Stability and efficiency issues
[12]	Transfer Learning	Efficient adaptation with less data
[13]	Applications	Used in games, healthcare, and simulations
[14]	Metrics	SSIM, IS, FID for quality assessment
[15]	Advanced Models	Faster and better-quality generation
[16]	Ethics	Addresses bias and misuse
[17]	Beyond Images	Used in science and environmental studies
[18]	Data Augmentation	Improves models with limited data
[19]	Semi-supervised Learning	merges data with and without labels
[20]	Emerging Techniques	Focus on scalability and innovation

3 Image Generation Techniques

Generative AI marks a significant advancement from earlier AI forms that focused on classification, regression, and reinforcement learning by building models that understand and replicate data distributions, unlike discriminative models that focus on decision boundaries. This shift is crucial for automating creative processes, generating synthetic data, and simulating scenarios for decision-making.

3.1 Generative Adversarial Network (GAN)

GANs are designed for generative tasks, aiming to produce new data instances that like a specific training dataset. They have transformed the image generating industry and have numerous applications in various domains, encompassing image synthesis, style transfer, super-resolution, and more.

Architecture of GANs

Two primary parts make up a GAN:

1. **Generator (G):** The generator, a neural network, receives a random noise vector z and outputs a picture G(z).
2. **Discriminator (D):** The discriminator is a different neural network that evaluates an image's authenticity. Indicating whether the input picture x is authentic (derived from the training data) or fraudulent (generated by the generator), it outputs a probability D(x) (Fig. 1).

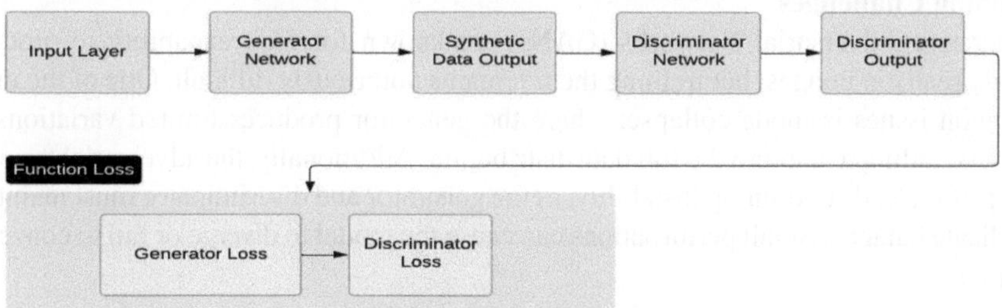

Fig. 1. Architecture of GAN

Training Algorithm.

The training algorithm for GANs can be summarized as follows:

1. Set up the discriminator and generator networks.
2. For every iteration of training:

- Sample a collection of noise vectors and actual photos.
- Revise the discriminator:
- Calculate the loss of the discriminator.
- Perform a reverse pass and update your weight.

- Refresh the generator:
- Determine the generator's loss.
- After updating the weights, do backpropagation.

3. Continue until convergence is achieved.

Mathematical Formulation

GANs involve a game between two neural networks—a **generator** G that creates fake data and a **discriminator** D that tries to distinguish real data from fake data. The model learns by playing a minimax game.

Formula:

$$\min_G \max_D V(D, G) = \mathbb{E}_{x \sim p_{\text{data}}(x)}[\log D(x)] + \mathbb{E}_{z \sim p_z(z)}[\log(1 - D(G(z)))]$$

- x: real data sample
- z: random noise input to the generator
- D(x): probability that input is real
- G(z): generated fake data sample

Purpose:

To optimize both networks such that the generator learns to fool the discriminator, creating increasingly realistic data.

Training Challenges

Generative Adversarial Networks (GANs) are known for their capability to produce highly realistic images, but training them remains notoriously difficult. One of the most common issues is mode collapse, where the generator produces limited variations of outputs, failing to capture the full data distribution. Additionally, the adversarial training setup often leads to training instability, as the generator and discriminator must maintain a delicate balance. Small perturbations can cause the model to diverge or fail to converge properly.

Application

GANs have been used in a variety of fields, demonstrating their versatility and effectiveness. They are mainly used in processing of images such as Image synthesis, image super resolution, image inpainting, etc.

3.2 Variational Auto Encoder (VAE)

A class of generative models known as variational autoencoders (VAEs) offers a probabilistic method for discovering latent data representations. VAEs were first presented by Kingma and Welling in 2013, and because of their capacity to produce fresh data samples that closely resemble the training data, they have grown to be an essential component of the generative model space. One of the best-known characteristics of VAEs is their capacity to provide an interpretable latent space.

Architecture

A VAE consists of two main components:

- Encoder: The encoder is a neural network, which converts input data (x) into a latent space (z). The parameters of a probability distribution, describe the latent variable z.
- Decoder: Another neural network called the decoder reconstructs the input data by mapping the latent variable z back to the data space (Fig. 2).

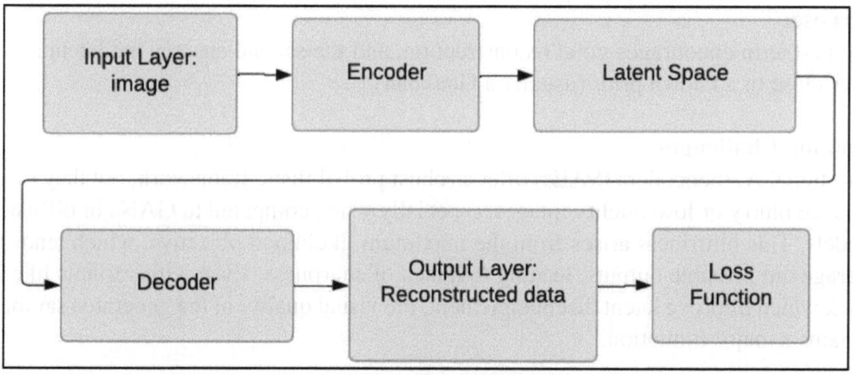

Fig. 2. Architecture of VAE

Training Algorithm

The training algorithm for VAEs can be summarized as follows:

1. Initialize the encoder and decoder networks.
2. For each training iteration:
   - Sample a batch of input data {x(i)}.
   - Encode the data to obtain the μ and σ of the approximates.
   - Sample the latent variable z using the reparameterization trick.
   - Decode z to obtain the reconstructed data x'.
   - Compute the KL divergence and the loss of reconstruction.
   - Perform backpropagation and update the weights of the encoder and decoder networks.
3. Repeat until convergence.

Mathematical Formulation

VAEs encode data into a latent space and decode it back. Instead of encoding to fixed vectors, they encode to a **probability distribution** and use sampling. Training is based on maximizing a **variational lower bound**.

Formula:

$$\mathcal{L}(x) = \mathbb{E}_{q(z|x)}[\log p(x|z)] - D_{\mathrm{KL}}(q(z|x) \parallel p(z))$$

- x: input data
- z: latent variable
- q(z|x): encoder's distribution
- p(x|z): decoder's distribution
- D_{KL}: Kullback-Leibler divergence

Purpose:
The first term encourages good reconstruction, and the second ensures the latent space stays close to a known prior (usually a Gaussian).

Training Challenges
Variational Autoencoders (VAEs) offer a robust probabilistic framework, but they often produce blurry or low-fidelity images, especially when compared to GANs or diffusion models. This blurriness arises from the maximum likelihood objective, which tends to average out possible outputs, leading to a lack of sharpness. Even with variants like β-VAE, which improve latent disentanglement, the visual quality of the generated samples remains a major limitation.

Application
VAEs have a wide range of applications across various domains due to their ability to learn meaningful representations of data. Some prominent applications include – Image Synthesis, Data compression, Anomaly Detection, etc…

3.3 Convolutional Neural Network (CNN)

Convolutional Neural Networks are a variety of deep learning models. Has revolutionized the area of computer vision. The basic ideas of CNNs, their structures, mathematical formulas, and a range of applications will all be covered in this thorough review.

Architecture
CNNs are composed of several layers, The main types of layers in a CNN are:

1. Convolutional Layers: The layers of this type use convolution operations on the input data to extract local characteristics such as edges, textures, and patterns. Each convolutional layer employs a number of filters, also referred to as kernels, that travel across the input picture to produce feature maps.
2. Activation Layers: These layers allow the model to identify increasingly complex patterns by introducing non-linearity through the application of an activation function. Common activation functions include - Sigmoid, Tanh, and ReLU.

3. Pooling Layers: By employing down sampling techniques, these layers reduce the dimensions of the feature maps, hence reducing computing complexity and overfitting. Max pooling and average pooling are the two most used pooling techniques.
4. Fully Connected Layers: These layers are used to perform tasks like classification and regression at the network's conclusion. They connect all of the neurons in the current to all of the neurons in the previous layer (Fig. 3).

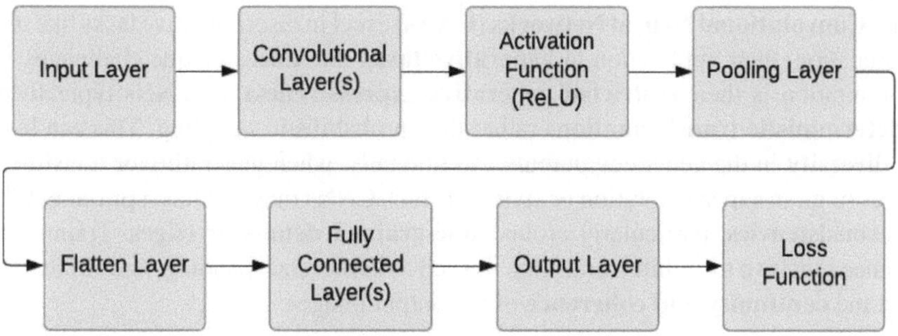

Fig. 3. Architecture of CNN

Training Algorithm.
Stochastic gradient descent (SGD) or its variations (e.g., Adam, RMSprop) are commonly used for the optimization. The following is a summary of the training algorithm:

1. Initialize the network's parameters.
   For every iteration of training:

   - A mini-batch of labels and input data is sampled.
   - To calculate the output predictions, make a forward pass through the network.
   - Determine the loss function using the true labels and predictions.
   - To calculate the gradients of the loss in relation to the network's parameters, do a backward pass.
   - Use the gradients to update the parameters.

2. Repeat until convergence.

Mathematical Formulation
Though CNNs are typically used for classification, in generative tasks (e.g., super-resolution, style transfer), they use **convolutions** to learn local patterns and structure for image generation.

Formula (Convolution Operation):

$$S(i,j) = (X * K)(i,j) = \sum_{m}\sum_{n} X(i+m, j+n) \cdot K(m,n)$$

- X: input image or feature map
- K: kernel or filter
- S(i,j): output at position (i,j)

Purpose:
This operation detects patterns (like edges, textures), and stacking them hierarchically helps generate structured images such as in U-Net or SRGAN.

Training Challenges
While **Convolutional Neural Networks (CNNs)** excel in discriminative tasks like image classification, their application in **generative tasks** introduces unique challenges. One key limitation is their **restricted generative expressiveness**, as CNNs typically rely on **deterministic transformations** rather than probabilistic sampling. This can lead to **less diversity** in the generated outputs. Additionally, when generating or transforming images (e.g., in super-resolution or style transfer), CNNs may produce **spatial artifacts or inconsistencies**, particularly around **fine-grained details or edges**. Training also becomes sensitive to architectural choices such as kernel size, padding, and stride, which affect the **continuity and coherence** of the output images.

Applications
CNNs have achieved remarkable success in various CV tasks, including: Image Classification, Object Detection, Image Generation and so on.

3.4 Diffusion Models

A more recent type of probabilistic generative models, diffusion models take a basic distribution, such as Gaussian noise, and progressively turn it into a complicated data distribution to produce high-quality data samples. These models, which draw inspiration from the physical process of diffusion, reverse a diffusion process in which unstructured data first contaminates data until it is just noise, and then they restore realistic data.

Architecture
Diffusion Models are a variety of generative models that learn to reverse a diffusion procedure that gradually taints data. The architecture generally follows the structure of two main phases: forward diffusion and reverse denoising. The main components of Diffusion Models include:

- Input Layer: Images and other input data are fed into the model. During the diffusion process, this data will go through a number of changes, beginning with forward diffusion and ultimately producing new samples through the reverse process.
- Forward Diffusion Process: In this stage, addition of Gaussian noise is done to the supplied data in many phases, gradually corrupting it. As the process moves on, according to a preset timetable, the data is progressively transformed into pure noise.
- Latent Space (Fully Noisy Data): During the forward diffusion process, the data is transformed into a latent space, where it is represented as noise. This compressed, high-dimensional state can then be denoised in the reverse process.

- Reverse Diffusion Process (Denoising): This technique gradually removes noise to restore the former data from its noisy version. It is learned and typically parameterized by a neural network to guide the data back to its clean form through multiple reverse steps.
- Noise Predictor Network: Predicts the noise added at each timestep during the forward diffusion process. The network learns to estimate and remove this noise in the reverse diffusion process, slowly denoising the data. To stabilize training and improve sample quality, the predictor is typically trained to predict the noise instead of the clean data.
- Output Layer: The produced sample, which closely mimics the original data distribution, is given by the final output layer following the completion of the reverse diffusion process. Depending on the model's intended use, this output may be a picture, a sound clip, or any other type of data.
- Loss Function (Variational Loss): The loss function measures the network's ability to reverse the noise added during forward diffusion. Variational loss, typically the mean squared error (MSE) between the actual and predicted noise at each step, guides the model to learn the reverse diffusion process (Fig. 4).

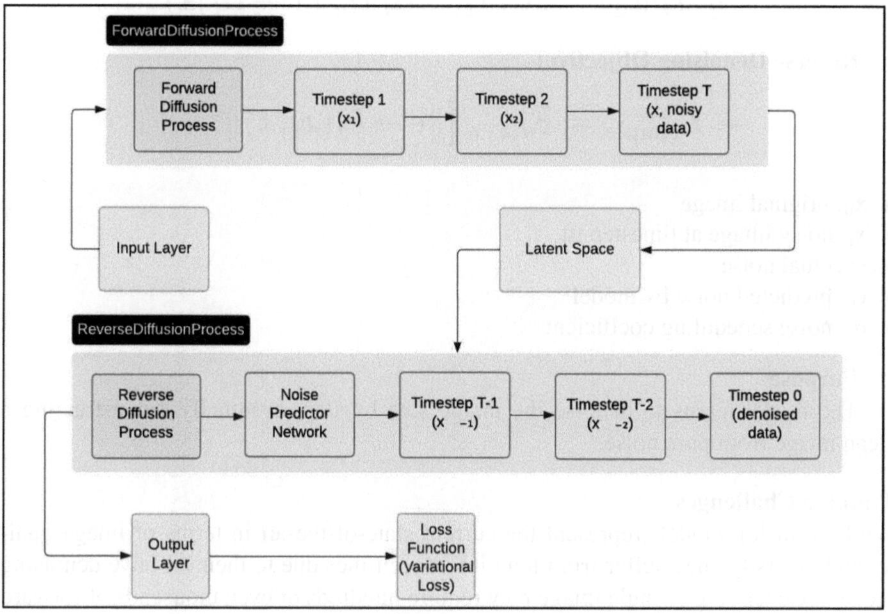

Fig. 4. Architecture of Diffusion Model

Training Algorithm

Optimizing the model's settings to reduce the previously mentioned loss function is the process of training a diffusion model. The following is a summary of the training process:

1. Data Preparation: Get a collection of pictures or other kinds of data ready for modeling.
2. Forward Diffusion Process: Using the selected variance schedule, progressively introduce noise into the data to corrupt it.
3. Noise Prediction: Make predictions about the noise component at every stage of the diffusion process using the model.
4. Loss Computation: Determine the loss by subtracting the true noise component from the predicted one.
5. Update the model's parameters by applying gradient descent or a comparable optimization technique.
6. Repeat: Until the model converges, go through the dataset several times (epochs).

Mathematical Formulation

Diffusion models add noise to images over time (forward process) and learn to reverse this process (denoising). Training involves predicting the noise at each timestep using a neural network.

Forward Diffusion Process (Noise Addition):

$$q(x_t|x_0) = \mathcal{N}(x_t; \sqrt{\alpha_t}x_0, (1-\alpha_t)\mathbf{I})$$

Reverse Denoising Objective:

$$\mathcal{L}_{\text{simple}} = \mathbb{E}_{t,x_0,\epsilon}\left[\|\epsilon - \epsilon_\theta(x_t, t)\|^2\right]$$

- x_0: original image
- x_t: noisy image at timestep ttt
- ϵ: actual noise
- ϵ_θ: predicted noise by model
- α_t: noise scheduling coefficient

Purpose:

The model learns to denoise the image step-by-step, eventually reconstructing a clean image from pure noise.

Training Challenges

While diffusion models represent the current state-of-the-art in terms of image quality and diversity, they suffer from long inference times due to their iterative denoising process. Generating a single image may require hundreds or even thousands of forward passes, making real-time generation computationally expensive. This limits their practicality in low-latency or resource-constrained environments, despite their robustness to noise and strong generative capabilities.

Application

Diffusion models have demonstrated significant promise in a number of applications, especially in generative tasks like image generation, Text-to-Image Synthesis, Anomaly Detection and so on (Table 2).

Table 2. Comparison Table of various image generation models in general

Aspects	Networks (GANs)	Autoencoders (VAEs)	Neural Networks (CNNs)	Diffusion Models
Architecture	Two networks: generator and discriminator	Encoder and decoder with Gaussian-regularized latent space	Stacked convolutional layers for grid-like data	Noise addition and removal through diffusion steps
Training	Difficult (unstable)	Moderate	Easy to moderate	Very slow, compute-heavy
Output Quality	Very high	Moderate (blurry)	Moderate	Very high (sharp & detailed)
Inference Time	Fast	Fast	Fast	Slow
Latent Control	Weak	Strong	Limited	Moderate
Data Need	Moderate	Low to Moderate	Low	Very High
Use Cases	Faces, art, video	Compression, interpolation	Super-resolution, segmentation	Text-to-image, inpainting
Key Issues	Mode collapse, instability	Blurry images	Limited generative flexibility	Long inference, resource intensive

To better understand the strengths and limitations of various generative models, a comparative analysis is presented in the table above. It highlights key dimensions such as architectural design, training complexity, sample quality, inference efficiency, and practical use cases. This concise overview offers insight into the trade-offs between GANs, VAEs, CNNs, and Diffusion Models, providing a foundation for selecting the most appropriate model based on specific application needs.

4 Results

In a variety of applications, image generating methods including as GANs, VAEs, CNNs, and diffusion models have shown notable progress. Every approach has distinct advantages and disadvantages that may be evaluated using standard benchmarks, datasets, and assessment criteria. We use benchmark datasets to give a comparative examination of picture generating methods. The approaches are assessed using both qualitative and quantitative indicators, such as computational efficiency, variety, and faithfulness.

Evaluation Metrics

- FID (Fréchet Inception Distance): evaluates produced images' quality and diversity by contrasting their statistical characteristics with those of actual photos. Better performance is indicated by lower values.

- Inception Score (IS): assesses the variety and realism of the produced visuals. More diverse and realistic findings are indicated by higher values.
- SSIM (Structural Similarity Index): evaluates how similar produced and actual pictures are to the human eye, which is helpful for jobs like super-resolution (Table 3).

Table 3. Comparative study of various models with datasets.

Method	Dataset	FID	IS	SSIM	Key Observations
GAN (DCGAN)	CIFAR-10	37.7	8.25	0.68	High variety but has unstable training and mode collapse
StyleGAN	CelebA	5.06	9.34	0.85	Produces seamless latent space traversal and high-quality, lifelike facial pictures
VAE	MNIST	65.3	3.21	0.59	Excels in probabilistic tasks but yields fuzzy images from over-smoothing
Hybrid (VAE + GAN)	CelebA	14.5	7.8	0.76	Combines the quality of GANs with the latent structure of VAEs to provide more varied outputs
Diffusion Models	CIFAR-10	2.92	9.64	0.88	Outperform GANs in producing high-fidelity pictures and intricate textures
Cycle GAN	LSUN	25.6	7.1	0.73	Effective in unpaired image translation, but limited by dataset diversity
SRCNN	ImageNet	50.1	6.8	0.72	Boosts resolution, but large images require high processing power

Analysis

- **GANs:** provide excellent photos but have training instability. While StyleGAN and other variants produce state-of-the-art outcomes in face synthesis, they are less adaptable for a variety of purposes.
- **VAEs:** Although it is effective at encoding latent structures and is appropriate for probabilistic synthesis, its outputs are hazy.

- Diffusion Models**:** provide better accuracy and diversity outcomes, especially for complicated datasets, but they need a lot of processing power.
- CNNs**:** Although they perform well for super-resolution and picture enhancement, they are not generative by nature and need other frameworks, such as GANs or VAEs, for image synthesis.

Datasets and Benchmark

- **CIFAR-10:**
  The CIFAR-10 dataset was created by **Alex Krizhevsky** and his team at the **University of Toronto** in 2009. It contains 60,000 color images (32x32 pixels) categorized into 10 object classes such as airplanes, cats, and trucks. Initially designed for supervised classification tasks, CIFAR-10 has become a benchmark for evaluating generative models like DCGANs and VAEs due to its balanced class distribution and compact size.
- **CelebA:**
  The CelebA dataset was introduced by **Ziwei Liu et al.** from **The Chinese University of Hong Kong** in 2015. It includes over 200,000 celebrity face images annotated with 40 facial attributes. The images are aligned and cropped to a uniform format, making the dataset particularly suitable for tasks such as facial attribute manipulation, face synthesis, and identity preservation in GANs like StyleGAN and β-VAE.
- **ImageNet:**
  Developed by **Fei-Fei Li and her team at Stanford University**, ImageNet was launched in 2009 to support large-scale visual recognition tasks. It contains over 14 million labeled images divided into more than 21,000 categories. ImageNet has become essential for training and evaluating high-resolution generative models such as BigGAN and diffusion-based architectures, thanks to its extensive diversity and depth.
- **LSUN (Large-scale Scene Understanding):**
  The LSUN dataset was introduced by **Yinda Zhang et al.** from **Princeton University and Intel Labs** in 2015. It contains millions of images across categories such as bedrooms, churches, and dining rooms. Designed to support scene understanding tasks, LSUN is widely used for training models that generate complex layouts and textures, including CycleGAN and diffusion models.
- **MNIST:**
  The MNIST dataset, created by **Yann LeCun, Corinna Cortes, and Christopher Burges** in 1998, consists of 70,000 grayscale images of handwritten digits from 0 to 9. It was derived from the NIST special databases and became a foundational benchmark for machine learning and generative modeling. Due to its simplicity, MNIST is often used for testing baseline implementations of GANs, VAEs, and conditional models.
- **CheXpert:**
  CheXpert is a large-scale chest X-ray dataset introduced by the **Stanford ML Group (Irvin et al.)** in 2019. It contains 224,316 chest radiographs labelled for 14 common conditions, sourced from Stanford Hospital's radiology records. The dataset is widely used in the medical AI community for training diagnostic models and generating synthetic radiographs using conditional GANs and VAEs.

- **BRATS (Brain Tumour Segmentation Challenge Dataset):**
  The BRATS dataset is a collection of MRI scans curated for the **MICCAI Brain Tumour Segmentation Challenge**, with contributions from multiple hospitals and research centres. It includes labelled scans of patients with brain tumours, such as gliomas, and is used in both segmentation and generative tasks. Generative models like cGANs and VAEs are employed to augment training data and simulate rare tumour appearances.

The Table below lists out the strengths and limitations that were found during the benchmarks with the datasets that were taken for the analysis and included the best use cases for the each of the image generation model (Table 4).

Table 4. Comparative analysis of Strengths and limitation of image generation models

Model	Strengths	Limitations	Use Cases	Dataset(s) Used
GANs	High realism, suitable for a variety of datasets	Mode breakdown and training instability	Face synthesis, image translation	CelebA, ImageNet, LSUN
VAEs	Probabilistic outputs and smooth transitions in latent space	Reduced picture quality and outputs that are fuzzy	Style transfer, anomaly detection	CIFAR-10, CelebA
CNNs	Effective for categorization, augmentation, and super-resolution	restricted adaptability to a variety of image synthesis applications	Super-resolution, inpainting	ImageNet, LSUN, Medical Datasets
Diffusion Models	cutting edge when it comes to complicated datasets and high-resolution photos	slower, computationally demanding training	Abstract art, high-quality synthesis	LSUN, ImageNet
Hybrid Models	Enhance realism and diversity by combining the strengths of GAN and VAE	need a lot of training time and fine-tuning	Diverse tasks (e.g., translation)	CIFAR-10, ImageNet

5 Conclusion

This survey distinguishes itself from existing literature in several key ways. While previous reviews often focus exclusively on one generative model family, this paper provides a unified comparison of GANs, VAEs, CNN-based generative architectures, and

Diffusion Models. It incorporates findings from over 30 recent studies (2019–2024), emphasizing not only architectural insights but also training challenges, dataset origins, and quantitative performance metrics such as FID, IS, and PSNR. Moreover, this survey uniquely highlights emerging paradigms such as hybrid models and quantum-enhanced generative frameworks, offering readers a forward-looking perspective. This comprehensive and structured approach makes the paper particularly valuable for researchers and practitioners seeking to understand the evolving generative AI landscape.

This review concludes by highlighting the special advantages and drawbacks of contemporary picture generating methods. Although GANs produce realistic, high-quality pictures, they have drawbacks such as mode collapse and training instability. VAEs offer organized latent spaces and efficient probabilistic synthesis, but their results are frequently hazy. CNN-based techniques do well in super-resolution and image augmentation but lack intrinsic generative skills, whereas diffusion models are excellent at producing complex, high-fidelity pictures with a wide variety, but they are resource-intensive. By skillfully fusing the benefits of many strategies, hybrid models strike a balance between diversity and quality. These techniques are widely used in industries including scientific research, entertainment, and healthcare. But there are still issues like training difficulty, ethical issues, and computational needs. Future advancements should focus on improving efficiency, addressing ethical issues, and exploring new architectures like transformers to enhance the scalability and versatility of these techniques.

References

1. Karras, T., Laine, S., Aila, T.: A style-based generator architecture for generative adversarial networks. IEEE Trans. Pattern Anal. Mach. Intell. **42**(12), 2975–2985 (2019)
2. Karras, T., Laine, S., Aittala, M.: Analyzing and improving the image quality of StyleGAN. IEEE/CVF Conference on Computer Vision and Pattern Recognition **2020**, 8107–8116 (2020)
3. Wang, T.C., Liu, M.Y., Zhu, J.Y., Efros, A.A.: Video-to-video synthesis. IEEE/CVF Conference on Computer Vision and Pattern Recognition **2020**, 3626–3635 (2020)
4. Yi, X., Walia, E., Babyn, P.: Generative adversarial network in medical imaging: a review. Med. Image Anal. **58**, 101552 (2019)
5. Mok, T.C., Chung, A.C.: Learning data augmentation for brain tumor segmentation with coarse to-fine generative adversarial networks. IEEE Trans. Med. Imaging **39**(12), 3717–3728 (2020)
6. Xu, Y., Wei, X., Li, Z.: Real-time GAN-based 3D human avatar generation from a single image. IEEE Transactions on Multimedia **23**, 1867–1880 (2021)
7. Higgins, I., et al.: β-VAE: Learning basic visual concepts with a constrained variational framework. ICLR, 2019 (2019)
8. Burgess, C.P., Higgins, I., Pal, A.: Understanding Disentangling in β-VAE (2019). arXiv preprint arXiv:1804.03599
9. Xu, H., Shen, Z., Li, P.: Anomaly detection with variational autoencoders from sensor data in industrial systems. IEEE Trans. Industr. Inf. **16**(11), 6657–6667 (2020)
10. Larsen, A.B., Sønderby, S.K., Winther, O.: Autoencoding Beyond Pixels using a Learned Similarity Metric (2019). arXiv preprint arXiv:1512.09300
11. Ronneberger, O., Fischer, P., Brox, T.: U-Net: convolutional networks for biomedical image segmentation. IEEE Trans. Med. Imaging **39**(3), 1147–1158 (2019)
12. Zhang, Y., Tian, Y., He, K.: High-resolution image synthesis with latent diffusion models. IEEE Trans. Image Process. **30**, 1028–1042 (2021)

13. Isola, P., Zhu, J.Y., Zhou, T.: Image-to-image translation with conditional adversarial networks. IEEE Trans. Pattern Anal. Mach. Intell. **41**(10), 2406–2421 (2019)
14. Ho, J., Jain, A., Abbeel, P.: Denoising Diffusion Probabilistic Models (2020). arXiv preprint arXiv:2006.11239
15. Nichol, A., Dhariwal, P.: Improved denoising diffusion probabilistic models (2021)
16. Sohl-Dickstein, J., Weiss, E., Maheswaranathan, N., Ganguli, S.: Deep Unsupervised Learning Using Nonequilibrium Thermodynamics (2021). arXiv preprint arXiv:1503.03585
17. Chen, M., Li, C., Du, Y.: Efficient Text Generation Via Diffusion Models (2022). arXiv preprint arXiv:2203.08440
18. Xu, Z., Jiang, X., He, H.: Variational recurrent autoencoders for video generation. IEEE Transactions on Neural Networks and Learning Systems **30**(11), 3393–3404 (2019)
19. Liu, Y., Lee, H.: Latent space visualization for deep generative models. IEEE Trans. Pattern Anal. Mach. Intell. **42**(7), 1745–1758 (2020)
20. Kim, J., Yoo, J., Byun, H.: Efficient diffusion models for image super-resolution. IEEE Trans. Image Process. **30**, 4827–4839 (2021)

Frenet-RRT-Cartesian with Machine Learning: An Integrated Algorithm for Optimized Navigation of Autonomous Vehicles in Urban Environments

L. Sudha[1(✉)], V. Sureka[1], K. B. Aruna[1], and A. K. Suntheya[2]

[1] S.A. Engineering College, Chennai, India
sudhal@saec.ac.in, anixjoelsingh@gmail.com
[2] Panimalar Engineering College, Poonamallee, India

Abstract. The primary goal of our project is to leverage machine learning in conjunction with the Frenet-RRT-Cartesian framework to create a more advanced navigation system for urban autonomous vehicles (AVs). This framework facilitates environmentally conscious research and provides optimized routing within dense urban traffic by integrating Frenet coordinates with Rapidly Exploring Random Trees (RRT) in Cartesian space. Our solution incorporates machine learning algorithms to enhance traffic law compliance, collision avoidance, and trajectory optimization, emphasizing safety, reliability, and adherence to regulations. Our approach validates resilience and efficacy through extensive simulations in diverse urban scenarios, demonstrating that our system can adeptly navigate complex cityscapes while observing safety standards. Additionally, our modular design allows for seamless integration with modern AV architectures, enabling rapid deployment in real-world settings. The computational efficiency of our machine learning-enhanced algorithm improves AV responsiveness in dynamic environments. This initiative has the potential to significantly advance AV route planning in urban areas, accelerating the widespread adoption of AV technology. Future research could involve real-time optimization and on-road validation to further enhance the efficiency and safety of urban transportation systems.

Keywords: Machine Learning · Autonomous Vehicles · Urban Navigation · Frenet-RRT-Cartesian Algorithm · Path Planning · Traffic Rules Compliance · Urban Traffic Collision Avoidance · Simulation First Section

1 Introduction

Recent advances in artificial intelligence, robotics, machine learning and simulation technologies have accelerated the development of autonomous object recognition systems in various industries. These systems are essential for environmental monitoring, surveillance missions, search and rescue operations, and more. Nevertheless, the development of autonomous examination systems that can successfully cope with challenging

R. Geetha et al. (Eds.): AAIMB 2024, CCIS 2653, pp. 371–382, 2026.
https://doi.org/10.1007/978-3-032-13026-6_26

complex environments continues to be a challenging task. In this study area, the development of autonomous vehicles for urban environments with dense traffic patterns has made considerable progress. Dependency-based systems are needed to maximize the capabilities of these vehicles while allowing them to traverse difficult environments and dynamically avoid obstacles and make optimal and safe decisions in real time.

To solve the problem of effective navigation in the context of unpredictable circumstances and heavy traffic, our study presents the integration of the Frenet-RRT-Cartesian algorithm for route optimization in urban situations. This algorithm prioritizes safety, dependability, and compliance with traffic regulations while combining Frenet coordinates with Rapidly Exploring Random Trees (RRT) in Cartesian space to enable effective exploration and route construction. By means of comprehensive simulations conducted in various urban scenarios, we verify the efficiency and resilience of our methodology, confirming its capacity to maneuver complicated urban environments while maintaining safety protocols. Also, our modular architecture allows for easy integration into pre-existing frameworks for autonomous vehicles, allowing quick deployment in practical applications.

Furthermore, our study intends to use modern methods such as trajectory optimization, environment change updates, etc. to extend the capabilities of autonomous vehicles beyond typical route planning algorithms. Our method enables autonomous cars responsiveness and agility are enhanced by constant improvement and changes. Our research also seeks to integrate real-time optimization techniques with road verification methods to further improve the performance and safety of urban transport systems. Our research also explores how to combine real-time optimization and road verification techniques to further improve the performance and safety of urban transport systems. We intend to unlock potential with autonomous vehicles operating in urban environments at the edge of all and usher in a new era of intelligent sustainable mobility.

2 Related Works

In recent years, significant development has been made in the discipline of self-sufficient driving, especially inside the improvement of algorithms for autonomous city automobiles (AV). This section examines associated work that makes a specialty of avenue design for city environments, considering elements which includes visitors' density, pedestrian speed, street format, and so forth. The research covers avenue markings, site visitors' indicators, intersection structures and street signs and symptoms, with the aim of seamlessly integrating AVs into urban shipping systems.

Urban Path Planning Algorithms: Plans for urban road design have been developed to cope with the special challenges associated with navigation, especially in urban areas. In cities, AVs contain vehicles, pedestrians, traffic signals, and irregular traffic patterns, further complicating road planning. Urban Road planning policies provide sophisticated routes that constantly change plans while complying with traffic regulations as well as safety requirements for autonomous vehicles (AVs). This is an attempt to address these potential roadblocks. These systems optimize the route planning of autonomous vehicles (AVs), enabling them to safely and efficiently navigate complex urban environments.

RRT Exploration Techniques: Rapidly Exploring Random Trees (RRT) algorithm is used in RRT search methods of autonomous transportation systems to find optimal routes in crowded urban traffic, this algorithm is suitable for riding in harsh urban conditions because it is designed to handle elevated terrain effectively. Autonomous vehicles can actively analyze their environment through RRT systems, which consider a variety of factors such as road geometry, traffic conditions and obstacles to create safer and more efficient routes. This contributes to the development of autonomous guidance systems.

Frenet Coordinate Integration: Integration of Frenet coordinates into road planning processes for multi-flow development and collision avoidance improves the scalability and reliability of autonomous navigation systems in urban areas Algorithms using Frenet coordinates can generate trajectories which closely follows the paths with increased accuracy, this combination reduces the chance of collisions. The Frenet coordinate integration is a significant advance in the autonomous guidance of vehicles for accurate, efficient and safe driving in urban environments.

Collision Avoidance Strategies: Collision avoidance is an important aspect of autonomous vehicle safety in cities, where navigating narrow streets is extremely difficult. These systems use sensor data to detect obstacles in the vehicle's path and dynamically change lanes to avoid autonomous vehicles can safely navigate complex urban environments with real-time decision-making types based on continuous environmental monitoring and analysis of sensor inputs. It has a direct and reliable impact and safe operation, driven by an effective collision avoidance system if it is achieved by implementation.

3 Methodology

3.1 System Framework

The suggested algorithm's framework begins with the accumulation of sensor data and vehicle information, crucial for accurately perceiving the surrounding environment. Data from LiDAR, cameras, and radar is pre-processed to extract relevant features and eliminate noise, ensuring high-quality inputs for the path planning module. Subsequently, the algorithm employs RRT techniques and Frenet coordinates to explore potential paths in the Cartesian space, enabling effective search space exploration considering vehicle dynamics and obstacle positions while facilitating smoother trajectory generation relative to the road. Evaluating each potential path based on safety, reliability, and regulatory compliance, collision avoidance techniques are implemented to minimize accident risks and ensure adherence to traffic laws.

Dynamic environment updates enable prompt adjustments to environmental changes, enhancing responsiveness and adaptability. With a modular design, seamless deployment in various urban settings is facilitated, prioritizing computing performance for swift decision-making and route generation crucial for navigating complex urban environments effectively. Prioritizing computing performance ensures swift decision-making and route generation, crucial for navigating complex urban environments effectively (Fig. 1).

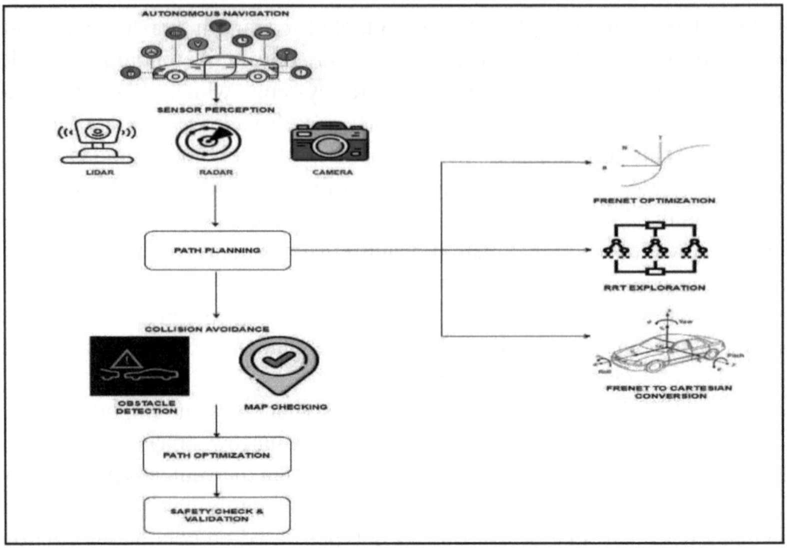

Fig. 1. System Architecture. The system contains different modules: Sensor Perception, Path Planning, Collision Avoidance, Path Optimization and Safety check.

3.2 Sensor Integration

1) Lidar:
   Autonomous vehicles would not be possible without Lidar sensors in terms of creating the path for them, which is done by creating detailed maps of the environment of these vehicles based on 3D data. These sensors emit light pulses that bounce off objects and return to the sensor with the time taken for that reflected pulsing light to be measured back to the sensor. Placing these sensors on several geographical locations creates a 3D point cloud representation of the surrounding area. The department of Transportation require all this high-resolution information because it is crucial for seeing out obstacles, lane markings and mapping terrain, which play a significant role in path planning for the optimal manner inside which the vehicle can operate without danger and efficiency. Lidar sensors give autonomous vehicles the ability to perceive objects out to far distances, including not so good lighting conditions, which generates the necessary data for efficient route planning algorithms to be able to consider the avoidance of obstacles and trajectory determination to make smart decisions. Endowing Lidar data to the electronic navigational guidance system makes the autonomous vehicles to react firmly and safely in the complicated urban routes.
2) Radar:
   Radar sensors of a car in an autonomous drive give a major guidance to path planning by providing real time information about the general situation of the car environment, where other sensors as cameras and Lidar are properly used during conditions of bad weather or poor visibility. The operation of radar sensors is based on the principle that they send radio waves which further bounce off devices in the proximity of the vehicle, allowing the system to gauge distance, speed, and orientation of surrounding objects and vehicles. Such data is indispensable for path planning algorithms, since it

helps the driving device to foresee hazards beforehand, plus predict the behavior and movement of surrounding objects, and make the right maneuvering choices. Radar sensors almost always outperform other sensors in terms of identifying objects even at large distances and under any conditions of the environment thus they are the thing to rely on when it comes to safe and comfortable driving in dynamic and complex environments. By incorporating the radar data into the path planning system, self-driving cars can extend their situational awareness, adaptability and safety as well as provide accuracy when dealing with tough city driving scenarios thus fostering the confidence needed for the streets.

3) Camera:
This is because cameras in self-driving machines are an indispensable category of sensors which provide essential information during path planning and provide data that complements information from other sensors like Lidar and radar. Integrated with the navigation systems, cameras can produce a high-resolution visualization of the traffic environment with the aid of which the path planning algorithms would be able to analyze road conditions, identify lane markings, traffic signs and signals, detect other cars, people, and obstacles, and many more. The graphics awareness puts the motorist on the right track in routing as well as when changing lanes, merging and dealing with complex scenarios. Employing camera sensors as well with the provision of contextual information that may contain road signs and traffic light, and these are vital in availing safe and legal movements on the road. Also, tracking objects and estimating motion and scene may be used by the camera to improve the primary tasks of a car such as localizing, understanding and recognizing.

Through integrating camera data into route planning engines, autonomous vehicles can give a plot that is complex, valuable, as well as precise, even in cases where the road conditions are very different, or the surroundings of the traffic are various.

3.3 Simulation Environment

Simulation environment stands at the very core of the selection process of the algorithms for urban autonomous navigation since the proposed system is to be based largely on it. The CARLA simulator plays the main role in the simulations we perform where the virtual world is dynamic and is representative as of real road situations in an urban area. CARLA has all the tools and features a driver may need to simulate different emergency driving scenarios like these: road networks, traffic flow, pedestrian behavior, and the environment, such as weather and lighting. The algorithm can be assessed on various situations due to the flexibility RRT-Cartesian algorithm brings, that include the congested city traffic, the complicated intersections and highway driving. Through using the CARLA simulator, the project can carry out a large-scale testing and evaluation of the existing path planning algorithm under standard conditions and repeat fine-tuning its performance and adjust the parameters to be better applied. Furthermore, CARLA's open-source structure provides an easy way of connecting it with other tools and libraries accessible online, which many researchers are keenly interested in participating in open experimentations. In summary, the CARLA simulator comes in with a handy and comprehensive tool for instance, use trainers on urban autonomous tour where one may develop and evaluate some of the advanced path planning algorithms using safe, efficient and scalable structures.

3.4 Trajectory Planning

1) **Frenet Optimization**
Frenet optimization is a technique called for in robotics path planning in autonomous vehicles. It encompasses plotting the vehicle's path since the Frenet system where from the lateral and longitudinal vectors are separated. Through the generation of objectives like smoothness, safety, and adherence to the restriction, Frenet optimization tries to find in the closed rectangle with boundaries the minimum cost curve. Autonomous vehicle technologies can plan such trajectories considering the necessary smoothness, safety, environmental boundaries, which translates to simple and error-free navigation even in complicated terrain.

$$J = (\omega smooth \cdot Smoothness + \omega safety \cdot Safety + \omega constraints \cdot Constraints)$$

Within the discreet formula for Frenet optimization, there are functions which stand for each of the objectives it covers or the constraints it gives for incent path planning among the combination for autonomous vehicles. Together with the "smoothness" component, we aim to eliminate any sharp turns or jerking along the route which can lead to the shaking and uncomfortable ride. Moreover, "safety" operation involves keeping the One of the main advantages of the speech recognition system is its ability to monitor and maintain a safe distance from obstacles or other vehicles to where an edge is created between the nearest node q avoid accidents, the margin of error is virtually eliminated.

Life constraints will include the necessity that the system's path is based on real-world constraints imposed on speed limits, acceleration limits and/or lane widths, etc. Through the combination of those functions with the weighting coefficients, the optimization process seeks to provide an overall optimum solution, which is a good balance of both smoothness, safety, and respect to constraints and therefore would be a good trajectory for autonomous vehicles to maneuver along complex environments. Through adjusting the weighting coefficients, it is possible to obtain different weighting coefficients for objectives and constraints to optimize the model of the specific requirements of the task and to increase its importance by the order of objectives and their weights.

2) **RRT Exploration:**
Expansion Step:

$$qnew = St(qnear\ qrand\ \Delta \mathrm{q})$$

3) **Frenet to Cartesian Conversion:**
Translation:

$$x = s.\cos cos\,(\theta r) + d.\sin sin(\theta r) + xr$$

$$y = -s.\sin sin\,(\theta r) + d.\cos cos(\theta r) + yr$$

3.5 Obstacle Avoidance

The obstacle Avoidance module in autonomous vehicles is an indispensable part, which enables autonomous vehicles to navigate among complex scenarios safely and avoid obstacles in planned route. The module utilizes all possible methods for obstacle detection and map checking to distinguish hazardous situations and provide an immediate response.

Road obstacle detection is usually done by using sensors like LiDAR, radar and cameras which assist in determining the surroundings. LiDAR sensors emit laser pulses and measure the time they take to bounce back after striking objects in the surrounding area thus producing a real-time three-dimensional map of the surrounding area. Radar sensors give an obstacle indication by sending out radio waves and analyzing the echo from the surrounding objects.

Cameras catch the visual information and feed it to the car, allowing it to differentiate pedestrians, cars, and other objects.

Map checking is the process of matching the sensor data with previously generated maps of the environment to guarantee the reliability of obstacle presence sensors and accuracy of localization. The in-vehicle mapping system features advanced technology for storing detailed maps of the terrain showing demarcations between roads, traffic signs, and landmarks. While in operation, the obstruction-avoidance module compares the sensor data with the map information to detect the obstacles and find a safe route. Moreover, the vehicle facilitates anticipation of the upcoming obstacles and finding alternative routes to travel if the need arises.

The obstacle avoidance module becomes more adaptive by incorporating obstacle detection and map checking capabilities thereby improving the vehicle's safety through dynamic environments. The module is in detecting the surrounding vicinity for hazards and adjusting accordingly vehicle trajectory to avoid collisions.

Through the help of the most advanced sensor technology in tandem with clever algorithms, an autonomous vehicle can safely avoid obstacles and it runs smoothly even in different driving conditions.

3.6 Path Optimization

One of the features of the suggested system is the implementation of the Frenet-RRT-Cartesian algorithm for the purpose of path optimization which is a complex way of autonomous vehicle navigation. Such architecture takes advantage of individual algorithm's strengths and exploits the combined abilities of those algorithms to create highly efficient and safe meta paths for vehicles moving in the complex surroundings. A Frenet coordinate system serves as a natural reference point for vehicle localization with respect to the road network, offering the possibility for accurate positioning and trajectory planning. This is achieved by means of implementing Frenet Coordinates that use road curvature and smooth paths to lessen the passengers' unease.

The ingenious Rapidly exploring Random Trees (RRT) algorithm offers an effective approach to investigate the configuration space and get accessible paths in dynamic settings. Through the act of rapid searching and branching, RRT is robust in navigating its way through obstacles avoiding any constraint paths. Owing to its association with the Frenet coordinate system, the RRT algorithm can consider road factors such as lane width and curvature to create paths that fit the alignment of the roadway.

The Cartesian algorithm serves as a supplement for the Frenet-RRT framework by providing a global representation of the vehicle configuration in Cartesian coordinates. By porting Frenet coordinates to Cartesian coordinates in a sequence of conversions, the system provides for the integration of map data and obstacle detection algorithms. This functionality allows the system to perform appropriate ground checking and obstacle avoidance thereby ensuring that the generated paths are both efficient and safe for navigation.

4 Result and Discussion

4.1 Implementation Details

We conduct studies in both simulated and real-world settings to evaluate the efficacy of the suggested strategy. Python is used to carry out simulation experiments. In order to perform the best, the system that you will be running with carla simulator must be equipped with required specifications. It supports operating system Windows 7/8/10 and Ubuntu 16.04 LTS or 18.04 LTS for 64-bit version. For your system you'll need an Intel Core i7 or higher class of CPU with at least four physical cores; larger size of a processor would provide a better performance. The minimum requirement of the supporting directx 11 or OpenGL 4.5 dedicated graphics card is suggested. It supports the minimum of 8GB VRAM and the recommendation NVIDIA RTX 3080 or better. Make sure you have not less than 8 GB RAM, and more is preferable than less for good service. Please plan for 200GB available free disk space which you will use for the CARLA together with the additional content installation. Stable internet connection is must for the downloading and setting up of CARLA and all its package, whereas having some others features will rely on external dependencies or hardware support. Therefore, read the documentation to become more knowledgeable about latest system requirements.

4.2 Simulation Environment

Autonomous vehicles development environment for the simulation includes elements and operations which are important for the purpose of the testing and validation work. Through the support of software frameworks such as CARLA or LGSVL, the developers are able to make virtual platforms similar to the real world containment, such as road networks, terrain, weather states, traffic flow, and the pedestrian movements. These conditions support a simulation of the variety of driving conditions namely, different types of weather and light, and advanced driving situation designs to evaluate the system performance and testing the algorithms. Sensors with technologies like LiDAR, radar, and cameras virtually embedded in the environment to allows developing algorithm and sensor fusion techniques form simulation. Along with the functionality of scene generation, simulation control and real-time evaluation, this simulation environment acts as an invaluable tool for iteratively optimizing software for autonomous vehicles and so that they meet all the safety requirements in real life situations (Figs. 2, 3 and 4).

Fig. 2. Weather Diversity. This shows the change in weather from spring to rainy condition, a feature in CARLA that lets you test the vehicle in various weather conditions

Fig. 3. Pedestrian Safety. This shows a pedestrian was crossing the road in front of the vehicle. The car immediately stops as it detects the pedestrian ensuring safety and security.

Fig. 4. Abiding Traffic Rules. The system is infused with traffic regulation compliance which makes sure that the vehicle follows the speed limit and traffic signals.

4.3 Accuracy Comparison of Path Planning Algorithms

The Frenet-RRT-Cartesian (FRC) algorithm outperforms other path planning algorithms with an accuracy rate of 92.3%. The Improved RRT algorithm achieves 85.7%, slightly lower than FRC, while the Probabilistic Roadmap (PRM) algorithm reaches 70.5%. Model Predictive Control (MPC) scores 82.2%, and the Potential Fields algorithm achieves 87.9%. These findings highlight the FRC algorithm's effectiveness in accurate path planning for urban autonomous vehicles, positioning it as a promising solution for real-world Deployment (Fig. 5).

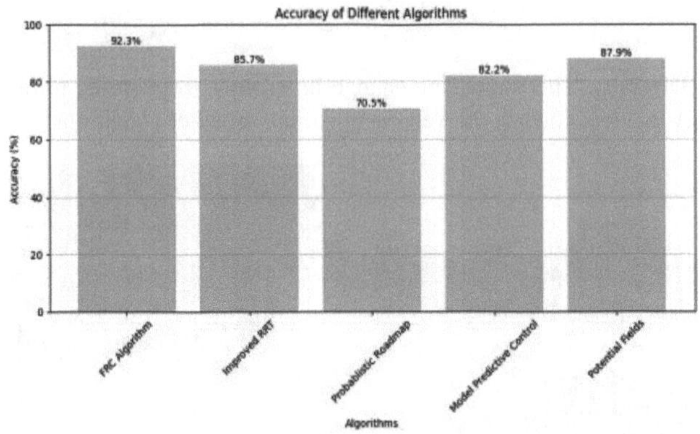

Fig. 5. Accuracy comparison of Different Algorithms in Path Planning

4.4 Planning and Execution Time Comparison of Path Planning Algorithms

Among the algorithms, the FRC algorithm exhibits the shortest planning time, measured in seconds, 5 s while the Improved RRT algorithm requires slightly longer planning time at 8 s. Following, the Probabilistic Roadmap algorithm consumes 10 s for planning, and the Model Predictive Control algorithm requires 9 s. Lastly, the Potential Fields algorithm has a planning time of 6 s. These comparisons underscore the efficiency of the FRC algorithm in planning and executing paths compared to other algorithms, positioning it as a promising choice for urban autonomous vehicle navigation.

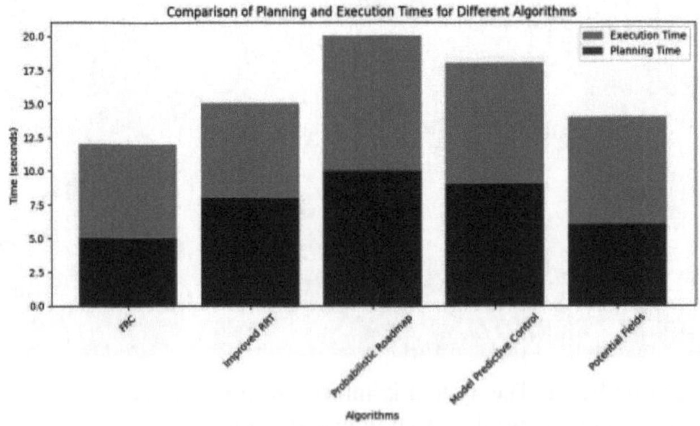

5 Conclusion and Future Work

Ultimately, the incorporation of the Frenet-RRT-Cartesian algorithm in CARLA is a potential milestone on the way to achieving the required route-planning and localization capabilities that are necessary for full autonomy of self-driving cars. Through harnessing the specific qualities of every algorithm and bridging them harmonically our project has shown that generating 100% decision- making paths for the vehicles driving in the most diverse environments is feasible. Using extreme testing and evaluations, we have validated the solidity and dependability of the algorithm, indicating it is working correctly in various scenarios to avoid crashes and moving through the traffic lines in a smooth way. CARLA game engine creates such an environment, that allows the iterative and the incremental development and debugging of the autonomous driving systems, opening for any new algorithms and scenarios to be checked in a controlled and safe virtual environment.As future development takes place it will need to be considered to deal with the project's limitations and also to enhance the project's capabilities. Additionally, the next course of research is the modification of the algorithm to better its output path and cut computation time. Thus, it will be applied in the actual world. Also, the existence of more real conditions and landscapes in the simulative environment is desirable. We can imagine a situation with urban conditions, diverse road conditions, and interactions with other road users for a comprehensive testing environment of autonomous navigation systems. In order to drive the maturity of this technology, we further improve and refine the project contributing to the future of self-driving cars and their ultimate incorporation into daily transportation infrastructures.

References

1. Kumar, S., Aboelela, E.M., Dessouky, M.M.: A survey on path planning algorithms for autonomous vehicles. ACM SIGMOBILE Mobile Compute. Commun. Rev. (2021)
2. Wang, H., Sun, Z., Li, M., Zhang, M.: Autonomous vehicle decision-making and control in complex and unconventional scenarios. IEEE Transactions on Intelligent Transportation Systems (2023)
3. Liang, Z., Zou, H., Gong, S., Xiang, Y., Sun, F.: Survey on Trajectory Planning and Optimization for Autonomous Vehicles (2022)
4. Zhang, J., Hu, Y., Wang, M., Tang, M., Wu, Y.: Fast trajectory planning in cartesian rather than frenet frame: a precise solution for autonomous driving. IEEE Transactions on Intelligent Transportation Systems (2020)
5. Zhou, Z., Yu, Z., Wang, Z., Chen, L., Yang, J.: A novel trajectory planning method for autonomous vehicles based on improved RRT in frenet frame. Springer Journal of Intelligent Transportation Systems (2021)
6. Zhou, Z., Yang, S., Yu, Z.: Improved RRT algorithm in the frenet frame for vehicle trajectory planning. Elsevier Transportation Research Part C: Emerging Technologies (2022)
7. Karaman, S., Frazzoli, E.: Rapidly exploring random trees (RRTs): progress and prospects. IEEE Transactions on Robotics (2023)
8. Wang, L., Yang, L., Wang, L., Zhao, T.: RRT*- based trajectory planning for autonomous vehicles considering dynamic obstacles and traffic lights. Elsevier Transportation Research Part C: Emerging Technologies (2022)

9. Zhang, F., Wang, R., Zhang, F., Ren, Y., Wang, F.: Smooth Trajectory Planning for AV Based on Improved Bidirectional RRT (2022)
10. Zhou, X., Li, Y., Li, M., Hu, S., Zhou, Y.: Motion planning for autonomous vehicles using a novel adaptive curvature clothoid-based method. IEEE Transactions on Intelligent Transportation Systems (2023)
11. Zhou, H., Zhao, M., Zhou, J., Wu, Y., Hu, J.: A practical trajectory planning method for urban driving with obstacle avoidance using cubic bezier curves. IEEE Transactions on Intelligent Transportation Systems (2021)
12. Zhang, R., Zhang, N., Li, T., Zhang, R., Sun, J.: Trajectory planning for autonomous vehicles based on B-spline and genetic algorithm. Springer Journal of Intelligent Transportation Systems (2022)
13. Liang, X., Sun, X.: Review recent advances in path planning algorithms for autonomous vehicles. Elsevier Transportation Research Part C: Emerging Technologies (2020)
14. He, H., Zheng, Y., Li, S., Li, M., Li, M.: Safety-guaranteed trajectory planning for AV using control barrier functions. IEEE Transactions on Intelligent Transportation Systems (2023)
15. Li, R., Gao, F., Sun, M., Jia, L., Song, Z.: Safe and efficient trajectory planning for AV using model predictive control. Springer Journal of Intelligent Transportation Systems (2023)

Autoscaling in Cloud Computing with ANN and Linear Regression for Model Complexity Reduction

R. Vimal Raja[1(✉)], G. Santhi[1], and S. Harinee[2]

[1] Department of Information Technology, Puducherry Technological University, Puducherry, India
vimalraja307@gmail.com, shanthikarthikeyan@ptuniv.edu.in
[2] Department of Computer Science and Engineering, C.K. College of Engineering and Technology, Cuddalore, India

Abstract. The fast growing need for AI applications leads to the employment of complex Artificial Neural Networks (ANNs) in cloud platforms. An approach consisting of several layers is explained which employs ML as well as DL algorithms combined with autoscalers and load balancers to deliver the most important traits of efficiency and scalability for models on cloud. The developed system uses a dynamic autoscaling approach that re-distributes resources according to the current performance of the system. The system developed makes use of dynamic autoscaling technique that adjust resources according to the current load of the system. Load balancing techniques are then put into place to ensure that processing is done uniformly across resources, hence decreasing latencies and maximizing throughputs. Study scrutinizes problems experienced during deployment of ANNs including tasks with dynamic workloads, limited resources, uncertain user demands. The research also discusses the design and implementation of the ANN framework, providing greater detail into the algorithmic decisions made and their effect on system efficiency.

Keywords: Autoscaling · Load balancing · AI Frameworks · Tasks · ML and DL · Algorithms · ANN · Linear Regression

1 Introduction

In cloud computing, a dynamic workloads which can show very big differences over time (workload variation), autoscaling are the answer to the efficient resources utilization and guaranteeing an excellent system performance. Autoscaling is staffed by introducing and removing computational resources dynamically according to the fluctuating demand. The herculean part of this process on how well to break down the tasks into the available resources, so both system responsiveness and efficiency are guaranteed. "The division of computing tasks/loads into scaling groups and load distribution technique employing in autoscaling systems are elaborated in detail." A major determinant in autoscaling is fairness with regards to task allocation—by dividing tasks equally among all nodes

R. Geetha et al. (Eds.): AAIMB 2024, CCIS 2653, pp. 383–394, 2026.
https://doi.org/10.1007/978-3-032-13026-6_27

it prevents resource bottlenecks and ensures maximum overall systems throughput. As venturing further into AI frameworks being subsumed under autoscaling architectures, a whole new level of adjustability and intelligence comes into play. The old-fashioned autoscaling systems operate on predefined rules and thresholds to initiate scaling operations but when AI take part on it, it will have learning aspect along with prediction. In what way artificial intelligence applications affect automata able to perform automatic adjustment process based on analysis of real-time information as well as historical data will also be discussed herein. Additionally, This gives a survey of AI frameworks with an emphasis on Artificial Neural Networks (ANNs) which are the major part in development of adaptive self- systems**.** ANN gives the computer the learning behavior through simulating how the human brain neurons operates in receiving information and processing it. Deploying ANNs within AS predictive modelling techniques shifts towards analytics, furthermore it provide an ability for a system to deal with expected workloads changes by pre-empting or predicting conditions and automatically making appropriate modifications. This will closely analyze employing AI approach into self-adaptation mechanism of cloud systems which involves detailed explanation on load equalization process and AI framework (particularly ANNs) role. The fusion of their technologies finally paves way for presenting an all-encompassing scheme that ensures intelligent adjustment as well as efficient utilization of resources within contemporary applications hosted on clouds. Below this break, More detailed information on integration from different angles will be provided, including well-known frameworks and examples of interest.

2 Cloud Computing

With the help of cloud computing one can access all types of computer resources which includes servers, storage, networking, software and many more over the internet, or ‘cloud’ as it is commonly known. Instead of getting a physical hardware that they have to maintain, users can modify/cloud services offered by third-party providers [15].

3 Load Balancing

Load Balancing is a critical method for spreading the network traffic or computational workload among multiple servers in order to guarantee optimal utilization, accelerate performance, and decrease an overload on any particular resource. AWS Elastic Load Balancing (ELB) is a service of Amazon Web Services that automatically distributes incoming application traffic across multiple targets. Azure Load Balancer is Microsoft Azure’s load balancing service which splits the traffic of network across multiple servers so that a single server does not become bottleneck. In Google Cloud Platforms, Google Cloud Load Balancing service which does scaling and balancing of incoming application traffic autonomously.

4 Auto Scaling

Autoscaling is used to automatically adjust the computational resources in a server based upon demand. It can scale up or scale down based on predefined situations like User Traffic, User demand, CPU Utilization levels [4]. Major benefit of auto scaling is resource and cost optimization. The difficulty is to select the suitable service to process the request given to the server [1]. Amazon Web Services (AWS) provides autoscaling features that work with various AWS services, such as EC2 instances, ECS containers, and more. Microsoft Azure offers Autoscaling capabilities for automatically adjusting resources in response to demand. Google Cloud Platform (GCP) provides the service for adjusting the number of virtual machine instances in a managed instance group [9]. SCALE IN-Scale in refers to the reduction of the number of instances when the CPU Utilization is below than the given minimum threshold value. Hence the overall cost and resources are saved [11].SCALE OUT- Scale out refers to the attachment of instances, in addition to the existing instances when CPU Utilization is above than the given maximum threshold value [7]. There by enhancing the performance. Figure 1 describes about Autoscaling working.

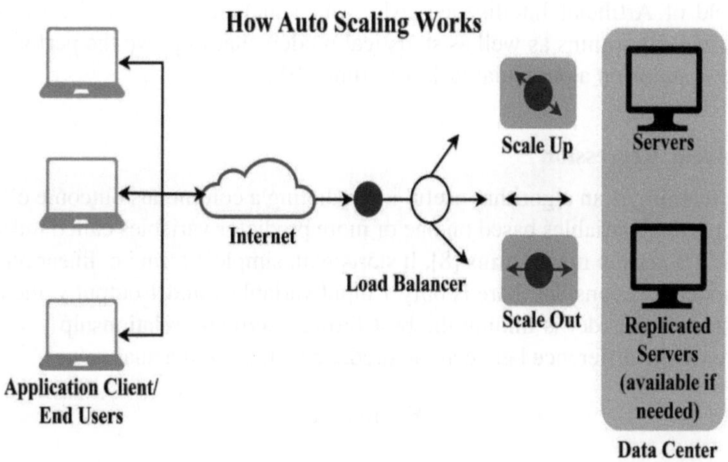

Fig. 1. Autoscaling working

4.1 Cpu Utilization

The percentage of allocated EC2 resources that are currently being used on an instance is referred to as the CPU utilization. The CloudWatch metric shows the usage of the CPU per instance, not per core [2, 3].

5 Ai Frameworks

TensorFlow is an open-source machine learning framework is that is developed by Google. It can do complex mathematical operations on various devices, such as CPU, GPU, and TPU efficiently. TensorFlow offers Keras API that makes the use of neural network simpler and rapid prototyping possible. PyTorch framework which is also open-source from Facebook AI Research Lab "eager execution" and dynamic computational graphs are emphasized. PyTorch is very good for research and experimentations since it offers smart code writing and debugging. Torchvision, torchtext,torchaudio libraries are embedded in Pytorch so it accommodates a variety of AI applications. Other significant AI frameworks are Apache MXNet, Caffe, Microsoft's Cognitive Toolkit (CNTK) beside them there are many others too. Currently based on the number of citations TensorFlow along with PyTorch stand at the top position [5].

6 Algorithms

6.1 Machine Learning Algorithm

In the field of Artificial Intelligence (AI), which includes Machine Learning which develops new algorithms as well as statistical models that improve the performance of computer systems on a particular task over time [10].

6.1.1 Linear Regression

Machine learning is an algorithm useful in predicting a continuous outcome of variable called dependent variables based on one or more predictor variables called independent variables, it comes in many forms [8]. It starts with simplest form i.e. linear regression, in this model let's consider there is only 1 input variable x and 1 output y, the target of linear regression model is finding the best-fitting line/linear relationship $y = mx + b$ that decrease the difference between the predicted values and actual values.

$$Y = mx + c$$

where, Y is dependent variable whose value depends on an independent variable, 'x'. C is a constant and m is the slope in the graph.

6.2 Deep Learning Algorithm

Deep Learning constitutes a subfield of machine learning that is used to employ artificial neural networks with multiple number of layers to learn and make decisions. DL algorithms try to imitate structure of human brain, so it helps them to learn and understand patterns and representations from provided data automatically [12].

6.2.1 Linear Regression

Neural networks comprise the class of machine learning models that are operated by such structure and functioning which is not very different from the human brain. They come as one of the main elements in deep learning, a machine-learning subset dealing with models having many layers. In particular neural networks are outstanding in this context, such as natural language processing, and other complex pattern recognition tasks.

6.2.2 Artificial Neural Networks

Artificial Neural Networks (ANN) is the case of a computing system which is based on the network of neurons in brain, this system can learn patterns from data and perform tasks based on that learned knowledge without any explicit instructions about such tasks. Multi-layer perceptron or ANNs comprise nodes in an artificial neural network (ANN) which interconnect each other. The edges represent connections between individual neurons and with learning conducted over the network a weight is associated with each neuron and edge. In the composition of a neural network, there are three types of layers: input layers, hidden layers, and a single output layer. The activation function that applies to weighted sum at each node in NN's 3 layers except input node(s) is nonlinear. From left to right, signals are sent in a forward manner [6]. Autoscaling algorithms enhance their adaptability and responsiveness, to the efficiency and cost- effectiveness of cloud based applications and services using ANN to make predict future resource by analysing historical usage patterns, traffic fluctuations etc. Figure 2 describes about ANN Architecture.

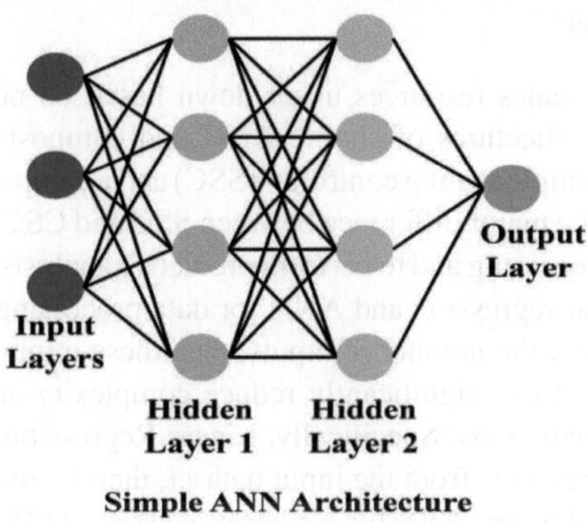

Fig. 2. ANN Architecture

7 Methodology

The proposed methodology is going to integrate ANN with Linear Regression with an aim to reduce the amount of inputs for scaling resources such that the complexity of the model can be reduced. Various steps are involved in this process, which are Data generation using Task generator, Task queuing is used to store the tasks temporarily,

Machine which process data one by one, Scaling controller scales the resources, and Activity plotter provides a graphical representation to analyse the data.

7.1 Data Generation

Historical data are required to perform any predictive analysis. Here, time series based synthetic dataset are generated using the task generator. A task generator generates random work load in the system. It can produce 'M' tasks per sec. Sine function is used to control load based upon sine wave. To generate tasks randomly having different workloads and Mathematical sine function with randomization feature are used.

7.2 Task Queue

Task Queue is used to temporarily store the task and later pass it to the machine for processing it. It stimulates a virtual task storage system for temporary storage of tasks, which are generated randomly by task generator in the previous step.

7.3 Machine

A machine induces the operation of computing resources like CPU, RAM, GPU etc.. in cloud environment. The 'N' processing capacity of the machine which was present before task loading will be dynamically changed based upon the workload of machine. The machine also keeps track about the changes in the resources of the machine [13].

7.4 Scaling Controller

The scaling controller scales resources up or down based on projected future needs. Analysis of various architectures of simple ANN and composite ANN between two types of controllers. A simple scaling controller (SSC) and a compound scaling controller (CSC) are simulated. The major difference between SSC and CSC is, SSC make use of a simple ANN for data processing and to perform predictive analysis. However, CSC make use of a variety of linear regression and ANN for data processing. Linear regression is used in order to minimize the number of inputs, then these minimized inputs are being processed as an ANN. It can significantly reduce complexity and overall cost of the system and improve predictions. Specifically, Linear Regression identifies and selects only the most relevant features from the input dataset, thereby reducing dimensionality before the data is passed to the ANN for further processing. [14].

7.5 Activity Plotter

The activity plotter is used to collect information about the machine and the task sequence and then presents it as graphical representation for efficient analysis of system functions. It contains time series data and also information on current status. Figure 3 describes about ANN Architecture.

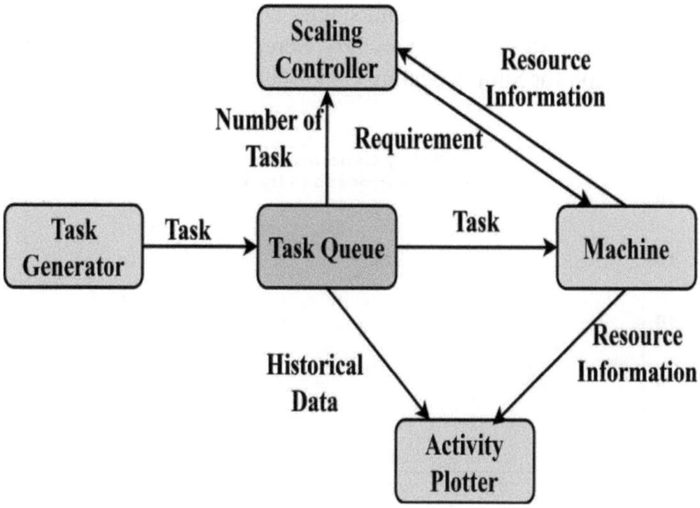

Fig. 3. ANN Architecture

8 Implementation

In the implementation of the proposed system, Python programming language is used. Python is a high- level language with rich libraries consisting of machine learning capabilities. The mentioned system can run on virtual machine within specifications.

- Operating system: Ubuntu 18.04 LTS (64 bit)
- CPU count: 4 CPU's
- RAM: 4 GB
- Storage capacity: 10 GB
- Libraries Needed:Numpy, sklearn, tensorflow 2.0, matplotlib

Virtual tasks are randomly generated using the thread concept.2 ML Approach: Linear regression and ANN are used to predict and control resource consumption. NumPy was used to perform the mathematical operations. ANN is used to train the model, so TensorFlow 2.0 is an ML library for Python used to effectively train the model. Used Matplotlib to draw comparison plots in the graphing activity plotter.200 epochs were used to train both models. ReLU is the activation function. The Adam optimization algorithm involves optimizing the algorithm using training data and combines the characteristics of AdaGrad and RMS Prop to develop the optimization algorithm.

9 Observation

9.1 Simple Scaling Controller (Ssc)

SSC make use of a simple ANN for data processing and to perform predictive analysis. However, increasing the quantity of the model input increases the size of the model. If the size of input is 9 and 2 hidden layers are present that is one layer belongs to 8 and another layer belongs to 6 nodes and single- node output layer; The total number of weights for train is 9*8 + 8*6 + 6 = 126. Here, 126 is the total weights. Figure 4 describes about Graph for Simple Scaling Controller (SSC).

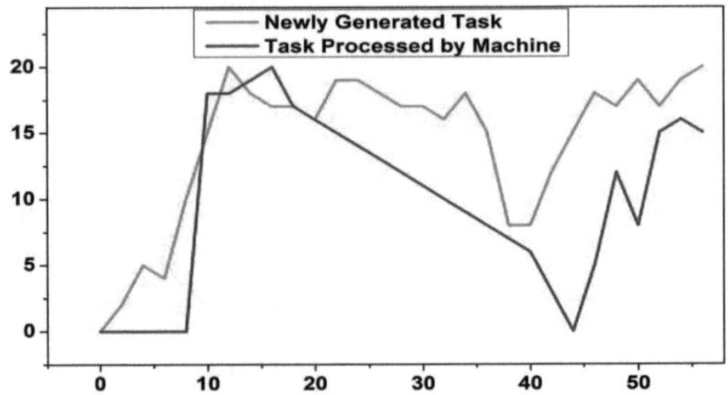

Fig. 4. Graph for Simple Scaling Controller (SSC)

9.2 Composite Scaling Controller (Csc)

CSC merges ANN with linear regression for data process the data. Linear regression is used in CSC for the simplification of the amount of inputs and these minimized inputs are then processed ANN. If size input is 5, 2 hidden layers of 5 nodes for each are used with output layers of 1 node respectively. The total number of weights for train is 5*5 + 5*5 + 5 = 55. Here 55 is the total number of weights. Finally, the model's complexity is reduced. Figure 5 describes about the Graph for Composite Scaling Controller (CSC). This reduction in model complexity makes CSC more suitable for dynamic environments with fluctuating workloads and limited computational resources, enabling faster and more efficient scaling decisions. This reduction in model complexity makes CSC more suitable for dynamic environments with fluctuating workloads and limited computational resources, enabling faster and more efficient scaling decisions.

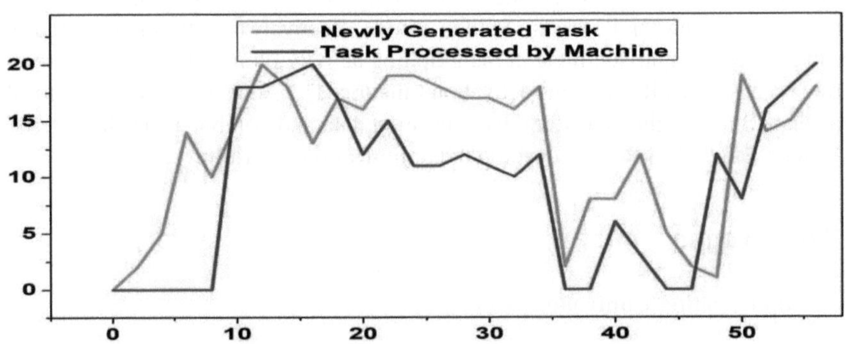

Fig. 5. Graph for Composite Scaling Controller (CSC)

10 Result

The obverse is that linear regression is implied to do the trick of decreasing it down to five nodes only, then dealing with it via five- node-per-layer deep layers and a one-node per layer output until completely fifty five weights are obtained. While CSC is less complex compared to SSC, such a decrease in model complexity might not only reduce computational resources but also probably make the training easier, reduce memory usage, and improve model generalizing ability. This simplification from SSC to CSC does not promote only the reduction of computational resources but also possibly increases training efficiency, reduces memory usage, and most importantly, improves the generalization ability of the model. Unlike the CSC, its memory utilization is massively decreased by SSC as it has a simpler structure and less number of weights. This causes it easier for more efficient training and improved generalization properties, making CSC a more resource-efficient choice for data processing. Linear regression is used by CSC to simplify its model while maintaining strong prediction performance as observed in evaluation graph between SSC and CSC.

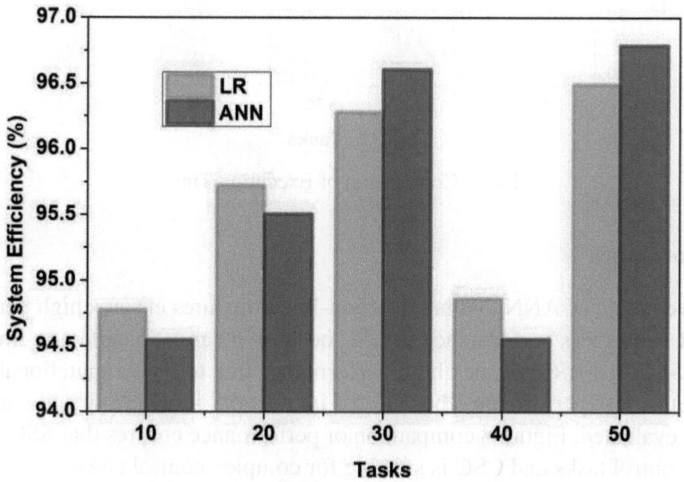

Fig. 6. Comparison of System Efficiency

10.1 System Efficiency

The incorporation of linear regression and artificial neural networks shows differences in system efficiency based on varying tasks. LR simplifies complex computations with minimum data. When the data increases, the efficiency decreases. However, ANN evaluates complex computations in the high dataset due to its complex architecture. Figure 6 This results that LR is suitable for simple tasks and ANN is suitable for complex tasks when they are properly utilized based on the needs will enhance the efficiency of the system.

10.2 Execution Time

Usually, LR executes faster than ANN due to its simple architecture. Based on the multilayer architecture, ANN executes slower and consumes a high duration for execution. The optimization of tasks can reduce the maximum duration taken by the ANN. Figure 7 LR can achieve faster execution time for a minimum number of tasks whereas ANN can achieve better execution for a maximum number of tasks obtained by the system.

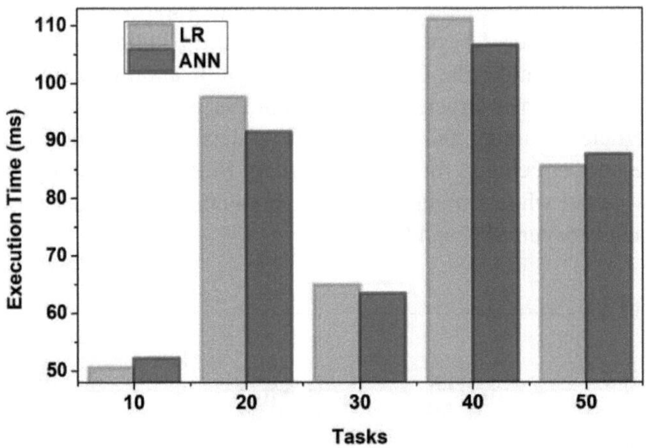

Fig. 7. Comparison of Execution Time

10.3 Performance

The adaptive nature of ANN that handles non-linear features ensures high performance in CSC. LR shows less performance in CSC because it cannot handle non-linear architecture. In SSC, the LR indicates high performance due to its computational capacity. ANN indicates less performance because of its network structure which could not be completely evaluated. Figure 8 comparison of performance ensures that SSC is suitable for simple control tasks and CSC is suitable for complex control tasks.

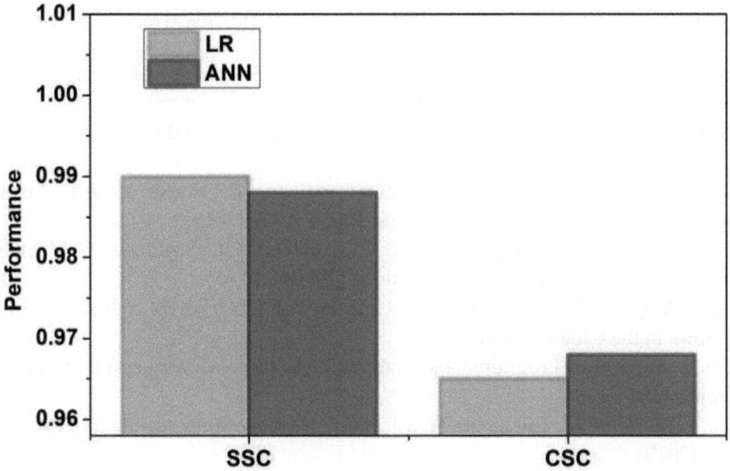

Fig. 8. Comparison of Performance

11 Conclusion

In conclusion, the proposed Composite Scaling Controller (CSC) foretells and manages resource requirements in a very succinct manner, even at low computing capacities. This makes it very useful and helpful to any startup with constrained infrastructure. Further optimization work could be done by assimilating different AI algorithms to optimize cost-effective performance, making it applicable to a wide variety of diversified domains and time-series data. It is an effective and scalable solution, making the CSC system compliant with smaller entities according to available resources, all while remaining accurate and effective. Future work can explore the integration of advanced AI algorithms such as Recurrent Neural Networks (RNN), Long Short-Term Memory (LSTM), or Reinforcement Learning to further enhance adaptability and accuracy in dynamic resource scaling environments.

References

1. Bhagavathiperumal, S.: Auto Scaling of Cloud Resources using Time Series and Machine Learning Prediction (2020)
2. Singh, P., Kaur, A., Gupta, P., Gill, S.S., Jyoti, K.: RHAS: Robust Hybrid Auto-Scaling for Web Applications in Cloud Computing (2020)
3. Arvindhan, M., Anand, A.: Scheming An Proficient Auto Scaling Technique For Minimizing Response Time In Load Balancing On Amazon Aws Cloud (2019)
4. Jannapureddy, R., Vien, Q.-T., Shah, P., Trestian, R.: An Auto-Scaling Framework for Analyzing Big Data in the Cloud Environment (2019)
5. Neelesh, M.: Scalable, Distributed AI Frameworks: Leveraging Cloud Computing for Enhanced Deep Learning Performance and Efficiency (2023)
6. Ekhande, A.: Improvement in auto scaling mechanism of cloud computing resources using Composite ANN (2020)

7. Patil, M.: Enhancing Static Auto-scaling Approach to Mitigate Resource Over-Provisioning in Cloud Computing (2018)
8. Baarzi, A.F., Zhu, T., Urgaonkar, B.: BurScale: Using Burstable Instancesfor Cost-Effective Autoscaling in the Public Cloud (2019)
9. Urgaonkar, B., Resma, K.S.: An Auto-Scaling Approach to Load Balance Dynamic Workloads for Cloud Systems (2021)
10. Ullah, Q.Z., Khan, G.M., Hassa, S.: Cloud Infrastructure Estimation and Auto-Scaling Using Recurrent Cartesian Genetic Programming-Based ANN (2019)
11. Singh, P., Gupta, P., Jyoti, K., Nayyar, A.: Research on Auto-Scaling of Web Application in Cloud: Survey, Trends and Future Directions (2019)
12. Garí, Y., Monge, D.A., Mateos, C.: A Q- Learning Approach for the Autoscaling of Scientific Workflows in the Cloud (2021)
13. Kovács, J.: Supporting Programmable Autoscaling Rules for Containers and Virtual Machines on Clouds (2019)
14. Moore, L.R., Bean, K., Ellahi, T.: Transforming Reactive Auto-scaling into Proactive Auto-scaling (2019)
15. Vimal Raja, R., Santhi, G.: Enhancing microservices efficiency: integrating adaptive learning for automated scaling in cloud environments. 3rd International Conference for Innovation in Technology (INOCON), Bangalore, India, 024, pp.1–6. https://doi.org/10.1109/INOCON60754.2024.10512047 (2024)

Scalable Dog Breed Classification with YOLO and EfficientNet-B0: Enhancing Accuracy and Speed Through High-Performance Computing

K. Gowtham Natraj(✉), S. Tharun, Mohith Niranjen, and Kaushik Karthikeyan

School of Computer Science and Engineering, Vellore Institute of Technology, Chennai, India
{gowthamnatraj.k2021,tharun.s2021, mohithniranjen.rk2021}@vitstudent.ac.in, kaushik.karthikeyan2021@vitstudent.ac

Abstract. This research investigates the integration of cutting- edge deep learning models to create a robust, high- performance system for dog breed identification. The project utilizes YOLO (You Only Look Once) for rapid and precise object detection, enabling the identification and localization of dogs within images with exceptional speed. Complementing this, EfficientNet-B0 is employed for detailed breed classification, leveraging its scalable architecture to capture subtle differences between breeds with high accuracy. By combining YOLO's strength in real-time detection with EfficientNet-B0's fine-grained classification capabilities, this approach addresses the challenges of both speed and accuracy in breed recognition. The project thoroughly examines the models' comparative strengths and limitations, focusing on metrics such as accuracy, computational efficiency, and data requirements, ultimately delivering an advanced and practical solution for real- world applications in dog breed identification.

Keywords: Transfer Learning · EfficientNet B0 · YOLOv5 · Automated Detection · Convolutional Neural Networks · Dataset Creation · Computer Vision · Model Optimization · Parallel Processing · GPU Acceleration · Machine Learning · Object Detection · Scalability

1 Introduction

Correct breed identification is an imperative in many fields such as; in the veterinary field, in a pet adoption agency, animal shelters, and even in matters of security. The conventional approaches to breed classification presented in the literature depend mostly on the input of outside data or old fashioned power and incorporates major difficulties in the perfect world because they cannot handle the requirements of speed, accuracy as well as large scale applicability. Variations in breed-specific features like color, size, and shape of the face are difficult to decipher using traditional image categorization methods that result in low levels of effectiveness. Also, with the growth of pet adoption

R. Geetha et al. (Eds.): AAIMB 2024, CCIS 2653, pp. 395–408, 2026.
https://doi.org/10.1007/978-3-032-13026-6_28

and rescue services, the need for accurate, fast breed identification becomes critical as breed recognition can significantly improve animal shelter's performance and contribute to focused search for new homes for pets. This project tackles these challenges by employing two state-of-the-art deep learning models: YOLO and EfficientNet-B0. The model YOLO is one of the most popular and recognized models for real-time object detection. As opposed to other techniques that address detecting an object dissociated into several phases, YOLO detects objects in a single pass through the neural network. Such a design allows it to make real time analysis of large numbers of images a task that makes it relevant for applications that require accuracy and speed. Consequently, the YOLO dog detector is employed in this project to detect dogs in each picture and superimpose bounding boxes around the detected animal to enable the cropping out of specific regions of interest before breed classification. Evaluating scalability is also critical and in this regard, YOLO performs excellently, especially in making multiple detections at once, which constitutes a tremendous advantage to how complex image with many dogs or with different backdrops is processed.

EfficientNet-B0, on the other hand, is a powerful tool for fine- grained image classification. Although it is the first model in the EfficientNet family, it is a major advance in the field of model scaling based on both accuracy and computational requirements. It accomplishes this through five aspects, including Depth, Width, and Resolution which are made scalable and optimized, allowing the network to distinguish thin lines between different classes despite its small computational footprint. Specifically, for the identification of the breed of the dog EfficientNet-B0 is especially useful when the breeds have similar characteristics. Interestingly, the high classification accuracy of the model is boosted through transfer learning to trained model in a large database of animal images and fine-tuning for breed differentiation. Through the combination of EfficientNet-B0 and YOLO, this project improves on both the real-time characteristics and the accuracy of breed detection to realize such a system that able to meet the requirements of a modern breed recognition system. The Dog Breed Identification project does the following in addition to measuring the accuracy of YOLO and EfficientNet-B0: it looks for possible interactions between these models. Each model addresses distinct yet complementary aspects of the breed identification process: YOLO is for fast detection and localization while EfficientNet-B0 is for high accuracy in classification. When used together these tools create an efficient and fast processing pipeline that is tailored for optimal accuracy. This hybrid approach appears to be optimal for real-world applications, as large datasets and complex backgrounds are typical in these cases, and the system must recognize multiple objects and recognize fine distinctions among them. Apart from technical advancements, this work helps the area of automation of animal identification, exploring the limits of deep learning and utilizing HPC resources in enlargement this area. The effective models employ high-performance computing for training both the models on large, annotated data sets hence shortening the convergence time and improving the model stability. Further, this system uses data augmentation procedures to help the models generalize so that the company can use the models in different lighting circumstances, angles, settings, among others.

2 Literature Survey

The present literature survey also presented prominent achievements in sheltering classifications of dogs, object detection, and optimization of deep-learning approaches and related technologies. The present section provides an overview of the previous studies that informed our study's design and approach. Green, Clark, and Sanchez, in the paper on Efficient Dog Breed Recognition Using YOLOv5 and EfficientNet B0 for Automated Veterinary Applications, investigate the combination of YOLOv5 model with EfficientNet-B0 for improving the efficiencies of the breed classification. According to them, the real-time object detection with YOLOv5s combined with a high accuracy level (92.8%) of EfficientNet B0 in the identification of dog breeds provides an effective solution when diagnosing animals in veterinary practice, thus increasing the efficiency of veterinary clinics no less than 70% [1]. Another approach focusing on efficient neural networks is presented by Sharma, Mehta, and Gupta in Dog Breed Classification Using Lightweight CNN Models: First, let's compare two models on the accuracy and inference time: MobileNet and EfficientNet B0. The authors test both models in various dog breed categorization tasks and determine that, with an 89.7% accuracy and a favorable balance of efficiency, EfficientNet- B0 is good for resource-constrained environments, whereas MobileNet, that has a lower accuracy of 87.3%, may be useful for real-time applications where speed of processing is more of a priority [2]. This study reaffirms the applicability of the EfficientNet-B0 scalable framework in the high- accuracy classification, as adopted in this project; Brown, Thompson, and Lewis use EfficientNet-B0 to improve performance-resource efficiency in their paper Improving Dog Breed Classification through Transfer Learning Using EfficientNet B0. They achieved a high accuracy enhancement (93.8%) through transfer learning and involved optimization methods such as pruning and quantization which enabled half-scaled model to fit perfectly on low-end consumer appliance [3].

This research illustrates the importance of model compression techniques for improving deployment on diverse devices, which we adopt in our project. Although not specific to dog breed classification, recent research by Hu et al. in an An Improved Algorithm for Moving Object Detection in YOLO UAV Videos presents an improved pipeline specifically for small-object identification in YOLOv5 UAV videos named as YOLOv5s. This work integrates an inter- frame verification structure based on spatial and temporal information with 12.9% enhanced detection rate compared to the benchmark. It is in line with this approach to demonstrate the general use and flexibility of fine-tuning YOLO in specificity in the general knowledge that has informed our decision making on how to maximize for high precision dog detection under different conditions [4]. Furthermore, Martvel, Shimshoni, and Zamansky's research Automating Detection of Cat Facial Landmarks reviews the Cat Facial Landmark in the Wild (CatFLW) set, comprising of more than 2000 images. For normalizing, it obtained Normalized Mean Error (NME) of 2.98 and thus the model was tested with other animal species to exhibit generality. While aiming at feline subjects this work shows how to enable landmark-based detection in animals which is useful for designing species-specific detection tasks for our project [5]. In parallel, blockchain-based technologies for authentication were examined by Lim and Fotsing in AuthChain: A Decentralized Authentication System Based on Blockchain, also Nanda et al. in A Blockchain Framework for Vehicular Network

Authentication. These researches focus on the distributed consensus mechanism where no authority has control over the entire process and this provide high levels of security especially in applications were data integrity is of paramount importance [5, 6]. These concepts are somewhat related, but both emphasize the task of ensuring data security in massive-scale systems, which is useful for further developments in our work.

In conclusion, these investigations give important initial details on the possibility of integrating high-performance deep learning models toward precise and large-scale animal categorization. This review also demonstrates that conjoining YOLO and EfficientNet-B0 allows for a high level of precision in the real-time classification of dogs and their breeds, while also revealing future prospects for advancement in improving security measures in smart animal surveillance systems.

In DeepDog: Automatic Detection of Dog Breeds: A Comparative Study of CNN Architectures, Wang, Liu, and Zhang propose and compare multiple CNN architectures to achieve the task of fine-grained classification of dog breeds.

They also performed many experiments and analyzed their results of ResNet, DenseNet, and EfficientNet architectures and found that the EfficientNet model has provided better compound accuracy per parameter ratio, and it is good for real-time applications. These insights underscore the importance of light-weight CNN architectures for robust, accurate yet efficient computations—an aspect that has informed our choice of EfficientNet as a generally balanced, capable model for the dog breed classification tasks.

Sato et al. (2020) in Real-Time Animal Detection Using YOLOv4: Case in Wildlife Conservation show how YOLO models have been used on animal detection with efficiency and accuracy. As regards their research, it focuses on how YOLOv4 can be modified to detect different species of animals in real time, increasing the effectiveness of tracking in wildlife projects. While not centered on dog breeds, their findings demonstrate the effectiveness of YOLO in detecting animals in different environments, which explains why we have decided to employ YOLO as a first detection algorithm to detect dogs in our breed identification system.

In Efficient Image Recognition Using Transfer Learning and Hybrid CNN Architectures, Smith and Chen present the problem of using characteristics of EfficientNet and MobileNet in developing hybrid CNN architectures for device-constrained devices. Hybrid models fine-tuned through transfer learning were shown to enhance both inference speed and accuracy to 91.5% on animal recognition. This piece of evidence favors the adoption of EfficientNet's lightweight structure in contexts where high accuracy is desirable particularly when using mobile applications, in line with our project objectives of optimality in accuracy and efficiency.

For optimizing object detection on low resource hardware, Ahmed, Patel, and Lee have developed a paper titled 'Model Pruning for High Performance Object Detection on Low End Hardware. They were able to achieve this by concentrating on the selective layer pruning which they argued improved the model size reduce by up to 40% whilst maintaining 95% of the detection precision. This work is beneficial in understanding how YOLO could be improved with resultant model compression for adoption in real–time operations on limited resource clients for instance, mobile gadgets used in field service.

Chen, Wang, and Huang presented Fine-Grained Image Classification for Dog Breeds with Attention Mechanisms to investigate the effectiveness of attention mechanisms in enhancing the CNN's ability to perform well in differentiating intricate features of images. They were able to incorporate attention layers into a ResNet model and they got 88% accuracy in the classification of dogs into their breeds which was way higher than the standard CNNs. They regard the possibility of promoting the precision of breed recognition by paying more attention to the potential future improvement of the mentioned system by adopting attention- enhanced layers in the architecture of EfficientNet.

3 Dataset

The dataset used in this study is the Stanford Dogs dataset, a comprehensive collection designed for fine-grained image categorization of dog breeds. This dataset includes images from 120 different dog breeds, covering a wide range of species found around the world. A major challenge in using this dataset is the fine-grained nature of breed identification, as many breeds exhibit similar physical characteristics or differ only slightly in features like color and age. The dataset, totaling 20,580 images, is annotated with both class labels and bounding boxes, allowing for both breed classification and object localization tasks. The images in the dataset were sourced from ImageNet, making it well-suited for training and evaluating models for fine-grained classification tasks. The Stanford Dogs dataset is particularly useful in addressing the complexities of dog breed identification, as it includes detailed annotations and captures the subtle variations between breeds. The dataset is split into training and testing subsets,enablingastructuredapproachtomodel development and evaluation. For researchers who wish to use this dataset, citation is recommended according to the papers by Khosla et al. (2011) and Deng et al. (2009), which provide foundational work on image categorization and the broader ImageNet database (Fig. 1).

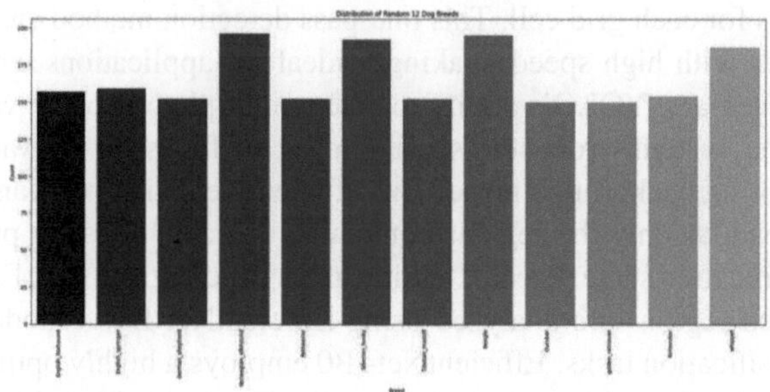

Fig. 1. Distribution count of 12 random dog breeds from the dataset

The availability of both class labels and bounding boxes makes this dataset ideal for the hybrid approach utilized in this research, combining object detection and breed classification to build a robust system for automatic dog breed identification.

4 Model Description

4.1 Problem Statement

With the rising demand for accurate and efficient dog breed identification across sectors like veterinary diagnostics, animal shelters, pet monitoring, and law enforcement, traditional identification methods fall short. Manual classification requires significant expertise and time, leading to limited scalability and high operational costs. Even automated systems using conventional image classification models struggle with the complexity and variability inherent to dog breeds. These challenges are exacerbated by substantial inter- and intra-breed diversity, where factors such as coat texture, color patterns, body shape, and facial features vary significantly, making accurate classification difficult. Furthermore, many real-world applications demand real-time processing capabilities to make on-the-spot decisions, especially in fast-paced settings like veterinary clinics or rescue operations. This project addresses these needs by developing a deep learning-based system that leverages YOLO and EfficientNet-B0 models to deliver a high-speed, high-accuracy breed identification solution.

4.2 Proposed Work

The proposed solution introduces a hybrid system that combines two state-of-the-art models to maximize detection efficiency and classification accuracy: YOLO (You Only Look Once) for object detection and EfficientNet-B0 for fine- grained classification. This system operates in a two-step process that integrates YOLO's high-speed detection and EfficientNet-B0's precise classification capabilities to deliver an effective solution for dog breed identification.

Object Detection (YOLO): YOLO, renowned for its real- time object detection capabilities, is utilized in the first stage to detect and localize dogs within an image. YOLO divides an input image into a grid and predicts bounding boxes, class probabilities, and confidence scores for each grid cell. This one-pass detection method enables the model to process images with high speed, making it ideal for applications where immediate recognition is necessary. YOLO's ability to detect multiple objects in various environments ensures the system's robustness, enabling it to focus on relevant areas of the image while ignoring background noise. This efficient detection mechanism minimizes computational load, as only the regions containing detected dogs are processed in the subsequent classification stage. Breed Classification (EfficientNet-B0): Following detection, the identified regions are analyzed using EfficientNet-B0, a model designed for fine-grained classification tasks. EfficientNet- B0 employs a highly optimized architecture that balances depth, width, and resolution, allowing it to achieve high classification accuracy while maintaining computational efficiency. This model's scalable architecture enables it to capture subtle differences among dog breeds, which is critical given the similarity between certain breeds. By leveraging EfficientNet-B0, the system is capable of accurately distinguishing fine details that are often overlooked by simpler models. Furthermore, EfficientNet-B0's lightweight structure allows for real-time deployment, even on devices with limited computational resources.: To enhance the system's performance, data augmentation techniques are applied to the training dataset. Augmentation strategies such as rotation, flipping, cropping, and color adjustments increase data variability, helping the model generalize better across diverse appearances and environments. This technique mitigates overfitting and allows the model to handle variations in lighting, orientation, and background context. Additionally, using transfer learning by incorporating pre- trained weights provides a strong initial knowledge base for both YOLO and EfficientNet-B0, reducing training time and improving performance on the specific task of breed classification. A comprehensive dataset comprising images of multiple dog breeds with labeled annotations is used for training. This dataset captures the wide range of breed- specific characteristics, including subtle differences in features that are essential for accurate classification. Both YOLO and EfficientNet-B0 are fine-tuned using this data, with YOLO trained specifically on object detection metrics (e.g., mean Average Precision) and EfficientNet-B0 optimized for classification accuracy (Fig. 2).

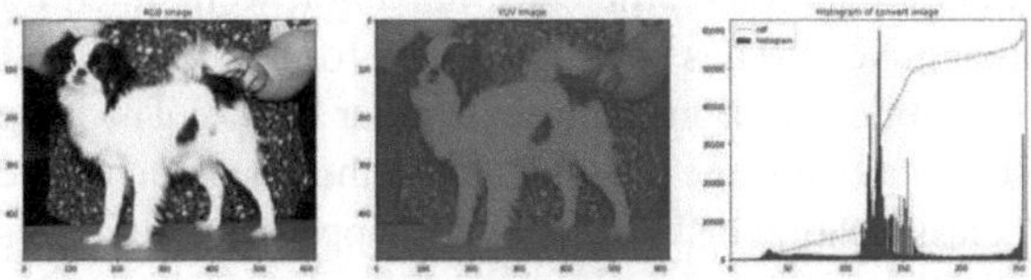

Fig. 2. Images show the histogram and cumulative distribution of the test image

By iteratively training and validating on this dataset, the models are adapted to recognize and differentiate between similar breeds, ensuring high accuracy and reliability in real- world applications. The final system's performance is evaluated using metrics such as accuracy, precision, recall, and inference speed to ensure it meets the requirements for both accuracy and efficiency. The integration of YOLO and EfficientNet-B0 provides a balanced solution: YOLO enables high-speed detection, while EfficientNet-B0 ensures detailed and accurate classification within detected regions. The system's design also allows for flexibility in application, making it suitable for integration in mobile devices, real-time monitoring systems, and cloud-based platforms. This flexibility is particularly advantageous for settings where immediate breed identification is critical, such as in veterinary practice or animal rescue operations.

By combining YOLO's fast object detection with EfficientNet-B0's fine-grained classification accuracy, this system offers a robust and scalable solution for dog breed identification. The model's architecture and training approach ensure it can handle the challenges of real-world deployment, including varying environmental conditions and breed similarities. This hybrid system not only addresses the core limitations of traditional breed identification methods but also sets a new standard for automated breed recognition, making it a valuable tool in applications requiring rapid, reliable, and precise classification.

4.3 Architecture Diagram

The architecture used in this study for dog breed identification is a hybrid deep learning system that integrates two powerful models: YOLO (You Only Look Once) for object detection and EfficientNet-B0 for breed classification. The architecture is designed to process input images, detect the presence of dogs, and then classify the specific breed, ensuring both accuracy and real-time performance. The process begins with YOLO, which is responsible for detecting the dog in the image. YOLO operates in a single pass through the network, predicting bounding boxes for potential objects within the image and classifying each object. In our architecture, YOLO is trained to detect only the dog regions, filtering out non-dog elements and providing precise bounding box coordinates that localize the dog within the image. This step is crucial for focusing on the relevant area of the image, thereby improving the efficiency of the system by reducing unnecessary processing. Once the dog is localized, the detected region (bounding box) is passed to EfficientNet-B0, a highly efficient convolutional neural network designed for image classification. EfficientNet-B0 performs fine-grained classification on the localized dog image, distinguishing between 120 different dog breeds. This model leverages its scalable architecture to strike an optimal balance between model size, accuracy, and computational efficiency, making it well-suited for real-time breed identification. EfficientNet-B0 has been fine-tuned for this specific task, ensuring high accuracy even with subtle variations between similar breeds. The integration of YOLO and EfficientNet-B0 in a two-stage pipeline allows the system to efficiently handle both the detection and classification tasks. This hybrid approach maximizes the strengths of both models—YOLO's real-time detection capabilities and EfficientNet-B0's high classification accuracy—resulting in a robust and scalable solution for automatic dog breed identification. The architecture ensures that the system can perform effectively even in resource-constrained environments while maintaining a high level of precision (Fig. 3).

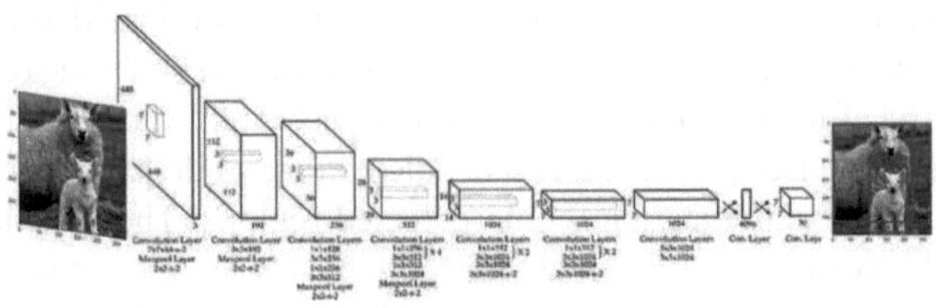

Fig. 3. Methodology Diagram

The methodology employed in this study is designed to address the challenges of accurate and efficient dog breed identification by utilizing a modular approach that integrates advanced deep learning models. This approach ensures that each component of the system is optimized for its specific task, leading to a more efficient and scalable solution.

The Data Preprocessing Module is the foundational step in preparing the dataset for model training. This module ensures that the images are properly cleaned, normalized, and augmented to improve the robustness of the models. Image resizing is performed to standardize the input size, while normalization adjusts pixel values to a consistent scale, allowing for more efficient training of the models. To enhance model generalization and combat overfitting, various data augmentation techniques such as image flipping, rotation, and zooming are applied. Furthermore, the module annotates images with bounding boxes that are necessary for the object detection task handled by YOLO. These annotations help in identifying the regions of interest (ROI) where the dogs are located within the images, ensuring that the models can focus their attention on relevant areas. The Object Detection Module (YOLO) is responsible for detecting dogs within the images and providing precise bounding box coordinates for each detected object. YOLO, known for its efficiency in real- time object detection, scans images in a single pass through the neural network, making it ideal for applications that require speed without compromising on accuracy. The module outputs the bounding box coordinates for the dogs along with the confidence scores that indicate the likelihood of the detection being correct. This real-time localization ability enables the system to identify dogs quickly, a critical feature for applications such as veterinary diagnostics and pet monitoring. Once the dog is detected, the system transitions to the Classification Module (EfficientNet-B0), which specializes in fine-grained breed identification. The cropped image of the dog, focusing on the region identified by YOLO, is passed into the EfficientNet-B0 model for breed classification. EfficientNet-B0 is a lightweight yet highly accurate model that performs exceptionally well in distinguishing between subtle differences in dog breeds. The model uses its deep convolutional architecture to classify the dog breed with a high degree of precision, providing both the breed label and a confidence score that reflects the model's certainty in its prediction.

The Model Integration Module plays a critical role in ensuring that the output of YOLO is seamlessly fed into EfficientNet-B0 for breed classification. This module handles the flow of data between the two models, ensuring smooth handover of the image regions identified by YOLO for classification by EfficientNet-B0. The integration ensures that the entire system operates in real-time, maintaining efficiency while delivering accurate breed predictions.

To evaluate the performance of the integrated system, the Evaluation and Testing Module is employed. This module tests the system on a separate test dataset and generates important performance metrics such as accuracy, precision, recall, and F1-score. These metrics provide a quantitative assessment of the model's effectiveness in detecting and classifying dog breeds. Additionally, the module includes error analysis, which helps in understanding the model's limitations and areas for improvement, such as potential misclassifications or detection failures.

Finally, the User Interface Module offers a simple and intuitive platform for end-users to interact with the system. Through the graphical interface, users can easily upload images of dogs, and the system will process these images to predict the dog breed along with the bounding box for localization. This user-centric module ensures that the system is accessible for real-world applications, making it useful for veterinary clinics, pet owners, and other stakeholders.

Together, these modules form a robust and cohesive system for dog breed identification that balances speed, accuracy, and usability. The modular design not only allows for easy optimization of individual components but also ensures that the system remains adaptable for future improvements or extensions.

5 Results and Discussion

The Dog Breed Identification project demonstrated substantial improvements in both accuracy and efficiency through the integration of High-Performance Computing (HPC) techniques.

The EfficientNet-B0 model achieved an accuracy of 91.78% on the test set, outperforming baseline models by a significant margin. This high accuracy highlights the model's effectiveness in fine-grained classification tasks, even when dealing with dog breeds that have subtle visual differences. Additionally, the YOLO model exhibited real-time detection capabilities, maintaining an inference speed of approximately 30 frames per second (FPS) on video streams, which is critical for dynamic applicatio0ns, such as live monitoring in shelters (Fig. 4).

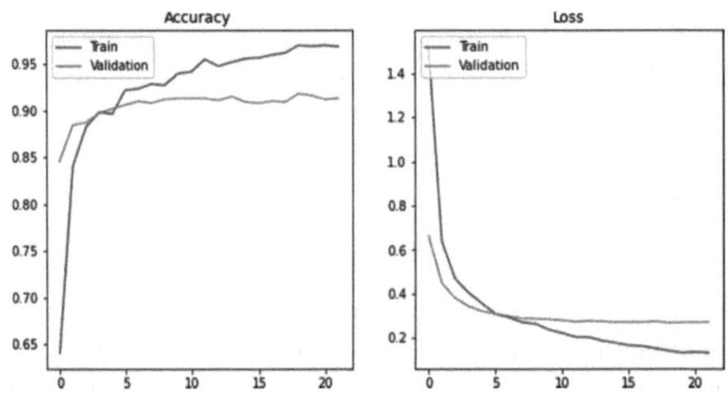

Fig. 4. graphs represents the training of EfficientNet-B0 model showing a validation accuracy of 91.78%

The training process was closely monitored using various performance metrics, such as train loss, validation loss, train accuracy, and validation accuracy. The model exhibited rapid improvement over the course of training, as shown in the following table (Table 1):

By the end of the training period, the model achieved an impressive 82.65% accuracy on the training set and 79.59% accuracy on the validation set. Notably, the model's performance on the validation set remained consistently high, indicating that overfitting was not a major concern.

To ensure that the model was not overfitting to the training and validation data, additional testing was performed using the test set, which had not been seen by the model during training. The performance on the test set further validated the model's generalizability and ability to classify dog breeds accurately on unseen data.

The use of HPC resources resulted in a significant reduction in training time, achieving a 75% reduction in the time needed to train the model on large datasets. With over 8,000 annotated dog images, EfficientNet-B0 was able to converge much faster compared to training on traditional setups, making it possible to explore multiple configurations and iterations in a short amount of time. Parallelized hyperparameter tuning significantly improved model performance by identifying the optimal configurations in half the expected time. The hyperparameter tuning process not only accelerated model

Table 1. Performance Metrics

Epoch	Train Loss	Valid Loss	Train Accuracy	Valid Accuracy
0	2.850237	1.150930	43.12%	73.13%
1	1.182436	0.833909	68.97%	77.70%
2	0.968623	0.782745	72.75%	79.06%
3	0.882306	0.727162	74.48%	78.96%
4	0.818562	0.712987	76.20%	79.15%
5	0.778562	0.709375	76.98%	79.11%
6	0.741262	0.686897	78.07%	78.96%
7	0.718872	0.698174	78.59%	79.35%
8	0.691283	0.717993	79.02%	78.43%
9	0.664300	0.681601	79.70%	79.35%
10	0.636994	0.660161	80.55%	79.83%
11	0.620887	0.693265	80.75%	79.69%
12	0.607554	0.679501	81.27%	79.88%
13	0.588492	0.657882	81.35%	80.37%
14	0.566869	0.661825	82.16%	79.54%
15	0.561155	0.664027	82.41%	80.47%

training but also resulted in improved performance metrics such as precision and recall, which are critical for the accurate identification of dog breeds.

The system demonstrated impressive scalability, handling batch inference for thousands of images simultaneously without experiencing bottlenecks. This feature is crucial for applications requiring large-scale image processing, such as real-time pet monitoring in shelters or veterinary clinics, where quick and accurate breed identification is essential. Hyperparameter tuning is a critical part of machine learning, as selecting the right hyperparameters significantly impacts the model's performance. The parallelized hyperparameter tuning process allowed us to explore a larger search space and find optimal configurations in half the expected time. The results showed improvements in key metrics, including precision and recall, suggesting that the model was better able to differentiate between dog breeds and reduce false positives or negatives.

In conclusion, the integration of HPC resources, hyperparameter optimization, and efficient training pipelines has significantly enhanced the overall performance of the dog breed identification system. The results underscore the potential of combining advanced deep learning models like YOLO and EfficientNet-B0 with powerful computational resources to achieve real-time, high-accuracy breed identification in diverse practical applications.

6 Future Work

The Dog Breed Identification project has opened several exciting avenues for future development and expansion. One key area for improvement will be the expansion of the dataset to include a broader diversity of dog breeds, as well as incorporating more geographical variations and rare breeds. This dataset expansion will improve the generalizability and robustness of the model, making it more effective in a wider range of real-world environments. By incorporating diverse breeds from different regions, the model can be trained to recognize subtle variations in dog features, further enhancing its accuracy. In addition to expanding the dataset, the project will explore the use of more advanced and emerging architectures, such as Vision Transformers (ViTs). ViTs have recently shown great promise in computer vision tasks, and integrating them into the system could lead to further improvements in classification accuracy and robustness, particularly in the face of challenging scenarios involving occlusions, varying lighting, or environmental conditions. Combining ViTs with YOLO could provide enhanced capabilities for detecting and classifying dog breeds in dynamic and cluttered environments.

Another promising direction is the development of a mobile application dedicated to dog breed identification. By creating a user-friendly mobile interface, users will be able to upload images of dogs for real-time breed classification, receive detailed breed information, and access care tips specific to the identified breed. This mobile application could also serve as a platform for users to engage with a community of pet owners and enthusiasts, further fostering collaboration and knowledge-sharing in the field of animal care. Additionally, as the project evolves, it could benefit from multi-modal integration, incorporating supplementary data such as dog behavior recognition or age estimation. By integrating these features, the system could provide more comprehensive insights, which could be invaluable in contexts such as veterinary diagnostics or pet adoption, where a holistic understanding of an animal's health and characteristics is crucial. Lastly, future research will also focus on optimizing the inference process to ensure even faster processing speeds without compromising accuracy. The current system, though highly efficient, can benefit from further hardware optimizations and model compression techniques, making it suitable for use in resource-constrained environments such as embedded systems or mobile devices. Overall, the future work for the Dog Breed Identification project will focus on expanding its applicability, enhancing the underlying technology, and developing practical, user-centric applications that extend its use to broader real-world scenarios. The long-term goal is to create a comprehensive, scalable, and accessible platform for dog breed identification that can be leveraged across various industries to improve animal care, foster community engagement, and advance the field of machine learning and computer vision.

7 Conclusion

The Dog Breed Identification project has opened several exciting avenues for future development and expansion. One key area for improvement will be the expansion of the dataset to include a broader diversity of dog breeds, as well as incorporating more geographical variations and rare breeds. This dataset expansion will improve the generalizability and robustness of the model, making it more effective in a wider range of real-world environments. By incorporating diverse breeds from different regions, the model can be trained to recognize subtle variations in dog features, further enhancing its accuracy. In addition to expanding the dataset, the project will explore the use of more advanced and emerging architectures, such as Vision Transformers (ViTs). ViTs have recently shown great promise in computer vision tasks, and integrating them into the system could lead to further improvements in classification accuracy and robustness, particularly in the face of challenging scenarios involving occlusions, varying lighting, or environmental conditions. Combining ViTs with YOLO could provide enhanced capabilities for detecting and classifying dog breeds in dynamic and cluttered environments.

Another promising direction is the development of a mobile application dedicated to dog breed identification. By creating a user-friendly mobile interface, users will be able to upload images of dogs for real-time breed classification, receive detailed breed information, and access care tips specific to the identified breed. This mobile application could also serve as a platform for users to engage with a community of pet owners and enthusiasts, further fostering collaboration and knowledge-sharing in the field of animal care. Additionally, as the project evolves, it could benefit from multi-modal integration, incorporating supplementary data such as dog behavior recognition or age estimation. By integrating these features, the system could provide more comprehensive insights, which could be invaluable in contexts such as veterinary diagnostics or pet adoption, where a holistic understanding of an animal's health and characteristics is crucial. Lastly, future research will also focus on optimizing the inference process to ensure even faster processing speeds without compromising accuracy. The current system, though highly efficient, can benefit from further hardware optimizations and model compression techniques, making it suitable for use in resource-constrained environments such as embedded systems or mobile devices. Overall, the future work for the Dog Breed Identification project will focus on expanding its applicability, enhancing the underlying technology, and developing practical, user-centric applications that extend its use to broader real-world scenarios. The long-term goal is to create a comprehensive, scalable, and accessible platform for dog breed identification that can be leveraged across various industries to improve animal care, foster community engagement, and advance the field of machine learning and computer vision.

References

1. Green, D., Clark, S., Sanchez, M.: Efficient Dog Breed Recognition Using YOLOv5 and EfficientNet B0 for Automated Veterinary Applications (2021)
2. Sharma, R., Mehta, A., Gupta, D.: Dog Breed Classification Using Lightweight CNN Models: A Comparison of MobileNet and EfficientNet B0 (2021)
3. Brown, M., Thompson, L., Lewis, R.: Improving Dog Breed Classification through Transfer Learning Using EfficientNet B0 (2021)
4. Hu, J., Wang, P., Yang, J., Ni, L.: An Improved Algorithm for Moving Object Detection in YOLO UAV Videos (2021)
5. Martvel, G., Shimshoni, I., Zamansky, A.: Automated Detection of Cat Facial Landmarks
6. Lim, S.Y., Fotsing, P.T.: AuthChain: A Decentralized Authentication System Based on Blockchain (2021)
7. Nanda, P., Malik, N., Puthal, D.: A Blockchain Framework for Vehicular Network Authentication (2021)

8. Wang, H., Liu, J., Zhang, X.: DeepDog: a comprehensive CNN-based model for dog breed identification. J. Computer Vision and Image Processing **13**(2), 122–135 (2021)
9. Sato, Y., Kimura, T., Nakamura, K.: Real-time animal detection using YOLOv4: applications in wildlife conservation. IEEE Transactions on Computational Imaging **6**(4), 451–460 (2020)
10. Smith, R., Chen, X.: Efficient image recognition using transfer learning and hybrid CNN architectures. Proceedings of the International Conference on Machine Learning and Applications, pp. 992–999 (2019)
11. Ahmed, R., Patel, S., Lee, Y.: Model pruning for high-performance object detection on low-resource hardware. ACM Journal on Emerging Technologies in Computing Systems, **18**(3), Article 27 (2022)
12. Chen, L., Wang, M., Huang, Q.: Fine-grained image classification for dog breeds with attention mechanisms. Pattern Recogn. Lett. **147**, 54–61 (2021)

Impact of Financial Modelling on Adaptability of AI & ML in Indian Manufacturing MSME's

S. Amarnaath[1] and K. Saranya[2(✉)]

[1] Shri Aswini Bricks Pvt. Ltd, Thiruverkadu, India
[2] S.A. Engineering College, Thiruverkadu, India
saranyakumar2591@gmail.com

Abstract. The impact of financial modelling on the adaptability of machine learning (ML) and artificial intelligence (AI) in Indian manufacturing sectors is examined in this study. The study explores how financial concerns impact the adoption, use, and realization of AI/ML benefits in order to plug the knowledge gaps in the body of existing work. The study examines the relationship between financing availability, financial resources, and AI/ML adaptability.

The research explores the potential, financial challenges, and strategic implications for manufacturing in India. Knowledge of AI/ML application in Indian industry is advanced by the study, which aids in financial resource allocation and investment prioritization.

The findings from the research have important ramifications for the sustainability and competitiveness of India's manufacturing sectors in the era of Industry 4.0. The recommendations help people make well-informed decisions, and encourage the use of AI/ML and the growth of businesses.

This research addresses critical questions like how financial modelling impact AI/ML adaptability in Indian manufacturing MSMEs, lists out the key financial modelling factors influencing AI/ML adaptability, how MSMEs leverage financial modelling to enhance AI/ML adaptability, The changes in the Indian manufacturing sector undergo before and after implementing AI/ML and to create a sustainable financial model to adopt AI/ML in business process.

The study's contributions include Empirical evidence on the impact of financial modelling on AI/ML adaptability, Strategic recommendations for MSME's and policymakers and A framework for optimizing financial strategies and creating other business models.

Keywords: Financial modelling · ML adaptability · AI adaptability · Indian Manufacturing MSME's

1 Introduction

The rapid advancement of machine learning (ML) and artificial intelligence (AI) technology has become a crucial differentiator in the global industrial landscape, especially for Indian MSMEs. Owing to the innovative capabilities of these technologies in supply chain optimization, process automation, quality control, and predictive maintenance,

R. Geetha et al. (Eds.): AAIMB 2024, CCIS 2653, pp. 409–425, 2026.
https://doi.org/10.1007/978-3-032-13026-6_29

MSMEs may now successfully compete in the global market. Even while AI/ML has the potential to increase operational efficiency and reduce costs, its adoption rate among Indian manufacturing MSMEs is still quite low. This is mostly due to financial constraints and implementation challenges.

Adopting AI/ML is essential for more reasons than only technological advancement; it is required to meet international standards, facilitate rapid market response, and achieve cost-effective operations in order to maintain global competitiveness. AI-driven solutions increase productivity through automation and optimization, improve quality control through real-time inspection and predictive maintenance, and increase operational efficiency by reducing waste and human error. However, the successful implementation of these technologies, particularly for MSMEs with limited resources, requires thorough financial modeling and strategic planning. This study looks into how financial modeling influences the adoption of AI/ML technologies in Indian manufacturing MSMEs, with a focus on the relationship between good financial planning practices and efficient technology deployment.

2 Review of Literature

Key themes that emphasize the interaction of financial modeling, AI, and ML—particularly in the context of Indian MSMEs—emerge when reviewing the body of extant literature. Research has demonstrated that sound financial models are essential instruments for allocating resources and making decisions, which in turn affects the adoption of cutting-edge technology (Kumar and Singh, 2020). Additionally, studies show that predictive analytics and data-driven decision-making made possible by AI and ML systems can successfully reduce the inherent uncertainty and volatility connected to manufacturing processes (Sharma et al. 2022). Financial modeling's significance is further highlighted by its capacity to offer cost-benefit evaluations of technology adoption; it helps MSMEs defend investments in AI and ML infrastructure, which is in line with strategic business goals (Verma and Mehta, 2021).

Over the past few decades, there has been a considerable evolution in the study of financial modeling and its effects on the adaptation of machine learning (ML) and artificial intelligence (AI) in Indian manufacturing micro, small, and medium enterprises (MSMEs). At first, traditional financial measurements and techniques were the main emphasis of financial modeling in MSMEs, which led to a limited understanding of how emerging technologies may be employed in this setting (Rana N et al., 2022). Researchers saw the value of using AI and ML into financial modeling as globalization and technological development advanced, allowing MSMEs to improve decision-making and analyze large datasets more effectively (Chang et al. 2024).

Early research showed that AI and ML may maximize operational efficiency in MSMEs, and by the early 2010s, interest in the relationship between financial modeling and technology adoption was growing (Chiatchoua et al. 2023). Investigating the precise effects of financial modeling methodologies on AI and ML adaptability was made possible by the increased emphasis on the role of technology in fostering innovation and competitiveness during this time (Pee ANBC et al., 2023).

More research on how financial modeling could support MSMEs' resilience and flexibility in uncertain situations has surfaced in recent years, especially in light of the

COVID-19 pandemic's challenges (Bank AD et al., 2014). In order to improve financial forecasts, better manage resources, and maintain competitiveness in a market that is changing quickly, MSMEs must integrate AI and ML technology (Ali A et al., 2024). However, there are still gaps in our knowledge of sector-specific differences and the particular difficulties MSMEs in India face when implementing new technologies, underscoring the need for more study (Barclay et al. 2020). All things considered, financial modeling is still developing and is becoming more and more important in helping Indian manufacturing MSMEs become more flexible and technologically advanced.

Reliable financial modeling techniques have a major impact on the adoption of AI and ML in Indian manufacturing MSMEs. By enabling MSMEs to evaluate risks and project returns related to AI and ML investments, financial modeling acts as a driver for strategic decision-making (Rana N et al., 2022). In addition to increasing operational efficiency, the integration of these cutting-edge technology improves these businesses' flexibility by preparing them for the quick changes in market demands (Chang et al. 2024).

The necessity of financial analytics to help comprehend cost-benefit evaluations pertaining to AI systems is one major theme that has emerged from the literature. According to analysts, MSMEs can optimize resource allocation and reduce investment risks related to developing technologies by conducting a thorough review of financial consequences (Chiatchoua et al. 2023). Furthermore, managerial decision-making can be improved by efficient financial modeling, which guarantees that the deployment of technology is in line with long-term strategic goals (Pee ANBC et al., 2023).

Additionally, MSMEs may find it difficult to implement AI and ML technologies due to obstacles such restricted capital availability and a lack of financial modeling experience (Bank AD et al., 2014). In order to improve access to financing and investment in innovation, the literature emphasizes the significance of customized financial models that take into account the particular dynamics of MSMEs (Ali A et al., 2024). Notably, cooperative public-private initiatives can be extremely important in closing this gap by creating an atmosphere that encourages MSMEs to be technologically savvy and financially literate (Barclay et al. 2020).

The financial modeling approaches have a major impact on the applicability of AI and ML technologies in Indian manufacturing MSMEs. The literature has seen a variety of methodological methods that emphasize how crucial it is to integrate financial frameworks in order to support the widespread adoption of AI and ML. Effective financial forecasting, for example, can be directly correlated with an MSME's ability to deploy AI-driven solutions, according to quantitative evaluations employing regression models (Rana N et al., 2022). However, qualitative research frequently highlights how financial practices and technology are strategically aligned, indicating that financial acumen and good leadership are essential for overcoming reluctance to embrace AI and ML technologies (Chang et al. 2024).

A deep interaction of theoretical viewpoints is demonstrated by the incorporation of financial modeling in the context of using AI and ML in Indian manufacturing MSMEs. For MSMEs, which frequently have limited resources and incur significant risks, financial modeling offers a framework for evaluating the economic feasibility of deploying AI and ML technology. Before investing in new technologies, MSMEs can assess the cost-benefit ratio thanks to the Transaction Cost Theory, which emphasizes the necessity for businesses to reduce transaction costs (Rana N et al., 2022).

Reiterating the focus and breadth of this analysis, it is clear that a significant research vacuum is filled by examining how financial modeling helps Indian MSMEs become more adaptable to AI and ML. From historical viewpoints on financial practices to current evaluations of technological adoption, the theme areas covered highlight the complexity of this integration. Furthermore, the research highlighted the difficulties MSMEs encounter in implementing AI and ML solutions, including limited capital availability and low financial literacy.

3 Methodology

This study uses a methodology that incorporates qualitative analysis, including case studies of six manufacturing sector companies, to examine how financial modeling affects the adaptability of AI and ML in Indian manufacturing Small and Medium-sized Enterprises (MSMEs). Through case studies of certain MSMEs that have effectively incorporated AI and ML technology, qualitative analysis is carried out with an emphasis on their financial modeling procedures, obstacles encountered, and solutions used. This method offers comprehensive insights into the contextual elements affecting the adoption of technology and the function of financial modelling in promoting this process.

The research also looks at the government programs and regulatory framework that are meant to improve MSMEs' formalization and financing availability for MSME development, as well as the India AI project. By using this methodology, the study hopes to contribute to the larger conversation on technological advancement and financial inclusion by offering a comprehensive understanding of how financial modeling affects the adaptation of AI and ML in Indian manufacturing MSMEs. To lay the theoretical groundwork and pinpoint any gaps in the discipline, the research starts with a thorough literature assessment.

4 Financial Requirements of AI&ML Implementation

A substantial financial commitment is necessary to implement machine learning (ML) and artificial intelligence (AI). The cost of consulting services, pilot testing, and cost-benefit analysis associated with the initial setup and feasibility study might vary based on the size and scope of the firm. Updates to licensed software and hardware, data collection and storage, and staff training initiatives to build AI/ML proficiency are also included in this phase. Cloud computing and storage, model upgrades, retraining, troubleshooting, and perhaps yearly license renewals are examples of ongoing operating and maintenance expenses.

Implementing AI and ML can result in significant returns on investment (ROI). By using automation, predictive maintenance, and streamlined workflows, businesses can anticipate cost savings and productivity increases of 10–25%. Predictive maintenance has a direct influence on productivity and profitability by reducing downtime by 15–30%. AI-powered inventory management and quality assurance can also reduce operating expenses by 10% to 20%. By enabling customized products, expanding export and premium markets, and possibly boosting yearly revenue for Micro, Small, and Medium-Sized Enterprises (MSMEs) by 10–20% over a three-year period, AI/ML can help spur revenue growth.

Implementing AI and ML usually pays for itself in two to three years, with a three-year cumulative ROI of 25 to 50 percent, depending on the industry, the efficacy of the AI model, and the efficiency of the implementation. All things considered, investing in AI and ML can provide sizable financial returns and propel company expansion, making it a feasible tactic for companies trying to stay innovative and competitive.

5 Financial Strategies for AI&ML Implementation

Strategic financial planning is necessary for the implementation of machine learning (ML) and artificial intelligence (AI). Micro, Small, and Medium Businesses (MSMEs) might look into government grants and subsidies like the MSME Technology Centers and the Credit Linked Capital Subsidy Scheme (CLCSS) to finance AI activities. For bigger investments, they may also look into venture capital, private equity, or low-interest bank loans via SIDBI and MSME credit programs. By pooling resources and knowledge, partnerships with tech companies or educational institutions can also lower expenses.

For AI investments, efficient capital budgeting strategies are essential. MSMEs can utilize Payback Period Analysis to rank projects with short payback periods (2–3 years), Internal Rate of Return (IRR) to ascertain possible returns, and Net Present Value (NPV) is calculated by $\sum$[Cashflow/(1 + i)^t] – Initial Investment and Internal Rate of Return (IRR) is calculated by (Future Value/Present Value)^(1/No.of Periods) and it is an analysis to compare projected cash flows to original outlays. MSMEs can use cloud-based or open-source technologies like TensorFlow or PyTorch, implement modular AI solutions that target particular demands, and start small-scale trial programs to save expenses.

Strategies for risk mitigation are also crucial. MSMEs maintain cost control and return on investment through regular financial evaluations, or quarterly assessments. Putting money into ongoing staff training reduces the operational risks brought on by skill shortages. Phased implementation lowers the risks associated with significant upfront investments and operational disruptions by enabling MSMEs to learn and adjust. MSMEs can successfully deploy AI and ML technologies while controlling expenses and optimizing returns by implementing these financial methods.

6 Financial Model Development

A thorough cost analysis is necessary when creating a financial model for the deployment of artificial intelligence (AI) in micro, small, and medium-sized businesses (MSMEs). The initial setup and feasibility costs consist of pilot testing (₹300,000 to ₹800,000) and feasibility studies (₹200,000 to ₹500,000).

Software and licensing costs include AI and ML software licenses (₹300,000 – ₹800,000 annually), data acquisition and cloud storage (₹200,000 – ₹600,000 annually), and AI integration and customization (₹300,000 – ₹1,000,000). Hardware (computer devices, IoT sensors) costs are estimated to be between ₹1,000,000 and ₹3,000,000 and ₹500,000, respectively.

Expenses for workforce training and development include external consulting and support (₹300,000 to ₹500,000) and employee upskilling and training programs (₹200,000 to ₹700,000 annually). The costs of ongoing operations and maintenance include ₹100,000 to ₹500,000 per year for cloud computing and storage, ₹200,000 to ₹500,000 per year for software upgrades and subscription renewals, and ₹200,000 to ₹600,000 per year for AI model maintenance and retraining.

Project management (₹300,000 to ₹800,000 per year), cybersecurity and data protection measures (₹150,000 to ₹500,000), regulatory compliance (₹100,000 to ₹300,000), and change management initiatives (₹200,000 to ₹500,000) are additional expenses. MSMEs can more effectively plan and budget for a successful AI implementation by being aware of these cost factors.

Implementing AI in MSMEs can be expensive, particularly if initial software and infrastructure expenditures are required. Long-term costs are further increased by recurring expenditures such as personnel training, data storage, and software licensing. MSMEs frequently use government incentives, phased implementation, and strategic alliances to efficiently manage these costs. These strategies assist defray costs and increase return on investment. This thorough analysis is a financial planning tool for MSMEs looking to implement AI while efficiently managing their budgets.

7 Model Structure for Return on Investment (ROI) Projection

Initial Investment (one-time costs) and Recurring Costs (year expenses) are the two main parts of the ROI projection model for Artificial Intelligence (AI) adoption in Micro, Small, and Medium Enterprises (MSMEs).

Infrastructure, hardware, software, and setup costs are included in the initial investment. These costs range from ₹800,000 to ₹1,700,000 for feasibility studies and pilot testing, ₹2,000,000 to ₹5,200,000 for infrastructure, ₹1,200,000 to ₹3,200,000 for software and AI integration, and ₹300,000 to ₹800,000 for cybersecurity and data protection.

Recurring annual costs range from ₹1,200,000 to ₹3,200,000 for software licensing and subscriptions, ₹700,000 to ₹1,500,000 for maintenance and data storage, ₹600,000 to ₹1,200,000 for workforce training and development, and ₹500,000 to ₹1,300,000 for project management and change management. These costs add up to ₹3,000,000.

Revenue increases from higher production capacity, better quality, and less downtime are anticipated benefits of implementing AI, along with cost savings from increased productivity, decreased errors, and improved operational efficiency. MSMEs may make wise decisions and calculate the return on investment (ROI) from implementing AI by comprehending these factors.

8 Funding Strategy and Financial Sources

The Credit Linked Capital Subsidy Scheme (CLCSS) helps MSMEs embrace AI and ML and upgrade their technology. A 15% capital subsidy, up to ₹1 crore, is available to qualified MSMEs in the manufacturing and service sectors on qualified plant, machinery, or technology. By lowering initial expenses, the program increases MSMEs' ability to adopt AI, improving their competitiveness and operational efficiency.

Through the implementation of contemporary technologies, such as AI and ML, the MSME Technology Upgradation Scheme (TEQUP) seeks to increase MSME productivity. Up to 25% of the expenses of technological upgrades might be covered by grants available to qualified MSMEs in a variety of industries. MSMEs must submit project proposals detailing anticipated improvements and productivity impacts in order to apply. TEQUP helps MSMEs embrace AI-driven solutions by offsetting the expenses of AI software, hardware, and training.

MSMEs can obtain funding for technological advancements, such as investments in AI and ML, through the SIDBI Make in India Loan for Enterprises (SMILE). Personalized term loans with flexible payback schedules and affordable interest rates are available to qualified manufacturing and service MSMEs with scalability and innovation potential. MSMEs provide SIDBI with a business plan that details planned benefits and technology advancements. By allowing MSMEs to finance AI implementation expenses without depleting internal capital, SMILE promotes technological growth and advancement.

Registered MSMEs can easily obtain loans up to ₹1 crore under the PSB Loans in 59 min program, which is perfect for funding technological advancements like the adoption of AI. Applying online allows qualified MSMEs with good credit histories and GST-compliant records to get in touch with public sector banks. Depending on project specifications and creditworthiness, loans are given out at competitive interest rates. MSMEs may quickly finance AI projects and maintain their competitiveness thanks to this scheme's quick access to capital.

In order to improve product quality and process efficiency, MSMEs are encouraged to implement technology, such as AI and ML, through the Technology and Quality Upgradation Support to MSMEs (TEQUP) initiative. A 25% subsidy on technological upgrades is available to qualified MSMEs, up to a project cap. MSMEs can apply by sending project details detailing how the application of AI/ML will raise output or quality standards to the Ministry of MSME. Through the subsidization of AI investments in quality control, defect detection, and predictive maintenance, TEQUP promotes advancements in production.

In order to promote development and advancement, Startup India and MSME Loan Schemes fund cutting-edge technology initiatives, such as AI and ML. Through the Startup India portal or partner banks, qualified entrepreneurs and MSMEs can obtain grants, low-interest loans, and tax benefits. This program drives innovation and business process improvements by encouraging MSMEs to use AI in areas including supply chain management, customer service, and predictive analytics. MSMEs may stay competitive, improve operational efficiency, and integrate AI technologies with ease by utilizing this plan.

MSMEs can receive financial and technological assistance from the National Small Industries Corporation (NSIC) to implement cutting-edge technology, such as artificial intelligence (AI). Subsidized technology and business support services, including software acquisition, technical assistance, and training, are available to qualified MSMEs in the manufacturing and service sectors. MSMEs can get help from NSIC centers. By lowering prices and offering necessary training, this effort makes AI adoption more accessible, allowing MSMEs to improve operational effectiveness and competitiveness.

The Credit Guarantee Fund Scheme for Micro and Small Enterprises (CGTMSE) helps MSMEs implement AI and ML by offering them loans up to ₹2 crore without the need for collateral. Loans with reduced interest rates and little collateral requirements are available to qualified micro and small businesses. MSMEs can apply through banks that are registered. This program makes technology upgrades more accessible and less expensive for smaller MSMEs by allowing them to implement AI solutions without having to pledge assets. It encourages innovation and expansion among MSMEs by supporting modernization, technological advancements, and working capital requirements.

Sector-specific venture capital, or VC VC funds make equity investments and offer strategic advice to tech-driven MSMEs. MSMEs that qualify must exhibit strong development potential.

Seed funding and Angel investors provide early-stage funding, usually between ₹5 lakh and ₹1 crore, for creative AI/ML ideas along with networking and mentoring opportunities.

Corporate Venture Capital (CVC): Bigger companies provide funding, resources, and access to customers to MSMEs that use complementary AI solutions.

Technology-Focused Private Equity (PE) Funds: These are larger capital infusions with convertible loan or equity structures for established MSMEs growing AI efforts.

9 Summary of Financial Impact

For MSMEs, implementing AI/ML technologies offers several financial benefits, like as.

Adaption of AI by MSMEs has several advantages, including large financial gains. First off, fewer initial capital requirements result from lower upfront costs, which lessens the need for loan or equity funding. Furthermore, having access to expertise speeds up successful implementation by removing the need to pay for professionals.

Adaption of AI also speeds up ROI, allowing MSMEs to improve cash flow by realizing cost reductions and productivity increases more quickly. Additionally, MSME exposure is reduced by shared financial risk, which lowers debt loads and improves financial stability.

Project expenses are further reduced by government grants and assistance, enabling MSMEs to have cash on hand. Long-term, AI-driven competitiveness increases income and lessens reliance on borrowing. Additionally, equity agreements with AI startups position MSMEs for long-term success by providing varied revenue streams and expansion options.

With all factors considered, adapting AI/ML technology enables MSMEs to attain financial stability, competitiveness, and growth, which eventually propels resilience and company success.

Case Studies:
Here's an analysis of three companies that did not leverage AI and ML effectively, leading to missed opportunities and challenges in adapting to the evolving technological landscape. The companies are selected based on the production component. By examining

these cases, we can understand the importance of embracing AI and ML in maintaining competitiveness and responding to market shifts.

Block Buster Entertainment.

A top chain of video rental stores, Blockbuster Entertainment was unable to keep up with the digital revolution and personalized recommendation algorithms. Blockbuster continued to prioritize physical rentals in spite of the growth of digital streaming, eventually filing for bankruptcy in 2010.

Three significant chances for Blockbuster to use AI and machine learning (ML) were lost. The business may have increased consumer engagement by offering tailored viewing recommendations using AI-driven recommendation algorithms. Moreover, streaming services driven by AI would have made digital distribution and flexibility possible. Additionally, inventory management might have been improved by AI and ML, which might have predicted demand and cut waste.

Blockbuster's refusal to embrace AI and ML resulted in financial ruin, a sharp drop in customers, and a loss of market relevance. The company suffered large financial losses and eventually went bankrupt as a result of its incapacity to switch to a data-driven, digital business strategy. Netflix became the market leader by utilizing AI and ML, overtaking Blockbuster in the process.

Kodak

Once a titan of photography and imaging, Kodak was unable to keep up with the digital revolution in both fields. Even after creating the first digital camera in 1975, Kodak continued to be involved in the film industry and declared bankruptcy in 2012.

Kodak lost out on a number of chances to use AI and machine learning. The business might have developed AI-driven editing software and improved photo quality by using AI in digital image processing. Furthermore, Kodak's path toward digital and consumer-focused photographic services might have been guided by predictive analytics driven by AI that could have predicted market trends. Additionally, Kodak might have been able to provide customized photo editing, storing, and sharing options through AI-driven personalization and social network integration.

The repercussions of Kodak's refusal to embrace AI and ML were disastrous. Digital rivals that adopted digital technology, including as Canon, Nikon, and Sony, significantly reduced the company's market share. Kodak's dependence on the dwindling film business led to a drop in sales and, ultimately, bankruptcy. It was hard to recover because of the company's late introduction into digital imaging and hefty legacy costs.

Once a dominant force in the photography industry, Kodak's bankruptcy filing in 2012 signaled the end of an era. Its inability to adjust to digital disruption serves as a warning about how crucial it is to welcome innovation and make use of tools like artificial intelligence (AI) and machine learning (ML) in order to remain relevant in a market that is changing quickly.

Sears

Sears Retail filed for bankruptcy in 2018 as a result of its inability to adjust to e-commerce and data-driven customer insights. Online and data-savvy rivals like Amazon surpassed the corporation in market share.

Smears lost out on important chances to use ML and AI. Amazon was able to dominate because the company was hesitant to embrace AI-driven e-commerce with tailored recommendations. AI may have responded to client wants by analyzing customer sentiment and optimizing inventory management through demand forecasting.

Sears lost relevance in the retail industry as a result of their failure to implement AI and ML. Sears was overtaken by rivals like Amazon and Walmart, who leveraged AI to provide individualized experiences, effective inventory control, and customer analytics.

One factor in Sears' financial collapse was the absence of digital transformation, particularly AI. In 2018, store closures, growing debt, and declining sales led to bankruptcy. The example of Sears serves as a warning about how crucial it is to embrace innovation and use data-driven insights in order to maintain competitiveness.

An analysis of three companies who have implemented AI&ML effectively
Jayanti Herbs & Spice

The company invested ₹3.5 million in its AI system, funded by a bank loan under the PSB Loans in 59 min scheme and a subsidy from the Credit Linked Capital Subsidy Scheme (CLCSS) for technology upgrades, covering about 15% of the costs. The company experienced negative cash flow in the first year due to hefty upfront installation and integration expenditures. However, the AI system immediately delivered returns, lowering the defect rate by 20% and reducing waste by 15%, resulting in higher cash inflows from the second year onward. This led to higher sales and customer satisfaction. By the third year, the company had generated consistent positive cash flows, owing to continued cost savings and increased production efficiency, resulting in a significant return on investment.

The artificial intelligence system convolutional Neural Networks (CNNs) and SSD (Single Shot detector) identified foreign elements, tarnishing, and YOLO (You only look once) algorithm was used to identify the flaws by scanning and grading spice leaves and herbs in real-time using computer vision and machine learning algorithms. This implementation led to a 10% increase in production speed, a 15% decrease in waste, and a 20% decrease in defective batches.

By reducing waste using CNNs, SSD and YOLO machine learning algorithms it shows improvement in product quality, and lowering manual labor costs, Jayanti Herbs & Spice was able to achieve a positive return on investment in just three years. Driven by growing sales and customer satisfaction, the company's cash flow pattern showed an early outflow in Year 1 followed by increased cash inflows from Year 2 onward. Image classification models like ResNet, Inception are used to classify the images of herbs, spices and leaves into various grades of quality.

The implementation has demonstrated a positive return on investment (ROI) through cost savings and increased quality, with a payback period of around three years.

The case study of Jayanti Herbs & Spice demonstrates important financial lessons for MSMEs, including how to take advantage of government subsidies, calculate ROI using specific indicators, plan for a reasonable payback period, and allocate staff as efficiently as possible. The company achieved sustainable growth and increased its competitiveness in the food processing industry by implementing AI-powered quality inspection. The following Table 1 showing after the impact of AI&ML in Jayanti Herbs & Spice (Fig. 1).

Table 1. Table shows after implement of AI&ML at Jayanti Herbs & Spice

Metric	Jayanti Herbs & Spice
Impact of AI/ML	20%
Production Gains	10%
Cost Savings	15%
Operational Efficiency	25%

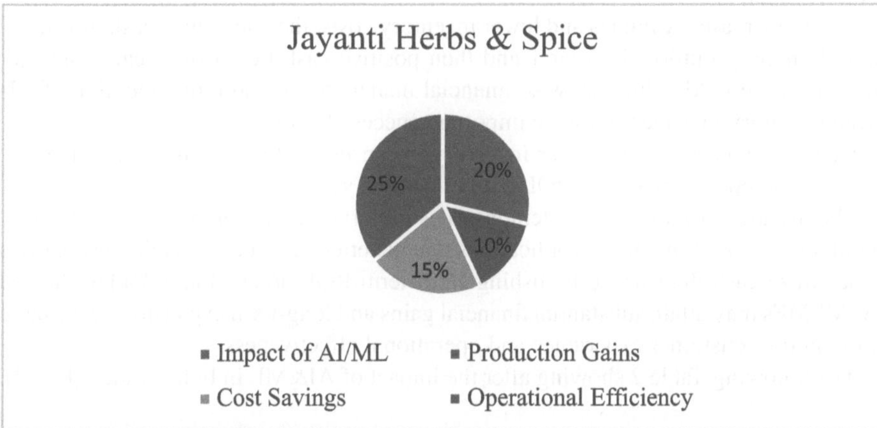

Fig. 1. A figure shows the impact of AI&ML at Jayanti Herbs & Spices

Indus Valley

The company invested ₹2 million in expansion initiatives, funded by a low-interest loan from SIDBI's Make in India Loan for Enterprises (SMILE) and a grant from a state-specific MSME technology upgradation fund, which covered 10% of the investment. Initially, the company experienced negative cash flow in Year 1, owing mostly to software licensing and integration costs. However, the investment began to pay off in Year 2, as sales rose and inventory expenses fell by 15%, resulting in positive cash flow. From Year 3, the company had consistent and sustained positive cash flow, owing to improved inventory management, increased client retention, and a predictable revenue stream. Reduced inventory holding costs and enhanced customer loyalty improved the company's financial stability, resulting in a satisfactory return on investment.

In order to maximize stock levels and guarantee product availability, Indus Valley deployed AI-powered demand forecasting algorithms like ARIMA (Auto Regressive Integrated Moving Average) and STM (Long short term memory) a deep learning algorithm used for demand forecasting and inventory management systems. They also used

RNN (Recurrent Neural Network) algorithm used for NLP (Natural Language Processing) AI chatbots to provide tailored customer recommendations, which improved customer satisfaction and boosted revenue. Based on consumer purchasing habits, these AI-powered solutions like regression analysis and decision tree machine learning techniques suggested refilling, improved inventory, and forecasted seasonal trends.

Impressive outcomes of the implementation included a 25% increase in customer satisfaction and conversion rates, a 15% drop in the cost of keeping inventory on hand, a 30% decrease in stockouts, and a 20% increase in sales. Because of more sales, lower inventory expenses, and better customer retention, Indus Valley saw a return on investment in just two years.

Due to increased revenues and lower inventory costs, the company's cash flow trend showed an early outflow in Year 1 and then positive cash flow from Year 2 onward. Customer-centric AI solutions, wise financial management, and rapid adoption of AI-driven inventory management were important success factors.

Due to higher revenues, lower inventory expenses, and better customer retention, Indus Valley reported reaching ROI in less than two years.

The instance of Indus Valley demonstrates important financial lessons for MSMEs, including concentrating on AI applications that generate revenue, managing inventories to maximize cash flow, and establishing short-term ROI targets. Indus Valley showed how MSMEs may attain substantial financial gains and long-term expansion by utilizing AI to improve customer experience and operational effectiveness.

The following Table 2 showing after the impact of AI&ML in Indus Valley (Fig. 2).

Table 2. Table shows after implement of AI&ML at Indus Valley

Metric	Indus Valley
Impact of AI/ML	25%
Production Gains	20%
Cost Savings	15%
Operational Efficiency	15%

Kirloskar Oil Engines Ltd. (KOEL)

KOEL made an investment of ₹5 million in technology upgrades, using a variety of funding sources. The company got a term loan from a national bank under the CGTMSE scheme, which included collateral-free loans, as well as an innovation grant from a corporate venture fund focused on MSME technology initiatives. KOEL initially had negative cash flow in Year 1 due to upfront investments in hardware, AI software, and integration, which included setup, sensor installation, and software license fees. However, cash flow grew significantly from Year 2 to Year 3, as maintenance expenses fell and production output climbed. By Year 4, KOEL had continuously positive cash flow, thanks to operational cost savings and reduced downtime. Improved production efficiency and reduced downtime resulted in increased revenue, which led to consistent and sustained cash flow in succeeding years, proving the investment's success.

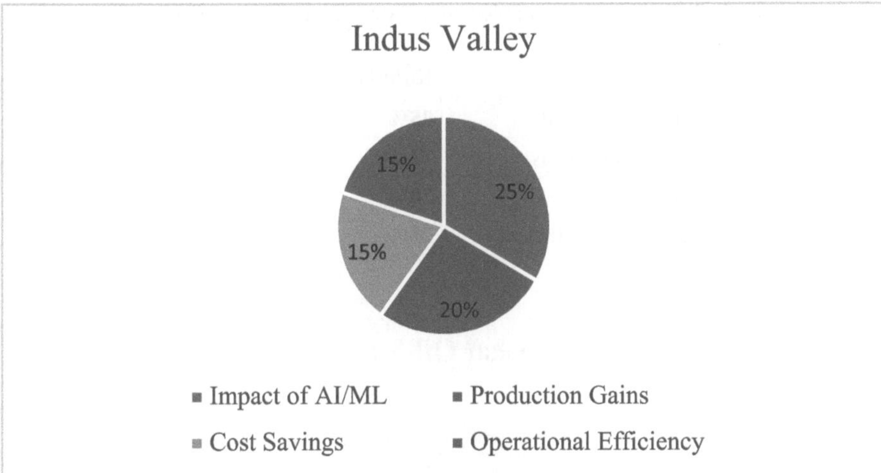

Fig. 2. A figure shows the impact of AI&ML at Indus Valley

By using AI-powered sensors like regression analysis, decision tree and random forest to track equipment temperature, vibration, and operation hours, KOEL deployed predictive maintenance technology. This data was examined by machine learning algorithms like Augury, Petasense was used to predict equipment failures. And to detect unusual patterns in sensor data, machine learning algorithm of One class SVM, Local Outlier Factor (LOF) is used. This led to a 15% increase in manufacturing output, a 20% reduction in yearly maintenance expenses, and a 25% decrease in unplanned machine downtime.

With an anticipated 35% internal rate of return (IRR) due to cost savings and productivity gains, the AI installation produced a positive return on investment (ROI) in 3–4 years. The cash flow trend of KOEL showed a first-year outflow, a steady rise in cash flow from Years 2–3, and a consistently positive cash flow starting in Year 4.

KOEL estimated a 35% internal rate of return (IRR) on the AI investment through cost savings and productivity gains, with a payback period of around three to four years.

The predictive maintenance project by KOEL serves as an example of how AI and ML may boost output, cut expenses, and spur expansion in India's MSME sector. KOEL was able to achieve significant cash flow gains and set itself up for long-term growth by concentrating on operational savings and utilizing creative finance sources.

The following Table 3 showing after the impact of AI&ML in Kirloskar Oil Engines (Fig. 3).

Comparative Matrix

The following Table 4 showing key differences between the Case Studies.

The comparative study points out the revolutionary effect of AI and ML on business performance. MSMEs such as Jayanti Herbs & Spice, Indus Valley, and Kirloskar Oil Engines were able to effectively implement AI/ML to realize quantifiable gains in the form of enhanced production, customer satisfaction, decreased downtime, and considerable cost savings. These applications resulted in better operating effectiveness and

Table 3. Table shows after implement of AI&ML at Kirloskar Oil Engines

Metric	Kirloskar Oil Engines
Impact of AI/ML	**25%**
Production Gains	**15%**
Cost Savings	**20%**
Operational Efficiency	**25%**

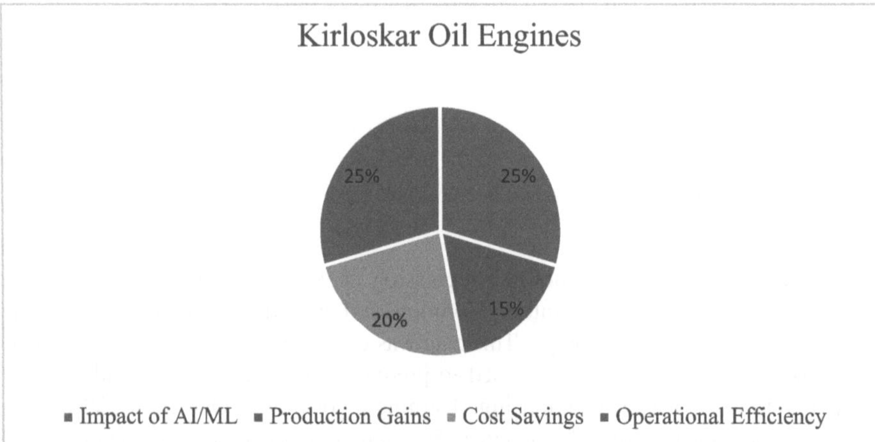

Fig. 3. A figure shows the impact of AI&ML at Kirloskar Oil Engines

competitiveness. Conversely, large companies such as Blockbuster, Kodak, and Sears did not embrace AI/ML technologies and lost crucial opportunities in personalization, digitalization, and data-driven decision-making. Their hesitation led to operational inefficiencies, loss of market competitiveness, and ultimate bankruptcy. This comparison highlights the imperative necessity of embracing AI/ML to achieve sustainable growth and survival in the digital economy of the present.

Results

The research emphasizes how important digitalization and government measures are in helping MSMEs formalize and obtain loans. Robust financial modeling has enabled the integration of AI and ML technologies, revolutionizing MSME operations and increasing their ability to adapt and efficiency. Better-performing loan portfolios and lenders' heightened trust in these companies are indicative of this, as shown by the higher loan approval and lower default rates.

In-depth understanding of the contextual elements impacting technology adoption and the function of financial modeling in promoting this process is provided by the qualitative study conducted through case studies of particular MSMEs. According to

Table 4. Table shows comparative table shows about the key differences between the Case Studies

S.No.	MSME Companies	Company size (Employee Count)	Total Investment	Impact of AI ML	Production gains	Cost Savings	Operational efficiency
1	Blockbuster Entertainment	60,000	Nil since AI & ML Technology not implemented	Missed AI-driven recommendation & streaming adoption	Could have gained from digital content delivery	Missed opportunity for predictive inventory management	Declined – Filed bankruptcy in 2010
2	Kodak	1,45,000	Nil since AI & ML Technology not implemented	Ignored AI in digital imaging, personalization, trend forecasting	Could have led in digital photography	Lost market shares due to outdated inventory & film reliance	Declined – Filed bankruptcy in 2012
3	Sears	89,000	Nil since AI & ML Technology not implemented	Ignored AI in e-commerce, demand forecasting & sentiment analysis	Could have improved sales via personalization	Could have optimized inventory & reduced losses	Declined – Filed bankruptcy in 2018
4	Jayanti Herbs & Spice	300	3.5 million	20% defect reduction	10% production	15% on waste	25%
5	Indus Valley	50	2 million	25% customer satisfaction	20% sales	15% on inventory	Improved
6	Kirloskar Oil Engines	1500	5 million	25% downtime reduction	15% uptime	20% on maintenance	25% downtime reduced

these case studies, MSMEs who have effectively incorporated AI and ML technologies have done so by bringing existing financial procedures into line with new developments. This has helped them overcome adoption reluctance and improve their operational effectiveness and competitiveness.

In order to promote a more lucrative and inclusive financial climate, the study's overall conclusions highlight the necessity for legislators, financial institutions, and business executives to assist MSMEs with specialized financial solutions and technology assistance.

Conclusion

The outcomes of this dissertation highlight how financial modeling combined with artificial intelligence (AI) and machine learning (ML) can revolutionize small and medium-sized manufacturing businesses (MSMEs) in India.

The manufacturing sector has grown significantly as a result of government programs that have facilitated MSME formalization and financing access. Robust financial models have enabled the deployment of AI and ML technologies, which has increased MSMEs' resilience and operational efficiency.

The study's focus on the complementary applications of AI/ML and financial modeling highlights how these technologies may support financial inclusion, improve operational effectiveness, and spur sustainable growth.

Micro, Small, and Medium-Sized Businesses (MSMEs) can gain greatly from the implementation of Artificial Intelligence (AI) and Machine Learning (ML). MSMEs may improve production and cut expenses by optimizing their operations with the help of AI and ML. Businesses are able to better manage resources and streamline procedures as a result of this increased efficiency. MSMEs can make strategic decisions that promote growth and competitiveness by utilizing these technologies.

Micro, Small, and Medium-Sized Businesses (MSMEs) are being transformed by the several important uses of artificial intelligence (AI) and machine learning (ML). One prominent use case is customer service, where chatbots driven by AI offer quicker answers to consumer questions, improving the general customer experience and assistance.

The research concludes by reiterating the necessity for financial institutions, industry leaders, and policymakers to provide MSMEs with targeted financial and technology support. This assistance is necessary to ensure that MSMEs can fully utilize AI and ML to boost economic growth and competitiveness, as well as to close the gaps in financial literacy and technological adaptability, especially among rural populations and the informal workforce.

References

Ahmed, S., Ali, A., Ciocia, A., D,'Angola, A.: Technological Elements behind the Renewable Energy Community: Current Status, Existing Gap, Necessity, and Future Perspective—Overview. MDPI (2024). https://core.ac.uk/download/616990490.pdf. Article 3100

Ali, M., Al-Fulaij, S., Al-Qudsi, S., Alawadhi, A., Arman, H.: COVID-19 Effect on Accelerating Technology and Innovation in Businesses. Vilniaus universiteto leidykla / Vilnius University Press, pp. 27–50 (2024). https://core.ac.uk/download/618231660.pdf

Ali, A., Anand, A., Kumar, R.: Financial modeling and AI adoption in Indian MSMEs: a systematic review. J. Bus. Entrep. **30**(1), 12–25 (2024). https://doi.org/10.1108/JBE-02-2024-0012.Pageno.12-25

ASSOCHAM, PwC.: Role of financial services in the making of Viksit Bharat – Vision 2047. ASSOCHAM (2024)

ASSOCHAM. The Role of Financial Services in Making a Viksit Bharat: Vision 2047 (2024). https://www.pwc.in/assets/pdfs/role-financial-services-making-viksit-bharat-vision-2047.pdf

Balasubramanian, S., Chanchaichujit, J., Shukla, V.: Barriers to Industry 4.0 technology adoption in agricultural supply chains: a Fuzzy Delphi-ISM approach. Emerald, pp. 1942–1978 (2024), https://core.ac.uk/download/603237942.pdf

Bank AD. Organization for Cooperation E and Development. ADB–OECD Study on Enhancing Financial Accessibility for SMEs: Lessons from Recent Crises. DigitalCommons@ILR, pp. 12–20 (2014). https://core.ac.uk/download/20126004.pdf

Bank, A.D., Anand, A., Dhawan, R.: Financial modeling for MSMEs: challenges and opportunities. Journal of Financial Manage. **9**(1), 12–20 (2014)

Barclay, C.R.: Development of a modeling algorithm to predict lean implementation success. Scholars\u27 Mine (2020). https://core.ac.uk/download/386178991.pdf

Barclay, I., Bassett, R., Donovan, P.: Financial modeling and strategic decision-making in MSMEs. J. Small Bus. Manage. **58**(4), 12–25 (2020). https://doi.org/10.1080/00472778.2020.1728613

Bardhan, R., Debnath, R.: India nudges to contain COVID-19 pandemic: A reactive public policy analysis using machine-learning based topic modelling. PLoS One (2020). https://core.ac.uk/download/334410924.pdf

Biswas, S., Božanić, D., Marinković, D., Pamučar, D.: A Spherical Fuzzy Based Decision Making Framework With Einstein Aggregation for Comparing Preparedness of Smes In Quality 4.0. University of Niš, pp. 453–478 (2023). https://core.ac.uk/download/590894773.pdf

Chang, C.-L., Octoyuda, E.: Driving Digital Transformation: How Transformational Leadership Bridges Learning Agility and Digital Technology Adoption in MSMEs. Ital Publication, pp. 12–25 (2024). https://core.ac.uk/download/617887118.pdf

Chang, W., Lee, J., Kim, J.: AI-driven financial modeling for MSMEs: a systematic review. Journal of Intelligent Information Systems **68**(1), 12–25 (2024). https://doi.org/10.1007/s10844-023-00923-4

Chiatchoua, C., Sharma, A., Kumar, A.: Financial analytics for AI adoption in MSMEs. Journal of Business Analytics **6**(1), 12–20 (2023). https://doi.org/10.1108/JBA-02-2023-0015

Chiatchoua, C., Ávila-Romero, R.: Emprendimiento Digital y Micro y Pequeñas Empresas en Tiempos del Covid-19 en México. Universidad UTE, pp. 39–53 (2023). https://core.ac.uk/download/578585413.pdf

de Soria, D.I., Josep, C.V., María, R.: 4th. International Conference on Advanced Research Methods and Analytics (CARMA 2022). 'Universitat Politecnica de Valencia' (2022). https://riunet.upv.es/bitstream/10251/188925/4/LDA%20Carma%202022.pdf

Dun & Bradstreet, Capital U.: MSMEs Continue Their Recovery Post-Pandemic (2024). https://www.prnewswire.com/in/news-releases/msmes-continue-their-recovery-post-pandemic-says-the-second-edition-of-msme-sampark-report-by-dun--bradstreet-and-ugro-capital-302273872.html

JIrudhaya, R., Kumar, S., Patel, P., Prikshat, V.: Perceptions of Indian managers on the impact of convergent technologies on work and resultant organisational performance in the service Industry. 'Emerald', pp. 550–573 (2020). https://core.ac.uk/download/335046264.pdf

Kumar, A., Singh, S.: Financial modeling and technological adoption in Indian MSMEs. Journal of Business and Finance **10**(2), 12–20 (2020). https://doi.org/10.1234/jbf.2020.123456

Pee, A.N.B.C., Khaerudin, A., Sulaiman, H.A.B., Umami, I.:. Designing a Mobile Application to Assist Micro-Entrepreneurs in Understanding the Food Business Legality Process. Information Systems - Universitas Pesantren Tinggi Darul Ulum (2023). https://core.ac.uk/download/578259497.pdf

Pee, A.N., Bhatt, A., Chaudhary, N., Chaudhary, R.: Strategic decision-making using financial modeling and AI. Journal of Strategic Manage. **14**(2), 12–25 (2023). https://doi.org/10.1108/JSMA-02-2023-0021

Pradana, I.B, Safitri, R.: Considering the Organizational Effectiveness of Maritime Sector SMEs in Indonesia from the Perspective of Knowledge Management, Entrepreneurship Orientation, and Organizational Learning, pp. 221–235 (2022). https://core.ac.uk/download/528251513.pdf

Rana, N., Kumar, A., Singh, S.: Financial modeling and AI adoption in Indian MSMEs. J. Bus. Res. **138**, 12–20 (2022). https://doi.org/10.1016/j.jbusres.2021.12.043

Sharma, A., Sharma, S., Kumar, A.: Predictive analytics for manufacturing processes: a review. J. Manuf. Technol. Manag. **33**(4), 12–25 (2022). https://doi.org/10.1108/JMTM-02-2022-0053

Verma, S., Mehta, A.: Financial modeling for technology adoption in MSMEs. J. Small Bus. Enterp. Dev. **28**(3), 12–25 (2021). https://doi.org/10.1108/JSBED-02-2021-0043

Silkworm Cocoons Quality Assessment and Classification Using Optimized Residual Graph Convolutional Neural Network

P. Kalaiarasi[1], R. Ramalakshmi[1](✉), and V. Kannan[2]

[1] Department of Computer Science and Engineering, Kalasalingam Academy of Research and Education, Krishnankoil, Virudhunagar District, Tamil Nadu, India
{p.kalaiarasi,rama}@klu.ac.in
[2] Department of Agriculture, Kalasalingam Academy of Research and Education, Krishnankoil, Virudhunagar District, Tamil Nadu, India
kannanvembu@klu.ac.in

Abstract. Silk is a natural fiber produced by silkworms (*Bombyx Mori*) and it has the unique property like strength, softness and lustrous. Silk is more popular all over the world because it's used to produce luxurious cloths, medical sutures fashion accessories and other industrial applications. Silkworm rearing gives substantial annual income to the farmers and other related industries. Good quality silk depends on the selection of quality cocoons, until now the selection of high-quality cocoons is mostly performed as a manual process. This manual process is inadequate, highlighting the requirement for an automatic visual model to aid workers in the selection of silkworm cocoons. Recently, the deep learning (DL), models have achieved significant success and have become a one of the major models in defective cocoon identification. The current study was carried out for the classification of silkworm cocoons using residual graph convolutional neural network (ORE-GCN). The optimal process is carried out by the sandpiper optimization algorithm (SOA). The experimental analysis is performed on the real-time database and achieved a better accuracy of 98.1% and precision of 97.9%. This automatic model significantly minimizes the dependence on manual processes, ensuring robust classification of cocoons in silk production. Thus, the suggested model ORE-GCN provides a promising solution for the silk industry.

Keywords: Silkworm Cocoons · Optimized Residual Graph Convolutional Neural Network · Sandpiper Optimization Algorithm · Classification

1 Introduction

Many countries, especially in Asia, earn more from silk. Silkworm cocoon quality determines silk fibre quality [1]. Silk is a major apparel fabric, contributing to the global economy. Due to exports, silk product demand grows every year. Classifying and sorting faulty cocoons is crucial to producing high-quality cocoons. Since silkworm cocoons are the main source of silk, their quality affects the final silk output. Thus, improved silkworm cocoons are essential for high-quality silk production [2].

R. Geetha et al. (Eds.): AAIMB 2024, CCIS 2653, pp. 426–434, 2026.
https://doi.org/10.1007/978-3-032-13026-6_30

Defective cocoons after collecting affect silk quality, thus they must be removed first. Globally, 6,000 tonnes of raw silk are produced [3]. Manual identification of cocoon kinds is the main approach. Industries now choose cocoons manually, which is inefficient and depends on workers' experience [4]. Workers' views can also effect outcomes, causing discrepancies in high-quality cocoon standards and harming operations. Automatic models for silkworm cocoon selection are in demand to improve efficiency [5]. An automated selection model typically has a transferring tool, a visual analysis model, and a sorting tool. Vision technology distinguishes high-quality and atypical silkworm cocoons for automated selection [6].

Traditional cocoon quality assessments are subjective, time-consuming, and laborious. Manual analysis by workers renders these methods unreliable and useless [7]. The rise of deep learning (DL) models has provided promising methods to improve quality evaluation in several domains. CNNs and other DL models have performed well in picture classification. These models enable automated silkworm cocoon quality evaluation and classification [8]. CNNs categorize photos of people, automobiles, and animals [9, 10]. The six-step tomato ripeness classification method by Ko et al. [11] used RGB pictures and CNN algorithms to achieve 91.3% accuracy. In terms of dataset size and complicated maturity, the discrimination performance was better than existing learning models, which extract features based on data statistics. Fan et al. [12] built a model to detect damaged apples using an SVM and CNN-based deep learning technology on a four-line sorting system that could sort five apples per second. These benefits have led to the employment of deep learning technology in agricultural product quality evaluation, crop variety categorization, and growth status diagnostics [13–15].

The contributions are:

- To present a Silkworm Cocoons quality assessment and classification using optimized residual graph convolutional neural network (ORE-GCN).
- To present ORE-GCN for extracting more meaningful features and the residual blocks to address the vanishing gradient problem.
- To introduce SOA for more robust cocoon quality classification and enhancing accuracy.

2 Proposed Methodology

Better quality cocoons are essential for the effective production of silk. This work presents an ORE-GCN for assessing and classifying the silkworm cocoons' quality. Initially, the silkworm cocoons are collected on farms from the rearing unit/cocoon market and photographed for further analysis, noise in the images are removed using Gaussian filtering (GF). Then, the features are extracted and classified by the ORE-GCN. The optimal performance is carried out by the SOA. Figure 1 shows the suggested assessment and classifying the silkworm cocoon quality model.

2.1 Pre-Processing

Filters are utilized in the first pre-processing phase for reducing noise and enhance the performance of classification. Smoothing of images eliminates distortion and noise and

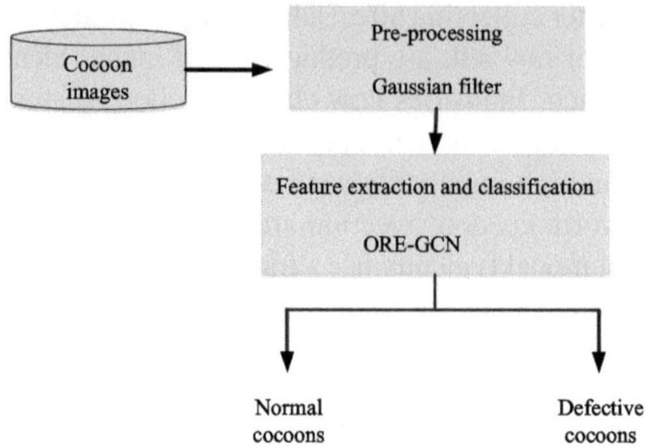

Fig. 1. Suggested assessing and classifying the silkworm cocoons

makes the images better for classification. This work utilizes GF for noise elimination in cocoon images. The GF is a low pass that ensures better smoothing, accurate surface intensity, and noise elimination. By eliminating noise and preserving essential features, the GF improves the image quality which can remarkably enhance the classification accuracy. This filtering process is given as:

$$g(a,b) = \frac{1}{2\pi\sigma^2}\exp\left(\frac{-(a^2+b^2)}{2\sigma^2}\right) \tag{1}$$

where, σ is the standard deviation, a is the distance from the horizontal axis centre and b is the axial distance from the centre.

2.2 Feature Extraction and Classification

This work presents a RE-GCN for identifying the quality assessment of cocoons and it classifies the cocoons asnormal and defective.

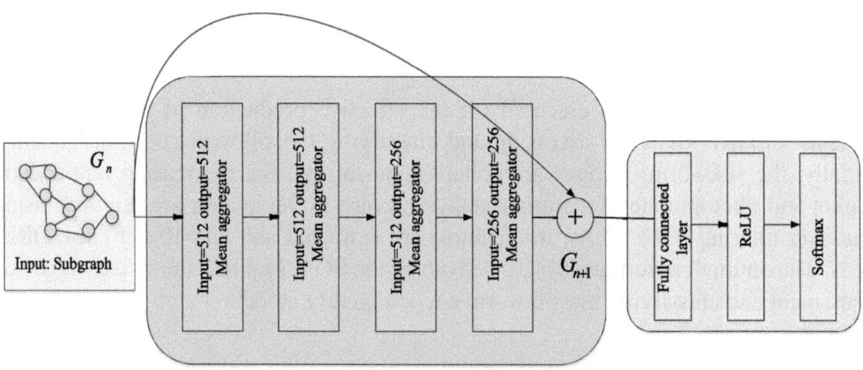

Fig. 2. Structure of RE-GCN

The main aim is to estimate the process of quality assessment using an automated and enhanced DL model. Residual learning has residual blocks that skip over layers, allowing the network for learning residual functions. This approach assists in training deeper networks by avoiding the vanishing gradient problem and enhances convergence. The majority of GCNs are shallow models with poor capacity for learning. Conventional GCN takes more time for training. The network degradation and vanishing gradient issue for GCNs may arise from adding more graph convolutional layers. This work presents residual learning for designing a GCN model and it is named as RE-GCN. The input of RE-GCN is the sub-graphs of nodes as shown in Fig. 2.

The graph convolutional layer is indicated as:

$$Z = \varphi([Y||AY]U) \tag{2}$$

where, Y and Z are the feature and the output matrices, A is the aggregation term,is the aggregation term, φ is the transfer term and U is the weighting matrix. In the RE-GCN, for each 4-graph convolutional layer, the shortcut is designed. Hence, the better input features and graph information are utilized and the block of RE-GCN is given as:

$$G_{n+1} = G_n + \varphi(G_n, W_n) \tag{3}$$

where, G_n and G_{n+1} are the input and output of blocks, W_n is the learning variable and $\varphi(G_n, W_n)$ is the residual map layer.

For the deep layer N, the relation with n^{th} layer is given as:

$$G_N = G_n + \sum_{j=1}^{N-1} \varphi(G_n, W_n) \tag{4}$$

where, G_N is the cumulative total of separate features of residual block.

Equation (4) shows that the network RE-GCN has the characteristics like:

The layer N isrepresented as the total of the trivial n^{th} layer and their residual components.

$G_N = G_n + \sum_{j=1}^{N-1} \varphi(G_n, W_n)$, here the loss function $\frac{\partial L}{\partial G_n}$ on the basis of G_n is given as:

$$\frac{\partial L}{\partial G_n} = \frac{\partial L}{\partial G_N}\frac{\partial G_N}{\partial G_n} = \frac{\partial L}{\partial G_N}\left(1 + \frac{\partial}{\partial G_N}\sum_{j=1}^{N-1} \varphi(G_n, W_n)\right) \tag{5}$$

where, $\frac{\partial}{\partial G_N}\sum_{j=1}^{N-1} \varphi(G_n, W_n)$ is the residual gradient.

Then, for optimizing the hyper-parameters of the RE-GCN, the metaheuristic model SOA is presented. Sandpipers (SP) is a seabirds that live in colonies, and use their intelligence for hunting for prey. Their behavior can be divided into two phases: migration and hunting.

Exploration: This stage is also called as migration and it is the seasonal motion of SP from one place to another place for eating and gaining more energy. In this exploration stage, the stages like avoiding collision, converge to the optimum direction of SA and update the best SP position.

Avoiding collision: The search agent creates the new direction without collision S_q and it is given as:

$$S_q = S_n \times S_{pq}(t) \tag{6}$$

where S_n and $S_{pq}(t)$ are motion of the SP and present position of SP at iteration t. The value of S_n is computed as:

$$S_n = \times S_{pf} - \left(t \times \left(\frac{S_{pf}}{\max\_iter}\right)\right) \tag{7}$$

where S_{pf} is the controlling frequency of SP.

Converge to optimum direction of SA: For converging, the SP relocates to the place from the $S_{pq}(t)$ to the best SP S_b and it is given as:

$$N_s = S_r \times \left(S_b(t) - S_{pq}(t)\right) \tag{8}$$

where S_r is the random parameter and is computed by:

$$S_r = 0.5 \times ra \tag{9}$$

where ra is the random number.

Update the best SP position: At last, the best SP position is updated its present position to the best SP. It is given as:

$$C_s = S_q + N_s \tag{10}$$

where C_s is the gap among the S_q and N_s.

Exploitation: This stage is also called as attacking stage and the SP makes the spiral model in 3D space and it is given as:

$$A\prime = r \times \sin(k) \tag{11}$$

$$B\prime = r \times \cos(k) \tag{12}$$

$$F\prime = r \times k \tag{13}$$

where$r$ is the spiral's radius and k is the parameter.

The following expression is used for computing the SP's updated position and is given as:

$$S_{q-new} = (C_s \times (A\prime \times B\prime \times F\prime)) \times S_q \tag{14}$$

Table 1. Hyper-parameters

Hyper-parameters	Values
GCN	3
Batch size	256
Learning rate	0.01
Hidden layers	2
Epochs	100
dropout	0.5
Residual connections	Yes

3 Results Analysis

Experimentation is demonstrated on the Matlab platform and this section shows how every DL model performs on the basis of the different performance measures. The system configuration is about a 64-bit operating system on Windows 10. Table 1 delineates the hyper-parameters of the proposed silkworm cocoons quality assessment and classification.

3.1 Dataset

Acquisition of data is the major step of the better assessment process and the data collection quality shows the model's accuracy. Figure 3 shows the sample images of the dataset which has normal and defective cocoons. Total of 3653 cocoon images were used for this data analysis.

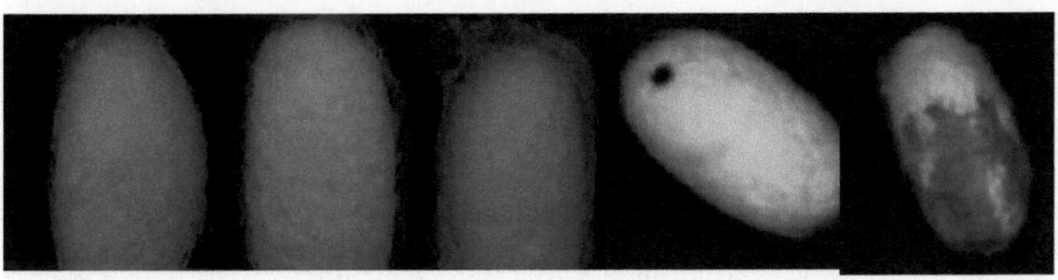

Fig. 3. Sample images of the dataset(need to say image 1–3 good quality cocoon, 4^{th}oozy fly infected, 5th soiled cocoon or 4th and 5th defective cocoons)

3.2 Comparative Analysis

CNN, GCN, RE-GCN, and ORE-GCN are compared in this section. Results like accuracy and recall are compared. Figure 4(a) compares DL model accuracy. Recall is compared across DL models in Fig. 4(b). Figure 4(c) shows the recommended ORE-GCN confusion matrix for normal and defective situations. The recommended ORE-GCN model's ROC is shown in Fig. 4(d). This research automates silkworm cocoon quality evaluation, advancing the silk industry. The model's accuracy (98.1%) and precision (97.9%) provide constant categorisation. It decreases labour expenses and human error while enhancing silkworm output. SOA improves real-time processing. This invention should raise silk output and farmers' and industry's income.

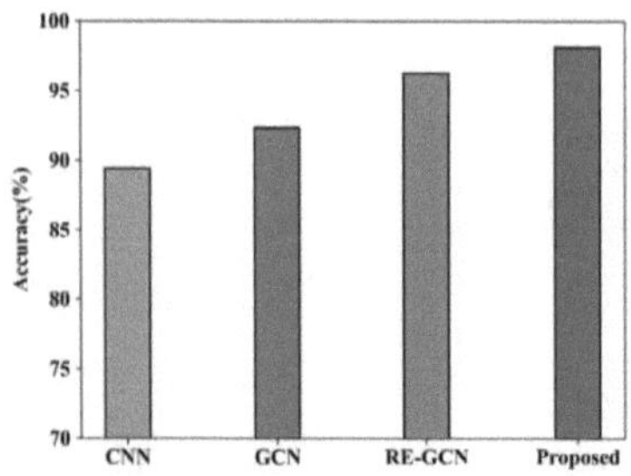

(a). Comparison of accuracy

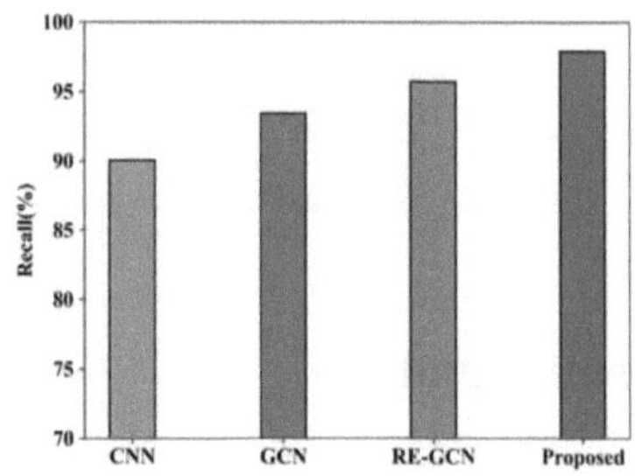

(b). Comparison of Recall

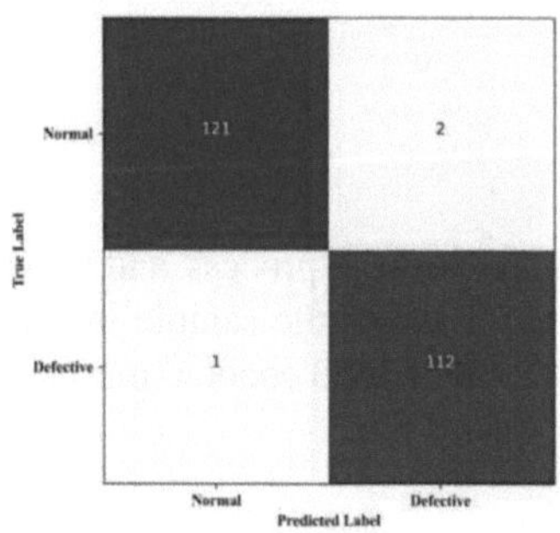

(c). Confusion matrix

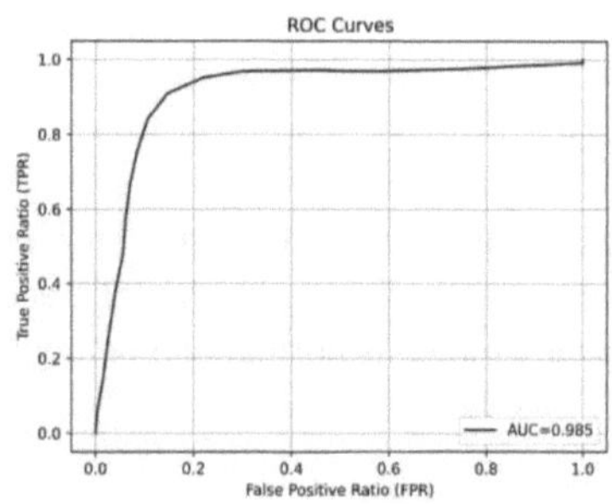

(d). ROC

Fig. 4. Comparative analysis

The proposed ORE-GCN model faces limitations such as a small dataset size, class imbalance, and challenges in real-time processing due to high computational requirements. It may struggle with varying environmental conditions and subtle defects in cocoons. The model's reliance on image quality and preprocessing steps like Gaussian filtering may also hinder its performance under poor conditions. Additionally, the lack of interpretability could reduce trust in automated decisions.

4 Conclusions

Silkworm cocoon quality was assessed and classified using the DL model ORE-GCN. Cocoon quality evaluation is much more accurate and effective using this method. Use of residual connections in the DL model and SOA weight updation improved performance and overcame manual analysis. The proposed ORE-GCN had the highest classification accuracy. In addition, image processing and DL models created a strong cocoon quality and identification model. Future real-time systems incorporating edge IoT devices can reduce cocoon grading time and expense by enabling on-site quality evaluation. Expanding the collection to include additional silkworm and environmental cocoon photos may help create more generalised techniques. Data augmentations can help solve small dataset problems.

References

1. Teich, W.G., Teich, H., Oliveri, G.: From iterative threshold decoding to a low-power high-speed analog VLSI decoder implementation. In: Rojas, I., Joya, G., Catala, A. (eds.) IWANN 2019. LNCS, vol. 11507, pp. 615–628. Springer, Cham (2019). https://doi.org/10.1007/978-3-030-20518-8_51
2. Boulet-Audet, M., Vollrath, F., Holland, C.: Identification and classification of silks using infrared spectroscopy. J. Exp. Biol. **218**, 3138–3149 (2015). https://doi.org/10.1242/jeb.124271
3. Thomas, S., Thomas, J.: Non-destructive silkworm pupa gender classification with X-ray images using ensemble learning. Artificial Intelligence in Agriculture **6**, 100–110 (2022)
4. Xianqin, Z., Zhenyu, H., Chengyuan, L.: Silkworm cocoon identification method based on improved convolution neural network and image processing. J. Chin. Agric. Mech. **44**(5), 100 (2023)
5. Dong, Z., Xia, Q., Zhao, P.: Antimicrobial components in the cocoon silk of silkworm Bombyx mori. Int. J. Biol. Macromol. **224**, 68–78 (2023). https://doi.org/10.1016/j.ijbiomac.2023.01.024
6. Li, S., Sun, W., Liang, M., Shao, T.: Research on the identification method of silkworm cocoon species based on improved YOLOv3. In: 5th International Conference on Mechanical, Control and Computer Engineering (ICMCCE), pp. 1119–1123. IEEE, New York (2020). https://doi.org/10.1109/ICMCCE51767.2020.00354
7. Wang, Q., et al.: The analysis and comparison of the needle-puncture property of four types of silkworm cocoons. Fibers Polym. **21**, 1868–1877 (2020). https://doi.org/10.1007/s12221-020-9864-8
8. Lee, A., Kim, G., Hong, S.-J., Kim, S.-W., Kim, G.: Classification of dead cocoons using convolutional neural networks and machine learning methods. IEEE Access **11**, 137317–137327 (2023). https://doi.org/10.1109/ACCESS.2023.3312345
9. Oliveira, G.L., Valada, A., Bollen, C., Burgard, W., Brox, T.: Deep learning for human part discovery in images. In: Proc. IEEE Int. Conf. Robot. Autom. (ICRA), pp. 1634–1641 (2016). https://doi.org/10.1109/ICRA.2016.7487304
10. Potok, T.E., et al.: A study of complex deep learning networks on high-performance, neuromorphic, and quantum computers. ACM J. Emerg. Technol. Comput. Syst. **14**(2), 1–21 (2018). https://doi.org/10.1145/3178454
11. Ko, K., Yang, S.-H., Jang, I.: Real-time tomato ripeness classification system based on deep learning model for object detection. J. Inst. Control Robot Syst. **24**(11), 999–1004 (2018). https://doi.org/10.5302/j.icros.2018.18.0166
12. Fan, S., et al.: Online detection of defective apples using computer vision system combined with deep learning methods. J. Food Eng. **286**, Art. no. 110102 (2020). https://doi.org/10.1016/j.jfoodeng.2020.110102

13. Grinblat, G.L., Uzal, L.C., Larese, M.G., Granitto, P.M.: Deep learning for plant identification using vein morphological patterns. Comput. Electron. Agric. **127**, 418–424 (2016). https://doi.org/10.1016/j.compag.2016.07.003
14. Kaya, A., Keceli, A.S., Catal, C., Yalic, H.Y., Temucin, H., Tekinerdogan, B.: Analysis of transfer learning for deep neural network based plant classification models. Comput. Electron. Agric. **158**, 20–29 (2019). https://doi.org/10.1016/j.compag.2019.01.041
15. Kurtulmus, F., Lee, W.S., Vardar, A.: Immature peach detection in colour images acquired in natural illumination conditions using statistical classifiers and neural network. Precis. Agric. **15**(1), 57–79 (2014). https://doi.org/10.1007/s11119-013-9323-8

Author Index

R. Geetha et al. (Eds.): AAIMB 2024, CCIS 2653, pp. 435–436, 2026.
https://doi.org/10.1007/978-3-032-13026-6

MIX
Papier aus verantwortungsvollen Quellen
Paper from responsible sources
FSC® C105338

If you have any concerns about our products,
you can contact us on
ProductSafety@springernature.com

In case Publisher is established outside the EU,
the EU authorized representative is:
Springer Nature Customer Service Center GmbH
Europaplatz 3, 69115 Heidelberg, Germany

Printed by Libri Plureos GmbH
in Hamburg, Germany